Simon Haykin

Computer Algorithms

Sara Baase

San Diego State University

Computer Algorithms

Introduction to Design and Analysis

SECOND EDITION

 Addison-Wesley Publishing Company

Reading, Massachusetts ▪ Menlo Park, California ▪ New York
Don Mills, Ontario ▪ Wokingham, England ▪ Amsterdam ▪ Bonn
Sydney ▪ Singapore ▪ Tokyo ▪ Madrid ▪ Bogotá
Santiago ▪ San Juan

This book is in the **Addison-Wesley Series in Computer Science**

Michael A. Harrison, *Consulting Editor*

James T. DeWolf, *Sponsoring Editor*
Bette J. Aaronson, *Production Supervisor*
Herb Caswell, *Text Design*
Stephanie Kaylin, *Copy Editor*
Marshall Henrichs, *Cover Design*
Hugh Crawford, *Manufacturing Supervisor*

Reprinted with corrections July, 1988
Library of Congress Cataloging-in-Publication Data

Baase, Sara.
  Computer algorithms.

  Includes index.
  Bibliography: p.
  1. Electronic digital computers--Programming.
2. Algorithms. I. Title.
QA76.6.B25 1988   519.7   87-9205
ISBN 0-201-06035-3

To Benjamin R. Tucker
for *Liberty*, 1881–1908

Preface

Purpose

This book is intended for a one-semester, upper-division course in algorithm analysis or data structures and algorithms where the emphasis is on algorithms. It has sufficient material to allow several choices of topics and algorithms for the latter part of the course.

The purpose of the book is threefold. It is intended to teach algorithms for solving real problems that arise frequently in computer applications, to teach basic principles and techniques of computational complexity (worst-case and average behavior, space usage, and lower bounds on the complexity of a problem), and to introduce the areas of *NP*-completeness and parallel algorithms.

Another of the book's aims that is at least as important as teaching the subject matter, is to develop in the reader the habit of always responding to a new algorithm with the questions: How good is it? Is there a better way? Therefore, instead of presenting a series of complete, "pulled-out-of-a-hat" algorithms with analysis, the text often discusses a problem first, considers one or more approaches to solving it (as a reader who sees the problem for the first time might), and then begins to develop an algorithm, analyzes it, and modifies or rejects it until a satisfactory result is produced. (Alternative approaches that are ultimately rejected are also considered in the exercises; it is useful for the reader to know why they were rejected.)

Questions such as: How can this be done more efficiently? What data structure would be useful here? Which operations should we focus on to analyze this algorithm? How must this variable (or data structure) be initialized? appear frequently throughout the text. Answers generally follow the questions, but I suggest the reader pause before reading the ensuing text and think up his or her own answers. Learning is not a passive process.

I hope the reader will also learn to be aware of how an algorithm actually behaves on various inputs — that is, Which branches are followed? What is the pattern of growth and shrinkage of stacks? How does presenting the input in different ways (e.g., listing the vertices or edges of a graph in a different order) affect the behavior? Such questions are raised in some of the exercises, but are not emphasized in the text because they require carefully going through the details of many examples.

Most of the algorithms presented are of practical use; I have chosen not to emphasize those with good asymptotic behavior that are poor for inputs of useful sizes (though some important ones are included). Specific algorithms were chosen for a variety of reasons including the importance of the problem, illustrating analysis techniques, illustrating techniques (e.g., depth-first search) that give rise to numerous algorithms, and illustrating the development and improvement of techniques and algorithms (e.g., Union-Find programs). (Section 2.9 on external sorting was included at the request of many of our working students who said it would be useful for their work.)

Prerequisites

Familiarity with data structures, such as linked lists, stacks, and trees, is assumed. The reader should also be comfortable with recursion.

Analysis of algorithms uses simple properties of logarithms and some calculus (differentiation to determine the order of a function and integration to approximate summations), though virtually no calculus is used beyond Chapter 2. I find many students intimidated when they see the first log or integral sign because a year or more has passed since they had a calculus course. Only a few properties of logs and a few integrals from first-semester calculus are needed. Section 1.2 reviews some of the necessary mathematics.

Changes from the First Edition

The major change is the addition of three new chapters: adversary arguments and selection, dynamic programming, and parallel algorithms.

Adversary arguments are an important technique in establishing lower bounds on the complexity of a problem. Students often find lower bounds difficult to understand, and many instructors have not covered them in an undergraduate course. The inclusion of this chapter, with several relatively easy adversary arguments, should bring the topic within the range of undergraduate students.

Dynamic programming is a classical algorithm design technique with a very large number of applications. The new chapter introduces this technique with a few examples.

In recent years, with highly parallel computers commercially available, there has been a tremendous amount of research in the field of parallel computation. It will continue to be a major focus of computational complexity in the future. The new chapter on parallel algorithms introduces this field.

Many students come into a course in algorithm analysis with a very vague idea of what "big oh" is, and they leave the course with a vague idea. The discussion of the order of functions in **Chapter 1** has been expanded in the hope that students will gain better mastery of the concepts and techniques for dealing with order. The section on mathematical tools has been expanded to provide a better review and reference for some of the mathematics used in the book.

In **Chapter 2**, Bubblesort has been replaced by Insertion Sort because Insertion Sort provides a good introductory example for worst-case and average analysis, and it is used in Shellsort. (Bubblesort was moved to the exercises.) Average analysis for Quicksort has been added. Mergesort was added because of its importance as an $O(n\lg n)$ sort and an example of Divide and Conquer, and because it is used as the basis of a parallel sort in Chapter 10.

Chapter 3 is the new chapter introducing adversary arguments and algorithms and lower bounds for selection problems.

In **Chapter 4**, the sections on the Dijkstra algorithms for shortest paths and minimum spanning trees have been rewritten. Since students should become adept at reading and writing recursive algorithms (and instruction in recursion seems to have improved in the past 10 years), the algorithms that use depth-first search are now written in recursive form. Many of the algorithms are more detailed.

The efficient and interesting Boyer-Moore algorithm was published while the first edition of this book was in production; it has been added to **Chapter 5**.

Chapter 6 is the new chapter introducing dynamic programming.

Chapters 7 and **8** have not had substantial changes; recent results on matrix multiplication are mentioned.

Several users of the first edition asked me to expand **Chapter 9** and I have done so. More *NP*-complete problems are described. Although Turing machines still are not explicitly mentioned, the definition of *NP* was changed to follow the Garey and Johnson definition of nondeterministic Turing machines. The notion of a polynomial transformation is defined more formally than in the first edition. I have added a discussion of two fundamental ideas: how to interpret the "length of the input" for number problems (e.g., prime testing), and the relation between the difficulty of decision problems and the related optimization problems. There is a new approximation algorithm for graph coloring.

Chapter 10 is the new chapter on parallel algorithms. It introduces the widely-used PRAM model for parallel computation (with some of the variant conventions for concurrent writes). This chapter includes the binary fan-in technique (for finding max, Boolean **or**, and addition, etc.), a sorting algorithm, an algorithm for finding connected components of a graph, and a lower bound proof for Boolean functions.

Improved Pedagogy

More than 100 new exercises have been added throughout the book. Algorithm techniques, such as Divide and Conquer, and greedy algorithms are explicitly mentioned. (In the first edition they were used but not identified as such.) To make the algorithms easier to read, they have been rewritten in a more modern language using features of Pascal and Modula-2.

I have been asked to mark the "hard" sections. Since the abilities, background, and mathematical sophistication of students at different universities vary considerably, the choice of hard sections certainly depends on the audience. I have starred sections that contain more complicated mathematics or more complex or sophisticated arguments than most others. I have also starred one or two sections that contain optional digressions.

I have starred exercises that use more than minimal mathematics, require substantial creativity, or require a long chain of reasoning. Some of the starred exercises have hints. Some exercises are somewhat open-ended. For example, one might ask for a good lower bound for the complexity of a problem, rather than asking the reader to show that a given function is a lower bound. I did this for two reasons. One is to make the form of the question more realistic; a solution must be discovered as well as verified. The other is that it may be hard for some students to prove the best known lower bound (or find the most efficient algorithm for a problem), but there is still a range of solutions they can offer to show their mastery of the techniques studied.

(Since I know that some instructors will skip starred sections, I have not starred a few sections that I consider essential to a course for which the book is used, even though they contain a lot of mathematics. For example, some, at least of the material in Section 1.4 on the asymptotic growth rate of functions, should be covered.)

Programs

The algorithms presented in this book are not programs; that is, many details not important to the method or the analysis are omitted. Of course, students should know how to implement efficient algorithms in efficient, debugged programs. I think that students taking an introductory course in computational complexity now have a stronger background in data structures and large programming projects than students of ten years ago. Many instructors may teach this course as a pure "theory" course without programming. Most chapters include a list of programming assignments, but they are only brief suggestions that should be amplified by instructors who choose to use them.

Course Outlines

Clearly the amount of material that can be covered in a course depends on the environment. The sample course outline that follows, without the topics shown in brackets, corresponds approximately to the course I taught at San Diego State University with three hours per week of lecture. At the University of California, Berkeley, the course has three lecture hours per week plus an hour per week of discussion section with a teaching assistant, and the students have a strong mathematical background. When I taught there, I covered most of the additional topics shown in brackets.

Chapter 1: I assign the whole chapter as reading but concentrate on Sections 1.4 and 1.5 in class.

Chapter 2: Sections 1–5 [Sections 6 and/or 7].

Chapter 3: Sections 1–2 [the remainder of the chapter].

Chapter 4: Sections 1–5 [Section 6 or a matching algorithm].

Chapter 5: The entire chapter.

Chapter 6: Section 1 and one other [several others].

Chapter 7: One or two topics from Sections 1–3 [include Strassen's algorithm].

Chapter 8: [Section 8.5].

Chapter 9: Sections 1, 2, and 4 [additional polynomial transformations, some discussion of Cook's theorem, Section 5 or 6]

Chapter 10: Sections 1 and 2 [one or more of the remaining sections].

Bibliography

The bibliography contains several kinds of entries: original sources for some of the material in the book, other papers related to the topics covered, and papers on additional topics or new results. Although the Bibliography is about twice as long as that in the first edition, it is far from complete. I have tried to include enough so that the motivated student will have a starting point for further reading.

Acknowledgments

I am happy to take this opportunity to thank the many people who helped me in big and small ways in the preparation of the second edition of this book.

I did most of the work on the second edition while I was a Visiting Professor in the Computer Science Division at the University of California, Berkeley. I appreciate the CS Division's hospitality, the access my visit provided to a large pool of helpful people and a good library, and the opportunity to sit in on courses taught by Dick Karp who always makes the subject of computational complexity exciting and beautiful in his superb lectures.

Some of the people who suggested topics and papers, contributed ideas and opinions, lent me papers, helped me to learn troff, suggested exercises, and reviewed some of the chapters are: Gilles Brassard, Doug Cooper, John Donald, Mike Harrison, Sampath Kannan, Dick Karp, Uzi Vishkin, Avi Wigderson, Yanjun Zhang, and others too numerous to mention but appreciated nonetheless. Of course, I'd have had nothing to write about without the many people who did the original research providing the material I so much enjoy learning and passing on to new generations of students. I thank them for their work.

My students at UC Berkeley and San Diego State used early versions of the manuscript as their textbook and helped find typos and obscurely worded exercises. (Particular thanks to Donald Chinn and Nels Olson for long lists of typos). In the years since the first edition appeared, several students and instructors who used the book sent me lists of typos and suggestions for changes. I don't have a complete list of names, but I appreciate the time and thought that went into their letters.

I found the reviews of the manuscript obtained by Addison-Wesley to be especially helpful and encouraging. My appreciation goes to the reviewers, Judith L. Gersting (Indiana University–Purdue University at Indianapolis), Jay L. Gischer (College of William and Mary), Steven Homer (Boston University), Michael J. Quinn (University of New Hampshire), and Diane M. Spresser (James Madison University); to Uzi Vishkin (Tel Aviv University and Courant Institute, New York University) and Avi Wigderson (Hebrew University, Jerusalem) for their comments on Chapter 10; and to Addison-Wesley's computer science editor, Jim DeWolf, for his work in overseeing the project and for not even mentioning how many times I stretched out the schedule. My thanks also to Bette Aaronson and Joe Vetere of Addison-Wesley for their efforts in the production of this book.

I thank Robert Cademy, Darrell Long, Jack Revelle, Roger Whitney, and Adobe Systems, Inc. for helping me solve the frustrating problem of getting drafts of the manuscript printed after I returned to San Diego State.

Finally, I thank my good friends, Jeannie and Corby, and my hiking buddies in the Bay Area and San Diego, for much needed breaks from work.

San Diego   S. B.

Contents

1

Analyzing Algorithms and Problems: Principles and Examples

1.1
Introduction

To say that a problem is solvable algorithmically means, informally, that a computer program can be written that will produce the correct answer for any input if we let it run long enough and allow it as much storage space as it needs. In the 1930s, before the advent of computers, mathematicians worked very actively to formalize and study the notion of an algorithm, which was then interpreted informally to mean a clearly specified set of simple instructions to be followed to solve a problem or compute a function. Various formal models of computation were devised and investigated. Much of the emphasis in the early work in this field, called *computability theory*, was on describing or characterizing those problems that could be solved algorithmically and on exhibiting some problems that could not be. One of the important negative results, established by Alan Turing, was the proof of the unsolvability of the "halting problem." The halting problem is to determine whether an arbitrary given algorithm (or computer program) will eventually halt (rather than, say, getting into an infinite loop) while working on a given input. There cannot exist a computer program that solves this problem.

Although computability theory has obvious and fundamental implications for computer science, the knowledge that a problem can theoretically be solved on a computer is not sufficient to tell us whether it is practical to do so. For example, a perfect chess-playing program could be written. This would not be a very difficult task. There is only a finite number of ways of arranging the chess pieces on the board, and under certain rules a game must terminate after a finite number of moves. The program could consider each of the computer's possible moves, each of its opponent's possible responses, each of its possible responses to those moves, and so on until each sequence of possible moves reaches an end. Then since it knows the ultimate result of each move, the computer can choose the best one. The number of distinct arrangements of pieces on the board that it is reasonable to consider (much less the number of sequences of moves) is roughly 10^{50} by some estimates. A program that examined them all would take several thousand years to run. Thus such a program has not been run.

There are numerous problems with practical applications that can be solved — that is, for which programs can be written — but for which the time and storage requirements are much too great for these programs to be of practical use. Clearly the time and space requirements of a program are of practical importance. They have become, therefore, the subject of theoretical study in the area of computer science called *computational complexity*. One branch of this study, which is not covered in this book, is concerned with setting up a formal and somewhat abstract theory of the complexity of computable functions. (Solving a problem is equivalent to computing a function from the set of inputs to the set of outputs.) Axioms for measures of complexity have been formulated; they are basic and general enough so that either the number of instructions executed or the number of storage bits used by a program can be taken as a complexity measure. Using these axioms, one can prove the existence of arbitrarily complex problems and of problems for which there is no best program.

The branch of computational complexity studied in this book is concerned with analyzing specific problems and specific algorithms. The book is intended to help the reader build up a repertoire of classic algorithms to solve common problems, some general design techniques, and some tools and principles for analyzing algorithms and problems. We will present, study, and analyze algorithms to solve a variety of problems for which computer programs are frequently used. We will analyze the amount of time the algorithms take to execute, and we will also often analyze the amount of space used by the algorithms. In the course of describing algorithms for a variety of problems, we will see that several algorithm design techniques often prove useful. Thus we will pause now and then to talk about some general techniques. We will also study the computational complexity of the problems themselves, that is, the time and space inherently required to solve the problem no matter what algorithm is used. We will study the class of *NP*-complete problems — problems for which no efficient algorithms are known — and consider some heuristics for getting useful results. And, finally, we will introduce the subject of algorithms for parallel computers.

In the following section we present some background and tools that will be used throughout the book, including a description of the algorithm language we will use.

1.2
The Algorithm Language, Mathematics, and Data Structures

1.2.1   The Algorithm Language

The main criterion for choosing an algorithm language for use in this book was that the algorithms should be easy to read. We want to focus on the strategy and techniques of an algorithm, not declarations and syntax details of concern to a compiler. The control structures of Pascal are fairly easy to understand, and Pascal has the advantage of being very widely known. However, Modula-2, though less widely known at this time, is similar to Pascal (at least in those parts of the language that would be used in this book) and has a few features that we prefer. Thus, the language used is closer to Modula-2 than Pascal, but any reader familiar with either language should be able to read the algorithms without difficulty. We have made modifications and extensions in the interests of clarity, simplicity, and readability. We will describe these, and, for the benefit of Pascal programmers, explain a few Modula-2 features that differ from Pascal.

Modula-2 and Pascal differ in the semantics of the evaluation of Boolean expressions. An expression like

$$(i \leq n) \textbf{ and } (A[i] \neq x)$$

is natural for controlling a **while** loop that searches an array. In Pascal all the terms of the expression are evaluated before determining if the expression is true or false. This causes an error if the array A has dimension $[1..n]$ because there will be an illegal reference to $A[n+1]$ if $i = n+1$. A similar problem occurs for loops that traverse a linked list; there may be an illegal use of a **nil** pointer. The solutions to this problem in Pascal are awkward. In Modula-2, if the first operand of an **and** or **or** determines the truth or falsity of the expression, the second operand is not evaluated. Thus if $i > n$ in the previous expression, the term $A[i] \neq x$ is not evaluated and there is no problem of an invalid reference to $A[n+1]$. We will assume the Modula-2 semantics in this book.

Although the parentheses in the expression

$$(i \leq n) \text{ and } (A[i] \neq x)$$

are required in Pascal and Modula-2, such expressions seem easier for people to read without the parentheses, so we will omit them and write, for example,

$$i \leq n \text{ and } A[i] \neq x.$$

The reserved word **begin**, required so often in Pascal, is almost always redundant because it follows another reserved word (e.g., **do** or **then**). In Modula-2 it is not needed, but we do need an **end** at the end of most control statements (**if**, **while**, and **for**).

Modula-2 allows us to specify an increment (besides 1 and -1) in a **for** statement. The form of the statement is:

for identifier := expression **to** expression **by** constantexpression **do**
    statement sequence
end

We will extend the **for** statement to allow notation such as

for $x \in S$ **do**

where S is a set. (We may use sets that would not be legal in Pascal or Modula-2.)

There are no line numbers in the language, but we will sometimes number the lines of an algorithm so that we can refer to them in the text.

There are two points where we prefer (and will use) the Pascal convention rather than Modula-2. Modula-2 uses the reserved word **procedure** in the header of both procedures and functions. We prefer to retain **function** for function headers. Also, we will use braces to enclose comments { like this }.

We will often omit explicit type declarations. For some algorithms, e.g., many sorting algorithms, the specific type of the data does not matter. In other cases, it is obvious from the context. We will occasionally make up new types, e.g., *BitString* or *Complex*, without defining them formally. In some such instances the implementation is clear; in others it is simply not very important.

The notation for defining and using records and pointers in both languages is almost identical and will be reviewed in the section on data structures.

We often write instructions in English to avoid cluttering the algorithm with details not relevant to the main problem. For example, we may say "Let x be the largest entry in A" where A is an array, or "Insert x at the end of L" where L is a linked list.

For improved readability, we often use mathematical notation instead of Pascal or Modula-2 operators. For example, we use $\leq$, $\geq$, and $\neq$ in place of <=, >=, and # (or <>). When computation is not the focus of a problem, we may omit multiplication and exponentiation operators and use ordinary mathematical notation instead, e.g.,

$$x := 3y^2$$

1.2.2 Some Mathematical Background

A variety of mathematical concepts, tools, and techniques are used in this book. Some should already be familiar to the reader; some will be new. The purpose of this section is to provide a brief review of and reference for some definitions and elementary properties about logarithms, probability, permutations, and summation formulas. We will introduce other mathematical tools, such as recurrence relations, in the text. The reader should be familiar with proofs by induction; there are some in the text.

Notation

For any real number x, $\lfloor x \rfloor$ (read "floor of x") is the largest integer less than or equal to x. $\lceil x \rceil$ (read "ceiling of x") is the smallest integer greater than or equal to x. For example, $\lfloor 2.9 \rfloor = 2$, and $\lceil 6.1 \rceil = 7$.

Logarithms

The mathematical tool used most extensively in this book is the logarithm function, usually to the base 2.

Definition   For $b > 1$ and $x > 0$, $\log_b x$ (read "log to the base b of x") is that number y such that $b^y = x$; i.e., $\log_b x$ is the power to which b must be raised to get x.

The following properties follow easily from the definition. In all cases, x's are arbitrary positive numbers and a is any real number.

1. $\log_b$ is a strictly increasing function, i.e., if $x_1 > x_2$, then $\log_b x_1 > \log_b x_2$.
2. $\log_b$ is a one-to-one function, i.e., if $\log_b x_1 = \log_b x_2$, then $x_1 = x_2$.
3. $\log_b 1 = 0$.
4. $\log_b b^a = a$.
5. $\log_b (x_1 x_2) = \log_b x_1 + \log_b x_2$.
6. $\log_b (x^a) = a \log_b x$.

7. $x_1^{\log_b x_2} = x_2^{\log_b x_1}$. (To prove this, show that the logs of both sides of the equation are equal.)

8. To convert from one base to another: $\log_{b_1} x = (\log_{b_2} x)/(\log_{b_2} b_1)$.

Since the log to the base 2 is used most often in computational complexity, there is a special notation for it: lg; i.e., $\lg x = \log_2 x$. The natural logarithm (log to the base e) is denoted by ln. Throughout the text we almost always take logs of integers, not arbitrary positive numbers, and we often need an integer value close to the log rather than its exact value. Let n be a positive integer. If n is a power of 2, say $n = 2^k$, for some integer k, then $\lg n = k$. If n is not a power of 2, then there is an integer k such that $2^k < n < 2^{k+1}$, and $\lfloor \lg n \rfloor = k$ and $\lceil \lg n \rceil = k+1$. $\lfloor \lg n \rfloor$ and $\lceil \lg n \rceil$ are used often. The reader should verify that

$$n \leq 2^{\lceil \lg n \rceil} < 2n$$

and

$$\frac{n}{2} < 2^{\lfloor \lg n \rfloor} \leq n.$$

Here are a few more useful facts.

1. $\lg e \approx 1.44$, $\ln 2 \approx 0.7$, and $\lg 10 \approx 3.32$.

2. The derivative of $\ln x$ is $1/x$ (and, using Property 8, the derivative of $\lg x$ is $\lg e / x$).

Probability

Suppose that in a given situation an event, or experiment, may have any one of, say, k outcomes $s_1, s_2, \ldots, s_k$. With each outcome s_i we associate a real number $p(s_i)$, called the probability of s_i, such that

1. $0 \leq p(s_i) \leq 1$ for $1 \leq i \leq k$, and

2. $p(s_1) + p(s_2) + \cdots + p(s_k) = 1$.

It is natural to interpret $p(s_i)$ as the ratio of the number of times s_i is expected to occur to the total number of times the experiment is repeated. If $p(s_i) = 0$, then s_i is impossible; if $p(s_i) = 1$, then s_i always occurs. (Note, however, that the definition does not require that the probabilities correspond to anything in the real world.) The examples most frequently used to illustrate the meaning of probability are flipping coins and throwing dice. If the "experiment" is the flip of a coin, then the coin may land with "heads" facing up or with "tails" facing up. We let $s_1 = $ 'heads' and $s_2 = $ 'tails' and assign $p(s_1) = 1/2$ and $p(s_2) = 1/2$. (If the reader objects because the coin could land on its edge, we may let $s_3 = $ 'edge' and define $p(s_3) = 0$.) If a six-sided die is thrown, there are six possible outcomes: for $1 \leq i \leq 6$, $s_i = $ "the die lands with side number i facing up," and $p(s_i) = 1/6$. In general, if there are k possible outcomes, each considered equally likely to occur, we let $p(s_i) = 1/k$ for each i.

We often need to consider the probability of any one of several specified outcomes occurring or the probability that the outcome has a particular property. Let S be a subset of $\{s_1, \ldots, s_k\}$. Then $p(S) = \sum_{s_i \in S} p(s_i)$. For example, the probability that when a die is thrown the number appearing is divisible by 3 is $p(\{s_3, s_6\}) = p(s_3) + p(s_6) = 1/3$.

Permutations

A permutation of n distinct objects is, informally speaking, a rearrangement of the objects. Let $S = \{s_1, s_2, \ldots, s_n\}$. Note that the elements of S are ordered by their indexes; i.e., s_1 is the first element, s_2 the second, and so on. A permutation of S is a one-to-one function π from the set $\{1, 2, \ldots, n\}$ onto itself. We think of π as rearranging S by moving the ith element, s_i, to the $\pi(i)$th position. We may describe π simply by listing its values, that is, $(\pi(1), \pi(2), \ldots, \pi(n))$. For example, for $n = 5$, $\pi = (4, 3, 1, 5, 2)$ rearranges the elements of S as follows: s_3, s_5, s_2, s_1, s_4.

The number of permutations of n distinct objects is $n!$. To see this, observe that the first element can be moved to any one of the n positions; then that position is filled and the second element can be moved to any of the $n-1$ remaining positions; the third element can be moved to any of the remaining $n-2$ positions, and so on. So the total number of possible rearrangements is $n \times (n-1) \times (n-2) \times \cdots \times 2 \times 1 = n!$.

Summation Formulas

There are several summations that occur frequently when analyzing algorithms. Formulas for some of them are listed here and in the next section, with brief hints that may help the reader to remember them.

1.   Sum of consecutive integers.

$$\sum_{i=1}^{n} i = \frac{n(n+1)}{2} \tag{1.1}$$

How to remember it. Write out the integers from 1 to n. Pair up the first and last, i.e., 1 and n; pair up the second and next to last, 2 and $n-1$, etc., as illustrated in Fig. 1.1. Each pair adds up to $n+1$. If n is even, there are $n/2$ pairs; hence the sum is $(n/2)(n+1)$. If n is odd, there are $(n-1)/2$ pairs and the extra number $(n+1)/2$ left in the middle; thus the sum is

Figure 1.1   Sum of consecutive integers.

$$\frac{n-1}{2}(n+1) + \frac{n+1}{2} = \frac{n}{2}(n+1).$$

2. Sum of squares.

$$\sum_{i=1}^{n} i^2 = \frac{2n^3 + 3n^2 + n}{6} \tag{1.2}$$

This can be proved by induction on n. The main thing to remember is that the sum of the first n squares is roughly $n^3/3$. Equation 1.2 is not used in the text, but it may be needed in some of the exercises.

3. Powers of 2.

$$\sum_{i=0}^{k} 2^i = 2^{k+1} - 1 \tag{1.3}$$

How to remember it. Think of each term 2^i as a 1-bit in a binary number:

$$\sum_{i=0}^{k} 2^i = 1\,1\,\cdots\,1.$$

There are $k+1$ 1s. If 1 is added to this number the result is

$$100\,\cdots\,0 = 2^{k+1}.$$

(This result can also be obtained by using the formula for the sum of a geometric progression given next.)

4. Geometric sums.

$$\sum_{i=0}^{k} \frac{1}{2^i} = 2 - \frac{1}{2^k} \tag{1.4}$$

This is a geometric progression. The general form and the formula for the sum are

$$\sum_{i=0}^{k} a^i = \frac{a^{k+1} - 1}{a - 1}.$$

To verify this, divide out the right-hand side.

5. Miscellaneous.

$$\sum_{i=1}^{k} i\,2^i = (k-1)2^{k+1} + 2 \tag{1.5}$$

Proof.

$$\sum_{i=1}^{k} i\,2^i = \sum_{i=1}^{k} i\,(2^{i+1} - 2^i) = \sum_{i=1}^{k} i\,2^{i+1} - \sum_{i=0}^{k-1} (i+1)2^{i+1}$$

$$= \sum_{i=1}^{k} i\,2^{i+1} - \sum_{i=0}^{k-1} i\,2^{i+1} - \sum_{i=0}^{k-1} 2^{i+1}$$

$$= k\,2^{k+1} - (2^{k+1} - 2) = (k-1)2^{k+1} + 2. \qquad \square$$

Summations Using Integration

Several summations that arise often in the analysis of algorithms can be approximated (or bounded from above or below) using integration. Suppose f is a continuous, decreasing function and that a and b are integers. Then, as Fig. 1.2 illustrates,

$$\int_{a}^{b+1} f(x)\,dx \le \sum_{i=a}^{b} f(i) \le \int_{a-1}^{b} f(x)\,dx. \qquad (1.6)$$

Similarly, if f is an increasing function,

$$\int_{a-1}^{b} f(x)\,dx \le \sum_{i=a}^{b} f(i) \le \int_{a}^{b+1} f(x)\,dx. \qquad (1.7)$$

Here are two examples that are used later in the text.

Example 1.1 An estimate for $\sum\limits_{i=2}^{n} \dfrac{1}{i}$.

$$\sum_{i=2}^{n} \frac{1}{i} \le \int_{1}^{n} \frac{dx}{x} = \ln x \Big|_{1}^{n} = \ln n - \ln 1 = \ln n$$

Similarly,

$$\sum_{i=2}^{n} \frac{1}{i} \ge \ln(n+1) - \ln 2.$$

Thus

$$\sum_{i=2}^{n} \frac{1}{i} \approx \ln n. \qquad (1.8)$$

$\blacksquare$

Example 1.2 A lower bound for $\sum\limits_{i=1}^{n} \lg i$.

$$\sum_{i=1}^{n} \lg i \ge n \lg n - 1.5n \qquad (1.9)$$

Proof. From Eq. 1.7 (and the observation that $\lg 1 = 0$) we have

$$\sum_{i=1}^{n} \lg i \ge \int_{1}^{n} \lg x\,dx.$$

$$\int_{1}^{n} \lg x\,dx = \int_{1}^{n} \lg e \ln x\,dx = \lg e \int_{1}^{n} \ln x\,dx$$

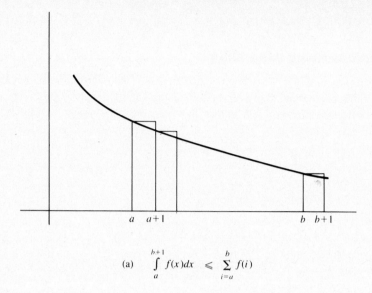

$$\text{(a)} \quad \int_a^{b+1} f(x)dx \;\leqslant\; \sum_{i=a}^{b} f(i)$$

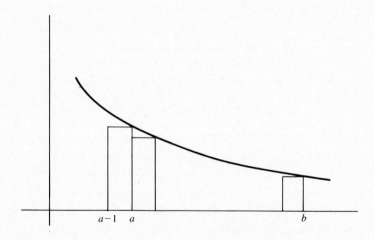

$$\text{(b)} \quad \sum_{i=a}^{b} f(i) \;\leqslant\; \int_{a-1}^{b} f(x)dx$$

Figure 1.2   Approximating a sum of values of a decreasing function.

$$= \lg e \,[x \ln x - x] \,\Big|_1^n = \lg e \,(n \ln n - n + 1)$$

$$= n \lg n - n \lg e + \lg e \geq n \lg n - n \lg e.$$

Since $\lg e \approx 1.44$, Eq. 1.9 follows.   ■

1.2.3 Data Structures

We assume that the reader is familiar with data structures such as linked lists, stacks, queues, and binary trees. The point of this section is to briefly review the elementary structures used in this book and some relevant terminology and notation.

The individual elements (also called *records* and sometimes *nodes*) of a data structure may be divided into several *fields*. Each field is given a name. For example, the record format for a linked list may be

> *ListNode* = **record**
> *name*: *CharacterString*;
> *age*: *integer*;
> *link*: *NodePointer*
> **end**

A *pointer* is a variable, array element, or field entry that "points to" a node; its value is an address (or array index) indicating the location of a node. Pointers are drawn as arrows in diagrams of data structures. The *null pointer* is a pointer value that indicates that there is no node pointed to; it is denoted by **nil**.

If *ptr* is a pointer, then *ptr*↑ is the node it points to. We refer to a particular field of the node by

> *ptr*↑.*fieldname*

where *fieldname* is the name of a field we wish to access. (This is the notation used by both Pascal and Modula-2.) If *ptr* = **nil**, then *ptr*↑ is undefined. More examples are given later in this section.

A *linked list* is a finite sequence of nodes such that each has a pointer field containing a pointer to the next node. A pointer to the list, i.e., to the first node, must be given. In a *simply linked* list the pointer field in the last node contains **nil**, and if the list is empty the pointer to the list has value **nil**. In the following example of a simply linked list, the node format, as described by *ListNode* earlier, is

and *first* is a pointer variable.

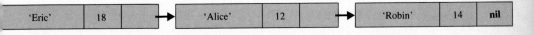

Using the notation just described for field entries, *first*↑.*name* = 'Eric', *first*↑.*link*↑.*age* = 12, and *first*↑.*link*↑.*link*↑.*link* = **nil**. If *p* is another pointer variable whose current value is the location of the third node, then *p*↑.*name* = 'Robin', and *p*↑.*link* = **nil**.

A *listhead* is a special node that may be included at the beginning of a linked list; it contains a pointer to the first data node and possibly some other information about the list (for example, the number of data nodes it contains). If a listhead is used, it is never deleted from the list; thus its use simplifies some of the details of inserting and deleting nodes by eliminating the special case of an empty list.

A *circular linked list* is a linked list in which the last node contains a pointer to the first node (or listhead, if there is one) instead of **nil**.

A *doubly linked list* is a linked list in which each node contains a pointer to the previous node, as well as to the next one. A doubly linked list may be circular and may have a listhead. The use of both of these features eliminates the special cases that occur when inserting or deleting nodes at either end of the list and thus simplifies algorithms. In the following example of a doubly linked circular list with listhead, the node format is

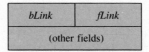

where *bLink* is the back link field and *fLink* is the forward link field. (The value of a pointer always indicates the location of the beginning of a node; to keep the picture clear the back links may not appear to do so.)

list

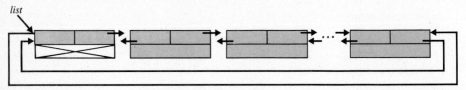

A *stack* is a list in which insertions and deletions are always made at one end, called the *top*. The top item in a stack is the one most recently inserted. A stack may be implemented or stored in an array or as a linked list. To *push* an item on a stack means to insert the item in the stack. To *pop* the stack means to remove the top entry. Stacks are sometimes called *lifo* lists ("last in, first out").

A *queue* is a list in which all insertions are done at one end, called the *rear* or *back*, and all deletions are done at the other end, called the *front*. A queue may be stored in an array or as a linked list. Queues are sometimes called *fifo* lists ("first in, first out").

A *binary tree* is a finite set of elements, called nodes, that is empty or else satisfies the following:

1. There is a distinguished node called the *root*, and
2. The remaining nodes are divided into two disjoint subsets, L and R, each of which is a binary tree. L is the *left subtree* of the root and R is the *right subtree* of the root.

Binary trees are represented on paper by diagrams such as the one in Fig. 1.3. If a node w is the root of the left (right) subtree of a node v, w is the *left (right) child*

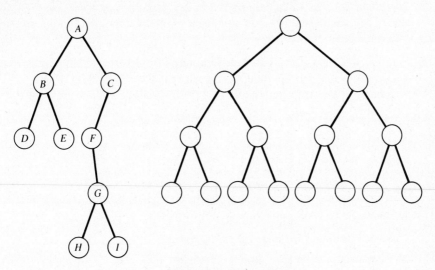

Figure 1.3 Binary trees.

of v and v the *parent* of w; there is a branch connecting v and w in the diagram. The *degree* of a node is the number of nonempty subtrees it has. A node with degree zero is a *leaf*. Nodes with positive degree are *internal nodes*. The *level* of the root is 0, and the level of any other node is one plus the level of its parent.[1] The *depth* (sometimes called the *height*) of a binary tree is the maximum of the levels of its leaves. A *complete binary tree* is a binary tree in which all internal nodes have degree 2 and all leaves are at the same level. The second binary tree in Fig. 1.3 is complete.

A binary tree is usually represented in a program as a linked structure: The location of the root is specified by a pointer variable and each node has two pointer fields, say *leftChild* and *rightChild*, which contain pointers to its left child and right child, respectively. Pointers to the parent of each node may be used also or instead if they are needed.

The following facts are used often in the text. The proofs are easy and are omitted.

Lemma 1.1   There are at most 2^l nodes at level l of a binary tree.

Lemma 1.2   A binary tree with depth d has at most $2^{d+1} - 1$ nodes.

Lemma 1.3   A binary tree with n nodes has depth at least $\lfloor \lg n \rfloor$.

---

[1] Beware: Some authors define *level* so that the level of the root is 1.

The standard traversal orderings for binary trees are *preorder*, *inorder*, and *postorder*. They are defined next; the argument to the traversal procedure is the root of the tree (or subtree) to be traversed.

Preorder(v):   Visit v; *Preorder*(left child of v); *Preorder*(right child of v).
Inorder(v):   *Inorder*(left child of v); Visit v; *Inorder*(right child of v).
Postorder(v): *Postorder*(left child of v); *Postorder*(right child of v); Visit v.

For the first binary tree in Fig. 1.3, the traversal orderings are:

Preorder: A B D E C F G H I
Inorder: D B E A F H G I C
Postorder: D E B H I G F C A

A *tree* (sometimes called a general tree to distinguish it from a binary tree) is a finite nonempty set of nodes satisfying the following:

1.   There is a distinguished node called the root, and
2.   The remaining nodes are partitioned into $m \geq 0$ disjoint subsets, $T_1, T_2, \ldots, T_m$, each of which is a tree. $T_1, T_2, \ldots, T_m$ are called the subtrees of the root.

The terms parent, child, leaf, degree, level, and depth are defined for trees as they are for binary trees with minor (if any) modifications.

It is also easy to extend the definitions of the preorder and postorder traversals to general trees. For preorder, for example, visit the root first, then (recursively) traverse each of the subtrees of the root, $T_1, T_2, \ldots, T_m$.

A *forest* is a finite (possibly empty) set of trees.

1.3
Analyzing Algorithms and Problems

1.3.1   Introduction

We analyze algorithms with the intention of improving them, if possible, and for choosing among several available for a problem. We will use the following criteria:

1.   Correctness
2.   Amount of work done
3.   Amount of space used
4.   Simplicity, clarity
5.   Optimality

We will discuss each of these criteria at length and give several examples of their application. When considering the optimality of algorithms, we will introduce techniques for establishing lower bounds on the complexity of problems.

1.3.2 Correctness

There are three major steps involved in establishing the correctness of an algorithm. First, before we can even attempt to determine whether an algorithm is correct, we must have a clear understanding of what "correct" means. We need a precise statement about what inputs it is expected to work on, and what result it is to produce for each input. Then we can try to prove this statement. There are two aspects to an algorithm: the solution method and the sequence of instructions for carrying it out, i.e., its implementation. Establishing the correctness of the method and/or formulas used may be easy or may require a long sequence of lemmas and theorems about the objects on which the algorithm works (e.g., graphs, permutations, matrices). For example, the validity of the Gauss elimination method for solving systems of linear equations depends on a number of theorems in linear algebra. Some of the methods used in algorithms in this book are not obviously correct; they must be justified by theorems.

Once the method is established, we implement it in a program. If an algorithm is fairly short and straightforward, we generally use some informal means of convincing ourselves that the various parts do what we expect them to do. We may check some details carefully (e.g., initial and final values of loop counters) and hand-simulate the algorithm on a few small examples. None of this proves that it is correct, but informal techniques may suffice for small programs. More formal techniques, such as loop invariants, may be used to verify correctness of parts of programs. Most programs written outside of classes are very large and very complex. To prove the correctness of a large program, we can try to break it down into smaller modules; show that, if all of the smaller modules do their jobs properly, then the whole program is correct; and then prove that each of the modules is correct. This task is made easier if (it may be more accurate to say "This task is possible only if") algorithms and programs are written in modules that are largely independent and can be verified separately. This is one of the many strong arguments for structured, modular programming. Most of the algorithms presented in this book are the small segments from which large programs are built, so we will not deal with the difficulties of proving the correctness of very long algorithms or programs.

One technique for rigorous proofs of correctness is to establish *loop invariants* by mathematical induction. Loop invariants are conditions and relationships that are satisfied by the variables and data structures at the end of each iteration of the loop. The loop invariants are carefully constructed so that it is easy to see that after the last iteration the algorithm will have accomplished what it is supposed to do. The loop invariants are established by induction on the number of passes through the loop. The details of the proof require carefully following the instructions in the loop.

We will illustrate this technique with a very simple example: the sequential search algorithm for finding the location, or index, of a given item x in a list, or array. The algorithm compares x to each entry in turn until a match is found or the list is exhausted. If x is not in the list, the algorithm returns 0 as its answer.

Algorithm 1.1   Sequential Search

Input: L, n, x, where L is an array with n entries (indexed $1..n$), and x is the item sought.

Output: *index*, the location of x in L (0 if x is not found).

1. *index* := 1;
2. **while** *index* $\leq n$ **and** $L[index] \neq x$ **do**[2]
3.   *index* := *index* + 1
4. **end** { while };
5. **if** *index* > n **then** *index* := 0 **end**

Before trying to show that the algorithm is correct, we should make a precise statement about what it is intended to do. This seems simple enough; using the preceding explanation, we may say that, if L is an array with n entries, the algorithm terminates with *index* equal to the index of the array entry containing x, if there is one, and *index* equal to 0 otherwise. This statement has two faults. It does not specify the result if x appears more than once in the list, and it does not indicate for which values of n the algorithm is expected to work. We may assume that n is non-negative, but what if the list is empty, i.e., if $n = 0$? A more precise statement is:

Given an array L containing n items $(n \geq 0)$ and given x, the sequential search algorithm terminates with *index* equal to the index of the first occurrence of x in L, if x is there, and equal to 0 otherwise.

Proof. First we establish the following statement by induction:

For $1 \leq k \leq n+1$, if and when control reaches the tests in line 2 for the kth time, the following conditions (loop invariants) are satisfied: *index* = k and for $1 \leq i < k, L[i] \neq x$.

Let $k = 1$. Then *index* = k from line 1, and the second condition is vacuously satisfied.

Now we assume that the conditions are satisfied for some $k < n+1$ and show that they hold for $k+1$. By the induction assumption, $L[i] \neq x$ for $1 \leq i < k$, and *index* = k when line 2 was executed the kth time. If the tests in line 2 are executed again, i.e., a $(k+1)$st time, we can conclude that the tests were satisfied the kth time, so $L[index] \neq x$; i.e., $L[k] \neq x$. Also, *index* is incremented in the loop, so the $(k+1)$st time the tests at line 2 are performed, the required conditions hold. This completes the induction proof.

Now suppose that the tests in line 2 are executed exactly k times. Clearly, $1 \leq k \leq n+1$. Consider the two possible cases when line 5 is executed after the loop. The output is 0 if and only if $k = n+1$. By the statement we just proved by induction, for $1 \leq i < n+1, L[i] \neq x$; i.e., x is not in the array, so the output 0 is correct.

---

[2] Reminder to Pascal programmers: By the semantics of our Boolean expressions, there is no reference to $L[index]$ if *index* > n.

(Note that this includes the case where $n = 0$ and the list is empty.) On the other hand, the output is *index* $= k \leq n$ if and only if the loop terminated because $L[k] = x$. Since for $1 \leq i < k$, $L[i] \neq x$, we can conclude that k is the index of the first occurrence of x in the array. Thus the algorithm is correct.   □

(Note that if we did not require the index of the *first* occurrence of x in L, but merely *any* occurrence, the algorithm could be shortened by searching the array backwards, starting with *index* := n, and omitting line 5.)

If this proof seems a bit tedious, imagine what a proof of correctness of a full-sized program with complex data and control structures would be like. But if one wants to verify rigorously that a program is correct, this is the sort of work that must be done. Programmers rarely write out such proofs in complete detail, but they should go through similar arguments to convince themselves that an algorithm or program works.

We will not do formal proofs of correctness in this book, though we will give arguments or explanations to justify complex or tricky parts of algorithms. The sample proof was presented here to give the reader a glimpse of what such proofs are like, and more basically, to show that correctness *can be proved*, though indeed for long and complex programs it is a formidable task.

1.3.3   Amount of Work Done

How shall we measure the amount of work done by an algorithm? The measure we choose should aid in comparing two algorithms for the same problem so that we can determine whether one is more efficient than the other. It would be handy if our measure of work gave some indication of how the actual execution times of the two algorithms compare, but we will not use execution time as a measure of work for a number of reasons. First, of course, it varies with the computer used, and we do not want to develop a theory for one particular computer. We may instead count all the instructions or statements executed by a program, but this measure still has several of the other faults of execution time. It is highly dependent on the programming language used and on the programmer's style. It would also require that we spend time and effort writing and debugging programs for each algorithm to be studied. We want a measure of work that tells us something about the efficiency of the *method* used by an algorithm independent not only of the computer, programming language, and programmer, but also of the many implementation details and over-head (or "bookkeeping" operations) such as incrementing loop indexes, computing array indexes, and setting pointers in data structures. Our measure of work should be both precise enough and general enough to develop a rich theory that is useful for many algorithms and applications.

A simple algorithm may consist of some initialization instructions and a loop. The number of passes made through the body of the loop is a fairly good indication of the work done by such an algorithm. Of course, the amount of work done in one pass through a loop may be much more than the amount done in another pass, and one algorithm may have longer loop bodies than another algorithm, but we are narrowing in on a good measure of work. Though some loops may have, say, five steps and some nine, for large inputs the number of passes through the loops will generally be large compared to the loop sizes. Thus, counting the passes through all the loops in the algorithm is a good idea.

In many cases, to analyze an algorithm we can isolate a particular operation fundamental to the problem under study (or to the types of algorithms being considered), ignore initialization, loop control, and other bookkeeping, and just count the chosen, or basic, operations performed by the algorithm. For many algorithms, exactly one of these operations is performed on each pass through the main loops of the algorithm, so this measure is similar to the one described in the previous paragraph.

Here are some examples of reasonable choices of basic operations for several problems.

Problem	*Operation*
1. Find x in a list of names.	Comparison of x with an entry in the list.
2. Multiply two matrices with real entries.	Multiplication of two real numbers (or multiplication and addition of real numbers)
3. Sort a list of numbers.	Comparison of two list entries.
4. Traverse a binary tree (represented as a linked structure where each node contains pointers to its left and right children).	Traversing a link. (Here, setting a pointer would be considered a basic operation rather than overhead.)

So long as the basic operation(s) is chosen well and the total number of operations performed is roughly proportional to the number of basic operations, we have a good measure of the work done by an algorithm and a good criterion for comparing several algorithms. This is the measure we will use in this chapter and in several other chapters in this book. The reader may not yet be entirely convinced that this is a good choice; we will be adding more justification for it in the next section. For now, we simply make a few points. First, in some situations, we may be intrinsically interested in the basic operation: It might be a very expensive operation compared to the others, or it might be of some theoretical interest. Second, we are often interested in the rate of growth of the time required for the algorithm as the inputs get larger. So long as the total number of operations is roughly proportional to the number of basic operations, just counting the latter can give us a pretty clear idea of how feasible it is to use the algorithm on large inputs. Finally, this choice of the measure of work allows a great deal of flexibility. Though we will often try to choose one, or at most two, specific operations to count, one can include some overhead operations, and, in the extreme, one could choose as the basic operations the set of machine instructions for a particular computer. At the other extreme, we could consider "one pass through a loop" as the basic operation. Thus by varying the choice of basic operations, we can vary the degree of precision and abstraction in our analysis to fit our needs.

What if we choose a basic operation for a problem and then find that the total number of operations performed by an algorithm is not proportional to the number of basic operations? What if it is substantially higher? In the extreme case, we might choose a basic operation for a certain problem and then discover that some algorithms for the problem use such different methods that they do not do *any* of the operations we are counting. In such a situation, we have two choices. We could abandon our focus on the particular operation and revert to counting passes through loops. Or, if we are especially interested in the particular operation chosen, we could restrict our study to a particular *class of algorithms*, one for which the chosen operation is appropriate. Algorithms using other techniques for which a different choice of basic operation is appropriate could be studied separately. A class of algorithms for a problem is usually defined by specifying the operations that may be performed on the data. (The degree of formality of the specifications will vary; usually informal descriptions will suffice in this book.)

In this section, we have often used the phrase "the amount of work done by an algorithm." It could be replaced by the term "the complexity of an algorithm." *Complexity* means the amount of work done, measured by some specified *complexity measure*, which in many of our examples is the number of specified basic operations performed. Note that, in this sense, complexity has nothing to do with how complicated or tricky an algorithm is; a very complicated algorithm may have low complexity. We will use the terms "complexity," "amount of work done," and "number of basic operations done" almost interchangeably in this book.

1.3.4   Average and Worst-Case Analysis

Now that we have a general approach to analyzing the amount of work done by an algorithm, we need a way of presenting the results of the analysis concisely. The amount of work done cannot be described by a single number because the number of steps performed is not the same for all inputs. We observe first that the amount of work done usually depends on the size of the input. For example, alphabetizing a list of 1000 names usually requires more operations than alphabetizing a list of 100 names, using the same algorithm. Solving a system of 12 linear equations in 12 unknowns generally takes more work than solving a system of 2 linear equations in 2 unknowns. We observe, secondly, that even if we consider only inputs of one size, the number of operations performed by an algorithm may depend on the particular input. An algorithm for alphabetizing a list of names may do very little work if only a few of the names are out of order, but it may have to do much more work on a list that is very scrambled. Solving a system of 12 linear equations may not require much work if most of the coefficients are zero.

The first observation indicates that we need a measure of the size of the input for a problem. It is usually easy to choose a reasonable measure of size. Here are some examples.

	Problem	Size of input
1.	Find x in a list of names.	The number of names in the list.
2.	Multiply two matrices.	The dimensions of the matrices.
3.	Sort a list of numbers.	The number of entries in the list.
4.	Traverse a binary tree.	The number of nodes in the tree.
5.	Solve a system of linear equations.	The number of equations, or the number of unknowns, or both.
6.	Solve a problem concerning a graph.	The number of nodes in the graph, or the number of edges, or both.

Even if the input size is fixed at, say, n, the number of operations performed may depend on the particular input. How, then, are the results of the analysis of an algorithm to be expressed? Most often we describe the behavior of an algorithm by stating its *worst-case complexity*. Let D_n be the set of inputs of size n for the problem under consideration, and let I be an element of D_n. Let $t(I)$ be the number of basic operations performed by the algorithm on input I. We define the function W by

$$W(n) = \max \{t(I) \mid I \in D_n\}.$$

$W(n)$ is the maximum number of basic operations performed by the algorithm on any input of size n. It is often not very difficult to compute. It is valuable because it gives an upper bound on the work done by the algorithm. The worst-case analysis could be used to help form an estimate for a time limit for a particular implementation of an algorithm. This is particularly useful in real-time applications. We will do worst-case analysis for most of the algorithms presented in this book. Unless otherwise stated, whenever we refer to the amount of work done by an algorithm, we will mean the amount of work done in the worst case.

It may seem that a more useful and natural way to describe the behavior of an algorithm is to tell how much work it does on the average; that is, to compute the number of operations performed for each input of size n and then take the average. In practice some inputs may occur much more frequently than others so a weighted average is more meaningful. Let $p(I)$ be the probability that input I occurs. Then the average behavior of the algorithm is defined as

$$A(n) = \sum_{I \in D_n} p(I)t(I).$$

We determine $t(I)$ by analyzing the algorithm, but $p(I)$ cannot be computed analytically. The function p is determined from experience and/or special information about the application for which the algorithm is to be used, or by making some simplifying assumption, e.g., that all inputs of size n are equally likely to occur. If p is complicated, the computation of average behavior is difficult. Also, of course, if p depends on a particular application of the algorithm, the function A describes the average behavior of the algorithm for only that application.

The following examples illustrate worst-case and average analysis.

Example 1.3   Sequential search

Problem: Let L be an array containing n entries. Find an index of a specified entry x, if x is in the array; return 0 as the answer if x is not in the array.

Algorithm: See Algorithm 1.1.

Basic operation: Comparison of x with a list entry.

Worst-case analysis: Clearly $W(n) = n$. The worst cases occur when x appears only in the last position in the list and when x is not in the list at all. In both of these cases x is compared to all n entries.

Average-behavior analysis: We will make several simplifying assumptions first to do an easy example, then do a slightly more complicated analysis with different assumptions. First let us assume that the elements in the list are distinct, that we know for sure that x is in the list, and that x is equally likely to be in any particular position. The inputs can be categorized according to where in the list x appears, so there are n inputs to consider. For $1 \le i \le n$, let I_i represent the case where x appears in the ith position in the list. Then, let $t(I)$ be the number of comparisons done (the number of times the condition $L[index] \ne x$ in line 2 of the algorithm is tested) by the algorithm on input I. Clearly, for $1 \le i \le n$, $t(I_i) = i$. Thus,

$$A(n) = \sum_{i=1}^{n} p(I_i) t(I_i) = \sum_{i=1}^{n} \frac{1}{n} i$$

$$= \frac{1}{n} \sum_{i=1}^{n} i = \frac{1}{n} \frac{n(n+1)}{2} = \frac{n+1}{2}.$$

This should satisfy our intuition that on the average, about half the list will be searched.

Now let us allow the case where x is not in the list at all. We will still assume the elements are distinct. Now there are $n+1$ inputs to consider. For $1 \le i \le n$, I_i represents the case where x appears in the ith position in the list, and I_{n+1} represents the case where x is not in the list. Let q be the probability that x is in the list, and let us assume that each position in the list is equally likely. Then for $1 \le i \le n$, $p(I_i) = q/n$ and $p(I_{n+1}) = 1-q$. As before, for $1 \le i \le n$, $t(I_i) = i$; also, $t(I_{n+1}) = n$. Thus,

$$A(n) = \sum_{i=1}^{n+1} p(I_i) t(I_i) = \sum_{i=1}^{n} \frac{q}{n} i + (1-q)n$$

$$= \frac{q}{n} \sum_{i=1}^{n} i + (1-q)n = \frac{q}{n} \frac{n(n+1)}{2} + (1-q)n$$

$$= q \frac{(n+1)}{2} + (1-q)n.$$

If $q = 1$, that is, if x is in the list, then $A(n) = (n+1)/2$, as before. If $q = 1/2$, that is, if there is a 50–50 chance that x is not in the list, then $A(n) = (n+1)/4 + n/2$; roughly three-fourths of the entries are examined. ∎

Example 1.3 illustrates how we should interpret D_n, the set of inputs of size n. Rather than consider all possible lists of names, numbers, or whatever, that could occur as inputs, we identify the properties of the inputs that affect the behavior of the algorithm; in this case, whether x is in the list at all and, if so, where it appears. An element I in D_n may be thought of as a set (or equivalence class) of all lists and values for x such that x occurs in the specified place in the list (or not at all). Then $t(I)$ is the number of operations done for any one of the inputs in I.

Observe also that the input for which an algorithm behaves worst depends on the particular algorithm, not on the problem. For Algorithm 1.1 a worst case occurs when the only position in the list containing x is the last. For an algorithm that searched the list from the end (i.e., beginning with $index = n$), a worst case would occur if x appeared only in the first position. (Another worst case would occur when x is not in the list at all.)

Finally, Example 1.3 illustrates an assumption we often make when doing average analysis of sorting and searching algorithms: that the elements are distinct. The average analysis for the case of distinct elements gives a fair approximation for the average behavior in cases with few duplicates. If there may be many duplicates, it is harder to make reasonable assumptions about the probability that x's first appearance in the list occurs at any particular position.

Example 1.4   Matrix multiplication

Problem: Let $A = (a_{ij})$ and $B = (b_{ij})$ be two $n \times n$ matrices with real entries. Compute the product matrix $C = AB$. (This problem will be discussed much more thoroughly in Chapter 7.)

Algorithm: Use the algorithm implied by the definition of the matrix product:

$$c_{ij} = \sum_{k=1}^{n} a_{ik} b_{kj} \quad \text{for } 1 \le i,j \le n.$$

Algorithm 1.2   Matrix Multiplication

```
for i := 1 to n do
    for j := 1 to n do
        cij := 0;
        for k := 1 to n do cij := cij + aik bkj end
    end { for j }
end { for i }
```

Basic operation: Multiplication of matrix entries.

Analysis: To compute each entry of C, n multiplications are done. C has n^2 entries so

$$A(n) = W(n) = n^3. \qquad \blacksquare$$

Example 1.4 illustrates that for some algorithms the instructions performed, hence the amount of work done, are independent of the input. Thus if n is the input size, $A(n) = W(n) = $ the number of basic operations done in all cases. In other algorithms for the same problem, this may not be true.

The concepts of worst-case and average-behavior analysis would be useful even if we had chosen a different measure of work (say, execution time). The observation that the amount of work done often depends on the size and properties of the input would lead to the study of average behavior and worst-case behavior, no matter what measures were used.

1.3.5   Space Usage

The number of memory cells used by a program, like the number of seconds required to execute a program, depends on the particular implementation. However, some conclusions about space usage can be made just by examining an algorithm. A program will require storage space for the instructions, the constants and variables used by the program, and the input data. It may also use some workspace for manipulating the data and storing information needed to carry out its computations. The input data themselves may be representable in several forms, some of which require more space than others. If the input data have one natural form (say, an array of numbers or a matrix), then we analyze the amount of *extra* space used, aside from the program and the input. If the amount of extra space is constant with respect to the input size, the algorithm is said to work *in place*. This term is used especially in reference to sorting algorithms. If the input can be represented in various forms then we will consider the space required for the input itself as well as any extra space used. In general, we will refer to the number of "cells" used without precisely defining cells. The reader may think of a cell as being large enough to hold one number. If the amount of space used depends on the particular input, worst-case and average-case analysis can be done.

1.3.6   Simplicity

It is often, though not always, the case that the simplest and most straightforward way of solving a problem is not the most efficient. Yet simplicity in an algorithm is a desirable feature. It may make verifying the correctness of the algorithm easier, and it makes writing, debugging, and modifying a program easier. The time needed to produce a debugged program should be considered when choosing an algorithm, but if the program is to be used very often, its efficiency will probably be the determining factor in the choice.

1.3.7   Optimality

No matter how clever we are, we cannot improve an algorithm for a problem beyond a certain point. Each problem has inherent complexity; that is, there is some minimum amount of work required to solve it. To analyze the complexity of a problem, as opposed to that of a specific algorithm, we choose a class of algorithms (often by specifying the types of operations the algorithms will be permitted to perform) and a measure of complexity, for example, the basic operation(s) to be counted. Then we may ask how many operations are actually *needed* to solve the problem. We say that an algorithm is *optimal* (in the worst case) if there is no algorithm in the class under study that performs fewer basic operations (in the worst case). Note that when we speak of algorithms in the class under study, we do not mean only those algorithms that people have thought of. We mean all possible algorithms, including those not yet discovered. "Optimal" does not mean "the best known"; it means "the best possible."

1.3.8   Lower Bounds and the Complexity of Problems

Then how can we show that an algorithm is optimal? Do we have to analyze individually every other possible algorithm (including the ones we have not even thought of)? Fortunately, no; we can prove theorems that establish a lower bound on the number of operations needed to solve a problem. Then any algorithm that performs that number of operations would be optimal. Thus there are two tasks to be carried out in order to find a good algorithm, or, from another point of view, to answer the question: How much work is necessary and sufficient to solve the problem?

1.   Devise what seems to be an efficient algorithm; call it A. Analyze A and find a function W such that, for inputs of size n, A does at most $W(n)$ steps in the worst case.

2.   For some function F, prove a theorem stating that, for any algorithm in the class under consideration, there is some input of size n for which the algorithm must perform at least $F(n)$ steps.

If the functions W and F are equal, then the algorithm A is optimal (for the worst case). If not, there may be a better algorithm or a better lower bound. Observe that analysis of a specific algorithm gives an *upper bound* on the number of steps necessary to solve a problem, and a theorem of the type described in (2) gives a *lower bound* on the number of steps necessary (in the worst case). In this book, we will see problems for which optimal algorithms are known, and other problems for which there is still a gap between the best-known lower bound and the best-known algorithm. Simple examples of each case follow.

The concept of a lower bound for the worst-case behavior of algorithms is very important in computational complexity. Example 1.5 and the problems studied in Section 1.5 and Chapters 2 and 3 will help to clarify the meaning of lower bounds and illustrate techniques for establishing them. The reader should keep in mind that the definition "F is a lower bound for a class of algorithms" means that for *any* algorithm in the class, and any input size n, there is *some* input of size n for which the algorithm must perform *at least* $F(n)$ basic operations.

Example 1.5   Finding the largest entry in a list

Problem: Find the largest entry in a list of n numbers.

Class of algorithms: Algorithms that can compare and copy list entries, but do no other operations on them.

Basic operation: Comparison of two list entries.

Upper bound: Suppose the numbers are in an array L. The following algorithm finds the maximum.

Algorithm 1.3   FindMax

Input: L, an array of numbers; $n \geq 1$, the number of entries.

Output: *max*, the largest entry in L.

```
1.  max := L[1];
2.  for index := 2 to n do
3.      if max < L[index] then
4.          max := L[index]
5.      end { if };
6.  end { for }
```

Comparisons of list entries are done in line 3, which is executed exactly $n-1$ times. Thus $n-1$ is an upper bound on the number of comparisons necessary to find the maximum in the worst case. Is there an algorithm that does fewer?

Lower bound: To establish a lower bound we may assume that the entries in the list are all distinct. This assumption is permissible because, if we can establish a lower bound on worst-case behavior for some subset of inputs (lists with distinct entries), it is a lower bound on worst-case behavior when all valid inputs are considered.

In a list with n distinct entries, $n-1$ entries are *not* the maximum. We can conclude that a particular entry is not the maximum only if it is smaller than at least one other entry in the list. Hence, $n-1$ entries must be "losers" in comparisons done by the algorithm. Each comparison has only one loser, so at least $n-1$ comparisons must be done. Thus $F(n) = n-1$ is a lower bound on the number of comparisons needed.

Conclusion: Algorithm 1.3 is optimal.   ∎

We could take a slightly different point of view to establish the lower bound in Example 1.5. If we are given an algorithm and a list of n numbers such that the algorithm halts and produces an answer after doing fewer than $n-1$ comparisons, then we can prove that the algorithm gives the *wrong* answer for some set of input data. If no more than $n-2$ comparisons are done, two entries are never losers; that is, they are not known to be smaller than any other entries. The algorithm can specify at most one of them as the maximum. We can simply replace the other with a larger number (if necessary). Since the results of all comparisons done will be the same as before, the algorithm will give the same answer as before and it will be wrong. This argument is a proof by contradiction. It illustrates a useful technique for establishing lower bounds, namely, to show that, if an algorithm does not do enough work, one can arrange the input so that the algorithm gives the wrong answer.

Example 1.6   Matrix multiplication

Problem: Let $A = (a_{ij})$ and $B = (b_{ij})$ be two $n \times n$ matrices with real entries. Compute the product matrix $C = AB$.

Class of algorithms: Algorithms that can perform multiplications, divisions, additions, and subtractions on the matrix entries and on the intermediate results obtained by performing these operations on the entries.

Basic operation: Multiplication.

Upper bound: The usual algorithm (see Example 1.4) does n^3 multiplications; hence at most n^3 multiplications are necessary.

Lower bound: It has been proved that at least n^2 multiplications are necessary.

Conclusions: There is no way to tell from the information given whether or not the usual algorithm is optimal. Some researchers have been trying to improve the lower bound, that is, to prove that more than n^2 multiplications are necessary, while others have looked for better algorithms. To date it has been shown that the usual algorithm is *not* optimal; there is a method that does approximately $n^{2.376}$ multiplications. Is this method optimal? The lower bound has not yet been improved, so we do not know if there are algorithms that do substantially fewer multiplications.   ∎

Up to now we have been discussing lower bounds and optimality of worst-case behavior. What about average behavior? We can use the same approach that we use with worst-case behavior. Choose what seems to be a good algorithm and figure out the function A such that the algorithm does $A(n)$ operations, on the average, for inputs of size n. Then prove a theorem stating that any algorithm in the class being studied must perform at least $G(n)$ operations on the average for inputs of size n. If $A = G$ (or if they are approximately equal), we can say that the average behavior of the algorithm is optimal. If not, look for a better algorithm or a better lower bound (or both).

We can use the same approach to investigate space usage. Analyze a particular algorithm to get an upper bound on the amount of space needed, and prove a theorem to establish a lower bound. Can we find one algorithm for a given problem that is optimal with respect to both the amount of work done and the amount of space used? The answer to this question is: sometimes. For some problems, there is a trade-off between time and space.

1.3.9 Implementation and Programming

Implementation is the task of turning an algorithm into a computer program. Algorithms may be described by detailed computer-language-like instructions for manipulating variables and data structures, or by very abstract, high-level explanations in English of solution methods for abstract problems, making no mention of computer representations of the objects involved. Thus the implementation of an algorithm may be a fairly straightforward translating job or it may be a very lengthy and difficult job requiring a number of important decisions on the part of the programmer, particularly concerning the choice of data structures. Where appropriate, we will discuss implementation in the general sense of choosing data structures and describing ways to carry out instructions given in an English description of an algorithm. Such discussion is included for two reasons. One, it is a natural and important part of the process of producing a (good) working program. Two, consideration of implementation details is often necessary for analyzing an algorithm; the amount of time required for performing various operations on abstract objects such as sets and graphs depends on how these objects are represented. For example, forming the union of two sets may require only one or two operations if the sets are represented as linked lists, but would require a large number of operations, proportional to the number of elements in one of the sets, if they are represented as arrays and one must be copied into the other.

In the narrow sense, implementation, or simply programming, means converting a fairly detailed description of an algorithm and the data structures it uses into a program for a particular computer. Our analysis will be implementation independent in this sense; in other words, it will be independent of the computer and programming language used and of many minor details of the algorithm or program.

A programmer can refine the analysis of algorithms under consideration using information about the particular computer to be used. For example, if more than one operation is counted, the operations can be weighted according to their execution times, or estimates of the actual number of seconds a program will use (in the worst or average case) can be made. Sometimes knowledge of the computer used will lead to a new analysis. For example, if the computer has any unusual, powerful instructions that can be used effectively in the problem at hand, then one can study the class of algorithms that make use of those instructions and count them as the basic operations. If the computer has a very limited instruction set that makes implementation of the basic operation awkward, a different class of algorithms may be considered. Generally, however, if the implementation-independent analysis has been done well, then the program-dependent analysis should serve mainly just to add more detail.

A detailed analysis of the amount of space used by the algorithms being studied is, of course, also appropriate when particular implementations are being considered.

Any special knowledge about the inputs to the problem for which an algorithm is sought can be used to refine the analysis. If, for example, the inputs will be restricted to a certain subset of all possible inputs, a worst-case analysis can be done for that subset. As we have noted, a good average-behavior analysis depends on knowing the probability of the various inputs occurring.

1.4
Classifying Functions by Their Growth Rates

1.4.1   Definitions and Notation

Just how good is our measure of work done by an algorithm? How precise a comparison can we make between two algorithms? Because we are not counting every step executed by an algorithm, our analysis necessarily has some imprecision. We have said that we will be content if the total number of steps is roughly proportional to the number of basic operations counted. This is good enough for separating algorithms that do drastically different amounts of work for large inputs.

Suppose one algorithm for a problem does $2n$ basic operations, hence roughly $2cn$ operations in total, for some constant c, and another algorithm does $4.5n$ basic operations, or $4.5c'n$ in total. Which one runs faster? We really don't know. The first algorithm may do many more overhead operations; i.e., its constant of proportionality may be a lot higher. Thus if the functions describing the behavior of two algorithms differ by a constant factor, it may be pointless to try to distinguish between them (unless we do a more refined analysis). We consider such algorithms to be in the same complexity class.

Suppose one algorithm for a problem does $n^3/2$ multiplications, and another algorithm does $5n^2$. Which algorithm will run faster? For small values of n the first does fewer multiplications, but for large values of n, the second is better — even if it does more overhead operations. The rate of growth of a cubic function is so much greater than that of a quadratic function that the constant of proportionality does not matter when n gets large.

As these examples suggest, we want a way to compare or classify functions that ignores constant factors and small inputs. We get just such a classification by studying what is called the *asymptotic growth rate*, or, simply, the *order* of functions.

We will use the usual notation for natural numbers and real numbers; i.e.:

$\mathbf{N} = \{0, 1, 2, 3, ...\}$ $\mathbf{N}^+ = \{1, 2, 3, ...\}$

$\mathbf{R}$ = the set of real numbers $\mathbf{R}^+$ = the set of positive reals $\mathbf{R}^* = \mathbf{R}^+ \cup \{0\}$

Let f and g be functions from $\mathbf{N}$ to $\mathbf{R}^*$. Figure 1.4 informally describes the sets we use to show the relationships between the orders of functions. Keeping the picture and the informal definitions in mind will help clarify the following formal definitions and properties.[3]

---

[3] Readers who plan to consult other books and papers should be aware that the definitions of O, Θ, Ω, and o are not yet standardized; variations may be encountered.

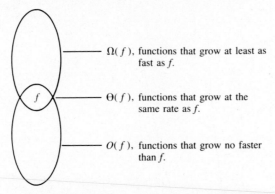

Figure 1.4 Big omega, big theta, and big oh.

Definition Let $f : \mathbf{N} \rightarrow \mathbf{R}^*$. $O(f)$ is the set of functions $g : \mathbf{N} \rightarrow \mathbf{R}^*$ such that for some $c \in \mathbf{R}^+$ and some $n_0 \in \mathbf{N}$, $g(n) \leq cf(n)$ for all $n \geq n_0$.

Notice that a function g may be in $O(f)$ even if $g(n) > f(n)$ for all n. The important point is that g is bounded by some constant multiple of f. Also, the relation between f and g for small values of n is not considered. Figure 1.5 shows the order relations for a few functions. (Note that the functions in Fig. 1.5 are drawn as continuous functions defined on $\mathbf{R}^+$ or $\mathbf{R}^*$. The functions that describe the behavior of most of the algorithms we will study have such natural extensions.)

The set $O(f)$ is usually called "big oh of f" or just "oh of f" although the "oh" is actually the Greek letter omicron. And, although we have defined $O(f)$ as a set, it is common practice to say "g is oh of f," rather than "g is a member of oh of f."

There is an alternative technique for showing that g is in $O(f)$:

$$g \in O(f) \text{ if } \lim_{n \to \infty} \frac{g(n)}{f(n)} = c, \text{ for some } c \in \mathbf{R}^*.$$

That is, if the limit of the ratio of g to f exists and is not ∞, then g grows no faster than f. If the limit is ∞, then g does grow faster than f.

The following theorem is useful for computing limits when f and g extend to continuous, differentiable functions on $\mathbf{R}^*$.

Theorem 1.4 L'Hôpital's Rule

If $\lim_{n \to \infty} f(n) = \lim_{n \to \infty} g(n) = \infty$, then $\lim_{n \to \infty} \frac{f(n)}{g(n)} = \lim_{n \to \infty} \frac{f'(n)}{g'(n)}$, if the derivatives f' and g' exist.

Example 1.7

Let $f(n) = n^3/2$ and $g(n) = 37n^2 + 120n + 17$. We will show that $g \in O(f)$, but $f \notin O(g)$.

Since for $n \geq 78$, $g(n) < 1f(n)$, it follows that $g \in O(f)$.

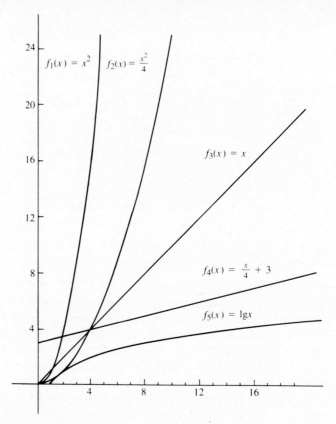

Figure 1.5   The orders of functions. $f_3 \in O(f_4)$, even though $f_3(x) > f_4(x)$ for $x > 4$, since both are linear. f_1 and f_2 are of the same order. They grow faster than the other three functions. f_5 is of the lowest order among the functions shown.

We could have come to the same conclusion from:

$$\lim_{n \to \infty} \frac{g(n)}{f(n)} = \lim_{n \to \infty} \frac{37n^2 + 120n + 17}{n^3/2} = \lim_{n \to \infty} (74/n + 240/n^2 + 34/n^3) = 0.$$

We can show that $f \notin O(g)$ by observing that the limit of f/g is infinity. Here is an alternative method. We assume $f \in O(g)$ and derive a contradiction. If $f \in O(g)$, then there exist c and n_0 such that for all $n \geq n_0$,

$$\frac{n^3}{2} \leq 37cn^2 + 120cn + 17c.$$

So

$$\frac{n}{2} \leq 37c + \frac{120c}{n} + \frac{17c}{n^2} \leq 174c.$$

Since c is a constant and n may be arbitrarily large, it is impossible to have $n/2 \leq 174c$. ∎

Example 1.8

Let $f(n) = n^2$ and $g(n) = n \lg n$. We will show that $g \in O(f)$, but $f \notin O(g)$.

$$\lim_{n \to \infty} \frac{g(n)}{f(n)} = \lim_{n \to \infty} \frac{n \lg n}{n^2} = \lim_{n \to \infty} \frac{\lg n}{n}$$

$$= \text{(using L'Hôpital's Rule)} \lim_{n \to \infty} \frac{\lg e / n}{1} = \lim_{n \to \infty} \frac{\lg e}{n} = 0$$

Therefore, $g \in O(f)$. However, since

$$\lim_{n \to \infty} \frac{f(n)}{g(n)} = \infty,$$

$f \notin O(g)$. ∎

The definition of $\Omega(f)$, the set of functions that grow at least as fast as f, is the dual of the definition of $O(f)$.

Definition Let $f : \mathbf{N} \to \mathbf{R}^*$. $\Omega(f)$ is the set of functions $g : \mathbf{N} \to \mathbf{R}^*$ such that for some $c \in \mathbf{R}^+$ and some $n_0 \in \mathbf{N}$, $g(n) \geq cf(n)$ for all $n \geq n_0$.

The alternative technique for showing that g is in $\Omega(f)$ is:

$$g \in \Omega(f) \text{ if } \lim_{n \to \infty} \frac{g(n)}{f(n)} = \infty \text{ or if } \lim_{n \to \infty} \frac{g(n)}{f(n)} = c > 0$$

(if the limit exists).

Definition Let $f : \mathbf{N} \to \mathbf{R}^*$. $\Theta(f) = O(f) \cap \Omega(f)$. (The most common way of reading "$g \in \Theta(f)$" is "g is order f.")

We also have:

$$g \in \Theta(f) \text{ if } \lim_{n \to \infty} \frac{g(n)}{f(n)} = c, \text{ for some } c \in \mathbf{R}^+.$$

That is, $c \neq 0$ and $c \neq \infty$.

Example 1.9

The worst-case complexities of Algorithm 1.1 (sequential search) and Algorithm 1.3 (finding the maximum element) are both in $\Theta(n)$. The complexity (worst case or average) of Algorithm 1.2 for matrix multiplication is in $\Theta(n^3)$. ∎

The terminology commonly used in talking about the order sets is imprecise. For example: "This is an order n^2 algorithm" really means that the function describing the behavior of the algorithm is in $\Theta(n^2)$.

Exercise   Show that $n(n-1)/2 \in \Theta(n^2)$.

Sometimes we wish to indicate that one function has strictly smaller asymptotic growth than another. We can use the following definition.

Definition $o(f)$ is the set of functions $g : \mathbf{N} \to \mathbf{R}^*$ such that $\lim\limits_{n \to \infty} g(n)/f(n) = 0$

1.4.2   How Important Is Order?

Table 1.1[4] shows the running times for several actual algorithms for the same problem. (The last column does not correspond to an algorithm for the problem; it is included to demonstrate how fast exponential functions grow, and hence how bad exponential algorithms are.) Look over the entries in the table to see how fast the running time increases with input size for the algorithms of higher complexities. One of the important lessons in the table is that the high constant factors on the $\Theta(n)$ and $\Theta(n\lg n)$ algorithms do not make them slower than the other algorithms except for very small inputs.

The second part of the table looks at the effect of asymptotic growth on the increase in the size of the input that can be handled with more computer time (or by using a faster computer). It is *not* true in general that if we multiply the time (or speed) by 60 we can handle an input 60 times as large; that is true only for algorithms whose complexity is in $O(n)$. The $\Theta(n^2)$ algorithm, for example, can handle an input only $\sqrt{60}$ times as large.

To further drive home the point that the order of the running time of an algorithm is more important than a constant factor (for large inputs), look at Table 1.2.[5]

Table 1.1
How functions grow.

	Algorithm	1	2	3	4	
Time (in μsecs.)		$33n$	$46n\lg n$	$13n^2$	$3.4n^3$	2^n
Time to solve for input size	$n=10$	.00033 sec.	.0015 sec.	.0013 sec.	.0034 sec.	.001 sec.
	$n=100$	.003 sec.	.03 sec.	.13 sec.	3.4 sec.	4×10^{14} centuries
	$n=1000$	.033 sec.	.45 sec.	13 sec.	.94 hrs.	
	$n=10,000$	.33 sec.	6.1 sec.	22 min.	39 days	
	$n=100,000$	3.3sec.	1.3 min.	1.5 days	108 years	
Approx. maximum input size in	1 sec. 1 min.	30,000 1,800,000	2000 82,000	280 2200	67 260	20 26

---

[4] This table (except the last column) is adapted from *Programming Pearls* by Jon Bentley (Addison-Wesley, Reading, Mass., 1986) and is reproduced here with permission.
[5] This table is also from *Programming Pearls* by Jon Bentley and is reproduced here with permission.

Table 1.2
Order wins out.

n	Cray-1 Fortran $3n^3$ nanosec.	TRS-80 Basic 19,500,000n nanosec.
10	3 microsec.	200 millisec.
100	3 millisec.	2 sec.
1000	3 sec.	20 sec.
2500	50 sec.	50 sec.
10,000	49 min.	3.2 min.
1,000,000	95 years	5.4 hours

Cray-1 is a trademark of Cray Research, Inc.
TRS-80 is a trademark of Tandy Corporation.

A program for the cubic algorithm from Table 1.1 was written for the Cray-1 super-computer; it ran in $3n^3$ nanoseconds for input of size n. The linear algorithm was programmed on a TRS-80; it ran in $19.5n$ milliseconds (which is $19,500,000n$ nanoseconds). Even though the constant on the linear algorithm is 6.5 million times as big as the constant on the cubic algorithm, the linear algorithm is faster for input sizes $n \geq 2500$. (Whether one considers this a large or small input size would depend on the context of the problem.)

If we focus on the order of functions (thus including, say, n and $1,000,000n$ in the same class), then when we can show that two functions are *not* of the same order, we are making a strong statement about the difference between the algorithms described by those functions. If two functions *are* of the same order, they may differ by a large constant factor. The constant, though, is irrelevant to the effects of improved computer speed on the maximum input size an algorithm can handle in a given amount of time. That is, the constant is irrelevant to the increase between the last two rows of Table 1.1. Let us look a little more closely at the meaning of those numbers.

Suppose we fix on a certain amount of time (one second, one minute — the specific choice is unimportant). Let s be the maximum input size a particular algorithm can handle within that amount of time. Now suppose we allow t times as much time (or our computer speed increases by a factor of t, either because technology has improved, or simply because we went out and bought a more expensive machine). Table 1.3 shows the effect of the speedup for several complexities.

The values in the third column are computed by observing that

$f(s_{new})$ = number of steps after speedup

$\quad = t$ times the number of steps before the speedup $= tf(s)$

and solving

$$f(s_{new}) = tf(s)$$

for s_{new}.

Table 1.3
Effect of increased computer speed on maximum input size.

Number of steps performed on input of size n $f(n)$	Maximum feasible input size s	Maximum feasible input size in t times as much time s_{new}
$\lg n$	s_1	s_1^t
n	s_2	ts_2
n^2	s_3	$\sqrt{t}s_3$
2^n	s_4	$s_4 + \lg t$

Now, if we multiply the functions in the first column by some constant c, the entries in the third column will not change! This is what we meant by saying that the constant is irrelevant to the effect of increased computer time (or speed) on the maximum input size an algorithm can handle.

1.4.3 Properties of O, Ω, and Θ

The order sets have a number of useful properties. Most of the proofs are left as exercises; they follow easily from the definitions. For all the properties, assume that $f, g, h : \mathbf{N} \rightarrow \mathbf{R}^*$.

1. Transitivity: if $f \in O(g)$ and $g \in O(h)$, then $f \in O(h)$.

    Proof. Let c_1 and n_1 be such that $f(n) \le c_1 g(n)$ for all $n \ge n_1$, and let c_2 and n_2 be such that $g(n) \le c_2 h(n)$ for all $n \ge n_2$. Then for all $n \ge \max(n_1, n_2)$, $f(n) \le c_1 c_2 h(n)$. So $f \in O(h)$. (The transitivity property also holds for Ω, Θ, and o.) □

2. $f \in O(g)$ if and only if $g \in \Omega(f)$.
3. If $f \in \Theta(g)$, then $g \in \Theta(f)$.
4. Θ defines an equivalence relation on the functions; each set $\Theta(f)$ is an equivalence class, which we call a complexity class.
5. $O(f+g) = O(\max\{f,g\})$. (This is also true for Ω and Θ. It is useful when analyzing complex algorithms, where f and g may describe the work done by different parts of the algorithm.)

Since Θ defines an equivalence relation, we can indicate the complexity class of an algorithm by specifying any function in the class. We usually choose the simplest representative. Thus if the number of steps carried out by an algorithm is described by the function $f(n) = n^3/6 + n^2 + 2\lg n + 12$, we say simply that the complexity of the algorithm is in $\Theta(n^3)$. If $f \in \Theta(n)$, we say that f is linear; if $f \in \Theta(n^2)$, we say f is quadratic; and if $f \in \Theta(n^3)$, f is cubic.[6] $O(1)$ denotes the set of functions bounded by a constant (for large n).

---

[6] Note that the terms *linear*, *quadratic*, and *cubic* are used somewhat more loosely here than they usually are used by mathematicians.

Here are two useful theorems. The proofs use the techniques presented in the section on definitions and notation, and are not hard; they are left for exercises.

Theorem 1.5 $\lg n$ is in $o(n^\alpha)$ for any $\alpha > 0$. That is, the log function grows more slowly than any power of n (including fractional powers).

Theorem 1.6 n^k is in $o(2^n)$ for any $k > 0$. That is, powers of n grow more slowly than the exponential function 2^n. (In fact, powers of n grow more slowly than any exponential function c^n where $c > 1$.)

1.5
Searching an Ordered List

1.5.1   The Problem and Some Solutions

To illustrate the ideas presented in the previous sections, we will study a familiar problem.

Given an array L containing n entries sorted in nondecreasing order, and given a value x, find an index of x in the list or, if x is not in the list, return 0 as the answer.

Let us pretend for the moment that we do not know the Binary Search algorithm; we approach the problem as if for the first time. We will consider various algorithms, analyze worst-case and average behavior, and finally consider Binary Search and show that it is optimal by establishing a lower bound on the number of comparisons needed.

Observe that the Sequential Search algorithm (Algorithm 1.1) solves the problem but makes no use of the fact that the entries in the list are in order. Can we modify that algorithm so that it uses the added information and does less work? The first improvement is prompted by the observation that, since the array is in nondecreasing order, as soon as an entry larger than x is encountered, the algorithm can terminate with the answer 0. How does this change affect the analysis? Clearly the modified algorithm is better in some cases; it will terminate sooner for some inputs. The worst-case complexity, however, remains unchanged. If x is the last entry in the list or if x is larger than all the entries, then the algorithm will do n comparisons. For the average analysis of the modified algorithm, we must know how likely it is that x is *between* any two list entries. Suppose we define a *gap*, g_i, to be the set of values y such that $L[i-1] < y < L[i]$ for $i = 2,\ldots,n$. Also, let g_1 be all values less than $L[1]$ and g_{n+1} all values greater than $L[n]$. We'll assume, as we did in Example 1.3, that there is a 50–50 chance that x is in the list; that, if it is, all positions in the list are equally likely (so have probability $1/2n$; and that if x is not in the list all gaps are equally likely (i.e., have probability $1/2(n+1)$). For $1 \le i \le n$, it takes i comparisons to determine that $x = L[i]$ or that x is in g_i, and it takes n comparisons to determine that x is in g_{n+1}. So we compute the average number of comparisons as follows:

$$A(n) = \sum_{i=1}^{n} \frac{1}{2n}i + \sum_{i=1}^{n} \frac{1}{2(n+1)}i + \frac{1}{2(n+1)}n.$$

The first term corresponds to cases in which x is in the list, and the other two terms correspond to cases in which x is not in the list. Evaluating the sums is easy and left as an exercise. The result is that $A(n)$ is roughly $n/2$. Algorithm 1.1 did $3n/4$ comparisons on the average, so the modified algorithm is an improvement, though its average behavior is still linear.

Let us try again. Can we find an algorithm that does substantially fewer than n comparisons in the worst case? Suppose we compare x to, say, every fourth entry in the list. If there is a match, we are done. If x is larger than the entry to which it is compared, say $L[i]$, then the three entries preceding $L[i]$ need not be examined explicitly. If $x < L[i]$, then x is between the last two entries to which it was compared. A few more comparisons (how many?) will suffice to determine the position of x if it is in the list or to determine that it is not there. The details of the algorithm and the analysis are left for the reader, but it is easy to see that only about one-fourth of the entries in the list are examined. Thus in the worst case approximately $n/4$ comparisons are done.

We could pursue the same scheme, choosing a large value for k and designing an algorithm that compares x to every kth entry, hence allowing us to eliminate from consideration $k-1$ keys at each comparison as we proceed through the list. Thus we do roughly n/k comparisons to locate a small section of L that may contain x. But for any fixed k the algorithm will still be linear. Can we do better?

The idea of the well-known Binary Search algorithm is to eliminate half the entries with each comparison. Instead of choosing a particular integer k and comparing x to every kth entry, we compare x first to the entry in the middle of the list. If x is larger, then, if it is in the list at all, it is in the second half; with one comparison the entire first half of the list is eliminated from consideration. Conversely, if x is smaller than the entry in the middle of the list, the second half of the list is eliminated from consideration. (Of course, if x is equal to the middle entry, there is nothing more to do.) Until x is found or it is determined that x is not in the list, compare x to the middle entry in the section of the list under consideration. After each comparison, the size of the section of the list that may contain x is cut in half.

Algorithm 1.4   Binary Search

Input: L, $n \geq 0$, and x, where L is an ordered array with n entries and x is the item sought.

Output: *index* such that $L[index] = x$ if x is in L and *index* = 0 if x is not in L.

1. *first* := 1; *last* := *n*;
   { *first* and *last* are the indexes of the first and last entries,
   respectively, of the section of the array currently being searched. }
2. *found* := *false*;

```
3.    while first ≤ last and not found do
4.        index := ⌊(first+last)/2⌋;      { index of middle entry }
5.        if x = L[index] then found := true
6.        elsif x < L[index] then last := index−1
7.        else first := index+1
8.        end { if }
9.    end { while };
10.   if not found then index := 0 end
```

1.5.2   Worst-Case Analysis of Binary Search

A reasonable choice of basic operation for the Binary Search algorithm is a comparison of x to a list entry. Let $W(n)$ be the number of such comparisons performed by the algorithm in the worst case on lists with n entries. It is usual to assume that one comparison with a three-way branch is done for the tests on x in lines 5 and 6; this is a reasonable assumption for an assembly language implementation. Thus $W(n)$ is also the number of passes through the **while** loop. Suppose $n > 1$. The first time line 3 is encountered, the task of the algorithm is to find x in a list of n entries indexed from $first = 1$ to $last = n$. It proceeds to line 5 and compares x to $L[\lfloor(1+n)/2\rfloor]$. In the worst case these keys are not equal and either $first$ or $last$ will be changed so that, on the next pass through the loop, the task is to find x in the section of the list indexed from $first$ to $last$, inclusive. How many entries are there in this section? If n is even, there are $n/2$ entries in the section of the list following $L[\lfloor(1+n)/2\rfloor]$ and $(n/2)−1$ entries in the section preceding it. If n is odd, there are $(n−1)/2$ entries in both sections. Hence, there are at most $\lfloor n/2 \rfloor$ entries in the section of the list in which the algorithm will look for x on the next pass through the loop. The number of comparisons done by the algorithm beginning with the second pass through the loop is at most the number it would do if the input were a list with $\lfloor n/2 \rfloor$ entries. Thus $W(n) = 1+W(\lfloor n/2 \rfloor)$. This is an example of a *recurrence relation*. A recurrence relation for a function is an equation relating the value of the function on the argument n to values of the function on smaller arguments. They occur often in the analysis of algorithms (especially recursive algorithms). The recurrence relation alone does not provide enough information to determine W, but W could be evaluated for arbitrary n if its value on the smallest argument in its domain were known. It is easy to see from the algorithm that $W(0)=0$. We will also compute $W(1)$.

If $n = 1$, then $first = last = index = 1$ and x is compared to $L[1]$. If $x \neq L[1]$, then either $last$ will be assigned 0 or $first$ will be assigned 2, and the instructions in the loop (in particular, the comparison) are not executed again. Thus, whether or not x is in the list, one comparison is done, so $W(1) = 1$. Thus what is known about W can be expressed in two equations:

$$W(n) = 1+W(\lfloor n/2 \rfloor) \quad \text{for } n > 1$$

$$W(1) = 1.$$

When a function is described by a recurrence relation, an equation giving the value of the function for a particular argument is called a *boundary condition*. The equation $W(1) = 1$ is a boundary condition for W.

Expanding the recurrence relation a few times should enable us to make a good guess at a formula describing $W(n)$ as a function of n without referring to other values of W. Such a formula is called a *closed form*. Expanding the recurrence relation gives

$$W(n) = 1 + W(\lfloor n/2 \rfloor) = 1 + 1 + W(\lfloor n/2^2 \rfloor)$$
$$= 1 + 1 + 1 + W(\lfloor n/2^3 \rfloor).$$

Each time the argument is divided by 2, 1 is added to the value of the function. Thus there will be approximately $\lg n$ terms in the sum, each equal to 1, but $\lg n$ may not be an integer, so a slight adjustment is necessary. The exact formula is established by induction on n.

Lemma 1.7 $W(n) = \lfloor \lg n \rfloor + 1$, for $n \geq 1$.

Proof, by induction on n. For the basis of the induction, let $n = 1$. Then we see that

$$\lfloor \lg n \rfloor + 1 = \lfloor \lg 1 \rfloor + 1 = 0 + 1 = 1 = W(1)$$

by the boundary condition. For the induction step, assume that $n > 1$ and that for $1 \leq k < n$, $W(k) = \lfloor \lg k \rfloor + 1$. Then

$$W(n) = 1 + W(\lfloor n/2 \rfloor) \quad \text{(by the recurrence relation)}$$
$$= 1 + \lfloor \lg \lfloor n/2 \rfloor \rfloor + 1 \quad \text{(by the induction hypothesis)}$$
$$= 2 + \lfloor \lg \lfloor n/2 \rfloor \rfloor$$
$$= \begin{cases} 2 + \lfloor \lg n - 1 \rfloor & \text{if } n \text{ is even, since } \lfloor n/2 \rfloor = n/2 \\ 2 + \lfloor \lg(n-1) - 1 \rfloor & \text{if } n \text{ is odd, since } \lfloor n/2 \rfloor = (n-1)/2 \end{cases}$$
$$= \begin{cases} 1 + \lfloor \lg n \rfloor & \text{if } n \text{ is even} \\ 1 + \lfloor \lg(n-1) \rfloor & \text{if } n \text{ is odd.} \end{cases}$$

If n is odd $\lfloor \lg n \rfloor = \lfloor \lg(n-1) \rfloor$, so in all cases $W(n) = 1 + \lfloor \lg n \rfloor$. ☐

Theorem 1.8 The Binary Search algorithm does $\lfloor \lg n \rfloor + 1$ comparisons of x with list entries in the worst case (where $n \geq 1$ is the number of list entries). Since one comparison is done on each loop iteration, the running time is in $\Theta(\lg n)$.

Binary Search does fewer comparisons in the worst case than a sequential search does on the average.

1.5.3 Average-Behavior Analysis

To simplify the analysis a little, we will assume that x appears in at most one place in the list. As we observed at the beginning of this section, there are $2n+1$ positions that x may occupy: the n positions in L and the $n+1$ gaps. For $1 \leq i \leq n$, let I_i represent all inputs for which $x = L[i]$. For $2 \leq i \leq n$, let I_{n+i} represent inputs for which $L[i-1] < x < L[i]$. I_{n+1} and I_{2n+1} represent inputs where $x < L[1]$ and $x > L[n]$, respectively. Let $t(I_i)$ be the number of comparisons of x with list entries done by Algorithm 1.4 on input I_i. Table 1.4 shows the values of t for $n = 25$. Observe that most inputs are worst cases; that is, it takes five comparisons to find x most of the time. So if we assume that all positions (including gaps) are equally likely, it is not unreasonable to expect the number of comparisons done on the average to be close to $\lg n$. Computation of the average yields $223/51$, or approximately 4.37, and $\lg 25 \approx 4.65$.

We will derive an approximate formula for the number of comparisons done on the average, given two assumptions:

1. All positions (including gaps) are equally likely, so for $1 \leq i \leq 2n+1$, $p(I_i) = 1/(2n+1)$.

2. $n = 2^k - 1$, for some integer $k \geq 1$.

The second assumption is made to simplify the analysis. The result for all values of n is very close to the result that we will obtain.

Observe that $k = \lfloor \lg n \rfloor + 1$, the number of comparisons done in the worst case. For $1 \leq t \leq k$, let s_t be the number of inputs for which the algorithm does t comparisons. For example, for $n = 25$, $s_3 = 4$ because three comparisons would be done for each of the four inputs I_3, I_9, I_{16}, and I_{22}. It is easy to see that $s_1 = 1 = 2^0$, $s_2 = 2 = 2^1$, $s_3 = 4 = 2^2$, and in general for $t < k$, $s_t = 2^{t-1}$. The algorithm does k comparisons if x is in any of 2^{k-1} positions in the list and if x is in any of the $n+1$ gaps,

Table 1.4
The number of comparisons done by Binary Search depending on the location of x; $n = 25$.

i	$t(I_i)$		i	$t(I_i)$
1	4		14	4
2	5		15	5
3	3		16	3
4	4		17	4
5	5		18	5
6	2		19	2
7	4		20	4
8	5		21	5
9	3		22	3
10	5		23	5
11	4		24	4
12	5		25	5
13	1		26,29,32,39,42,45	4
		gaps	all other gaps	5

so $s_k = 2^{k-1} + n + 1$. (If we did not assume $n = 2^k - 1$, only $k-1$ comparisons might be done for some of the gaps; see Table 1.4 for $n = 25$.) The average number of comparisons done is

$$A(n) = \frac{1}{2n+1} \sum_{t=1}^{k} t s_t = \frac{1}{2n+1} \left[\sum_{t=1}^{k} t 2^{t-1} + k(n+1) \right].$$

Using Eq. 1.5 we can conclude that

$$\sum_{t=1}^{k} t 2^{t-1} = \frac{1}{2} \sum_{t=1}^{k} t 2^{t} = (k-1)2^k + 1.$$

So, since $n = 2^k - 1$,

$$A(n) = \frac{(k-1)2^k + 1}{2n+1} + \frac{k(n+1)}{2n+1}$$

$$= \frac{(k-1)2^k + 1}{2^{k+1} - 1} + \frac{k 2^k}{2^{k+1} - 1}$$

$$= \frac{(2k-1)2^k + 1}{2^{k+1} - 1}$$

$$\approx \frac{2k-1}{2} = k - \frac{1}{2} = \lfloor \lg n \rfloor + \frac{1}{2}.$$

Thus we have proved the following theorem.

Theorem 1.9   Binary search (Algorithm 1.4) does approximately $\lfloor \lg n \rfloor + 1/2$ comparisons on the average for lists with n entries.

1.5.4   Optimality

We will show that the binary search algorithm is optimal in the class of algorithms that can do no other operations on the list entries except comparisons. We will establish a lower bound on the number of comparisons needed by examining *decision trees* for search algorithms in this class. Let A be such an algorithm. A decision tree for A and a given input size n is a binary tree whose nodes are labeled with numbers between 1 and n and are arranged according to the following rules:

1.   The root of the tree is labeled with the index of the first entry in the list to which the algorithm A compares x.

2.   Suppose the label on a particular node is i. Then the label on the left child of that node is the index of the entry to which the algorithm will compare x next if $x < L[i]$. The label on the right child is the index of the entry to which the algorithm will compare x next if $x > L[i]$. The node does not have a left (or right) child if the algorithm halts after comparing x to $L[i]$ and discovering that $x < L[i]$ (or $x > L[i]$). There is no branch for the case $x = L[i]$. A reasonable algorithm would do no more comparisons in that case.

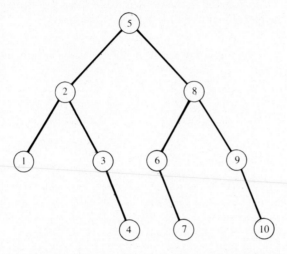

Figure 1.6   Decision tree for the binary search algorithm with $n = 10$.

The class of algorithms that can be modeled by such decision trees is very broad; it includes sequential search and the variations considered at the beginning of this section. Figure 1.6 shows the decision tree for the binary search algorithm with $n = 10$.

Given a particular input, algorithm A will perform the comparisons indicated along one path beginning at the root of its decision tree. The number of comparisons performed is the number of nodes on the path. The number of comparisons performed in the worst case is the number of nodes on the longest path from the root to a leaf; i.e., the depth of the tree plus one. Thus to establish a lower bound on the number of comparisons needed in the worst case, we establish a lower bound on the depth of the decision tree for A.

Let d be the depth of the decision tree for A, and let N be the number of nodes in the tree. We want to relate the depth to n, the input size. By Lemma 1.3, $d \geq \lfloor \lg N \rfloor$. We claim that $N \geq n$. Suppose, to the contrary, that there is no node labeled i, for some i between 1 and n. We can make up two input lists $L1$ and $L2$ such that for $1 \leq j \leq n$ and $j \neq i$, $L1[j] = L2[j] \neq x$ and $L1[i] = x$ but $L2[i] \neq x$. Since no node in the decision tree is labeled i, the algorithm A never compares x to $L1[i]$ or $L2[i]$. It behaves the same way on both inputs since their other entries are identical, and it must give the same output for both. Thus A gives the wrong output for at least one of the lists, and it is not a correct algorithm. We conclude that the tree has at least n nodes. So $d \geq \lfloor \lg N \rfloor \geq \lfloor \lg n \rfloor$. Recall that the number of comparisons done by A in the worst case is at least $d+1$, hence at least $\lfloor \lg n \rfloor +1$. Since A was an arbitrary algorithm from the class of algorithms considered, we have proved the following theorem.

Theorem 1.10   Any algorithm to find x in a list of n entries (by comparing x to list entries) must do at least $\lfloor \lg n \rfloor + 1$ comparisons for some input.

Corollary 1.11   Since Algorithm 1.4 does $\lfloor \lg n \rfloor + 1$ comparisons in the worst case, it is optimal.

Exercises

Section 1.2: The Algorithm Language, Mathematics, and Data Structures

1.1. Prove Property 7 of logarithms.

1.2. Prove Property 8 of logarithms.

1.3. Suppose three coins are lying on a table. One coin is chosen at random and flipped. What is the probability that after the flip the majority of the coins (that is, two or three of them) will have "heads" up if initially the sides facing up were

a) heads, tails, tails?
b) tails, tails, tails?
c) heads, heads, tails?

1.4. Consider four dice containing the numbers indicated below. For each pair of dice, say D_i and D_j with $1 \le i,\ j \le 4$ and $i \ne j$, compute the probability that on a fair toss of the two dice, the top face of D_i will show a higher number than the top face of D_j. (Show the results in a 4×4 matrix.)

$$D_1:\ 1, 2, 3, 9, 10, 11$$
$$D_2:\ 0, 1, 7, 8, 8, 9$$
$$D_3:\ 5, 5, 6, 6, 7, 7$$
$$D_4:\ 3, 4, 4, 5, 11, 12$$

(If you do the computation correctly and study the results carefully, you will discover that these dice have a surprising property. If you and another player were gambling on who throws the higher number, and you chose your die first, the other player could always choose a die with a high probability of beating yours. These dice are discussed in Gardner (1983), where their discovery is attributed to B. Efron.)

1.5. Give a formula for $\sum_{i=a}^{n} i$ where a is an integer between 1 and n.

1.6. Prove Eq. 1.2.

1.7. Prove Lemmas 1.1, 1.2, and 1.3.

Section 1.3: Analyzing Algorithms

1.8. Give a formula for the total number of operations done by the Sequential Search algorithm (Algorithm 1.1) in the worst case for a list with n entries. Count comparisons of x with list entries, comparisons with the variable *index*, additions, and assignments to *index*.

1.9. Write the sequential search algorithm (Algorithm 1.1) in an assembly language and find a formula for the exact number of statements executed in the worst case for a list with n entries.

1.10. a) Write an algorithm to find the median of three distinct integers a, b, and c.

b) Describe D, the set of inputs for your algorithm, in light of the discussion following Example 1.3.

c) How many comparisons does your algorithm do in the worst case? On the average?

d) How many comparisons are necessary in the worst case to find the median of three numbers? Justify your answer.

1.11. Write an algorithm to find the second-largest element in a list containing n entries. How many comparisons of list entries does your algorithm do in the worst case? (It is possible to do better than $2n-3$; we will consider this problem again.)

1.12. Write an algorithm to find both the smallest and largest elements in a list of n entries. Try to find a method that does at most roughly $1.5n$ comparisons of list entries.

1.13. Suppose the following algorithm is used to evaluate the polynomial

$$p(x) = a_n x^n + a_{n-1}x^{n-1} + \cdots + a_1 x + a_0.$$

$p := a_0;$
$xpower := 1;$
for $i := 1$ **to** n **do**
    $xpower := x*xpower;$
    $p := p + a_i *xpower$
end

a) How many multiplications are done in the worst case? How many additions?

b) How many multiplications are done on the average?

c) Can you improve on this algorithm? (We will consider this problem again.)

Section 1.4: Classifying Functions by Their Growth Rates

1.14. Let $p(n) = a_k n^k + a_{k-1}n^{k-1} + \cdots + a_1 n + a_0$ be a polynomial in n of degree k with $a_k > 0$. Prove that $p(n)$ is in $\Theta(n^k)$.

1.15. Let α and β be real numbers such that $0 < \alpha < \beta$. Show that n^α is in $O(n^\beta)$ but n^β is not in $O(n^\alpha)$.

*1.16. List the functions below from lowest order to highest order. If any two (or more) are of the same order, indicate which.

n	2^n	$n\lg n$	$\ln n$
$n - n^3 + 7n^5$	$\lg n$	$\sqrt{n}$	e^n
$n^2 + \lg n$	n^2	2^{n-1}	$\lg\lg n$
n^3	$(\lg n)^2$	$n!$	$n^{1+\epsilon}$ where $0 < \epsilon < 1$

*1.17. True or false: For any positive constant c, $f(cn) \in \Theta(f(n))$? (Hint: Consider some of the fast-growing functions listed in the preceding problem.)

*1.18. Prove Properties 2–5 in Section 1.4.3.

1.19. Prove Theorem 1.5.

*1.20. Prove Theorem 1.6.

*1.21. Show that the values in the third column of the speedup table (Table 1.3) are unchanged when we replace any function $f(n)$ in the first column by $cf(n)$ for any positive constant c.

*1.22. Give an example of two functions f, g: $\mathbf{N} \rightarrow \mathbf{R}^*$, such that $f \notin O(g)$ and $g \notin O(f)$.

1.23. Prove or disprove:

$$\sum_{i=1}^{n} i^2 \in \Theta(n^2)$$

Section 1.5: Searching an Ordered List

1.24. Write out the algorithm to find x in an ordered list by the method suggested in the text that compares x to every fourth entry until x itself or an entry larger than x is found, and then, in the latter case, searches for x among the preceding three. How many comparisons does your algorithm do in the worst case?

1.25. Draw a decision tree for the algorithm in the preceding exercise with $n = 17$.

1.26. Describe the decision tree for the sequential search algorithm (Algorithm 1.1) in Section 1.3 for an arbitrary n.

1.27. Show that $\lceil \lg(n+1) \rceil = \lfloor \lg n \rfloor + 1$.

1.28. Work out the exact formula for $A(n)$ given by a summation formula in Section 1.5.1.

1.29. How can you modify Binary Search (Algorithm 1.4) to eliminate unnecessary work if you are certain that x is in the list? Draw a decision tree for the modified algorithm for $n = 7$. Do worst-case and average-behavior analyses. (For the average, you may assume $n = 2^k - 1$ for some k.)

*1.30. Let S be a set of m integers and let L be a list of n integers ($n \leq m$) randomly chosen from the set S. Assume that the entries in L are sorted in ascending order. Let x be an element of S. On the average, how many comparisons will be done by binary search (Algorithm 1.4) given L and x as input? Express your answer as a function of n and m.

1.31. Suppose that the function Q is defined for all powers of 2 and is described by the following recurrence relation and boundary condition:

$$Q(n) = n - 1 + 2Q(n/2)$$
$$Q(1) = 0$$

Find a closed form for Q.

1.32. Suppose W satisfies the following recurrence relation and boundary condition (where c is a constant):

$$W(n) = cn + W(\lfloor n/2 \rfloor)$$
$$W(1) = 1.$$

What is the order of W?

Additional Problems

1.33. You have 50 coins that are all supposed to be gold coins of the same weight, but you know that one coin is fake and weighs less than the others. You have a balance scale; you can put any number of coins on each side of the scale at one time, and it will tell you if the two sides weigh the same, or which side is lighter if they do not weigh the same. Outline an algorithm for finding the fake coin. How many weighings will you do?

*1.34. The first n cells of the array L contain integers sorted in increasing order. The remaining cells all contain some very large integer that we may think of as infinity (e.g., *maxint* in Pascal). The array may be arbitrarily large (you may think of it as infinite), and *you don't know n.* Give an algorithm to find the position of a given integer *x* (*x* < *maxint*) in the array in *O*(lg*n*) time. (The technique used here is useful for certain arguments about *NP*-complete problems that we will see in Chapter 9.)

1.35. The Towers of Hanoi problem is often used as an example when teaching recursion. Six disks of different sizes are piled on a peg in order by size, with the largest at the bottom, as shown in Fig. 1.7. There are two empty pegs. The problem is to move all the disks to the third peg by moving only one at a time and never placing a disk on top of a smaller one. The second peg may be used for intermediate moves. The usual solution recursively moves all but the last disk from the starting peg to the spare peg, then moves the remaining disk on the start peg to the destination peg, and then recursively moves all the others from the spare peg to the destination peg. The three steps are illustrated in Fig. 1.8 and described in the following procedure.

> **procedure** *Hanoi* (*numberOfDisks, start, destination, spare*);
> { *start, destination,* and *spare* are peg numbers. }
> **begin**
>   **if** *numberOfDisks* > 0 **then**
>   *Hanoi* (*numberOfDisks*−1, *start, spare, destination*);
>   writeln ('Move top disk from peg ', *start,* ' to peg ', *destination,* '.');
>   *Hanoi* (*numberOfDisks*−1, *spare, destination, start*)
>   **end** { if }
> **end** { Hanoi }

Write a recurrence relation for the number of moves done. Then solve it.

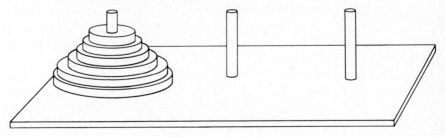

Figure 1.7   Towers of Hanoi.

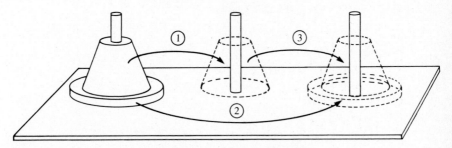

Figure 1.8   Moving the disks.

Notes and References

There are several texts on data structures that may be used for review and reference, for example, Kruse (1987), Tenenbaum and Augenstein (1986), and Aho, Hopcroft, and Ullman (1983). Several other texts on design and analysis of algorithms are listed in the Bibliography. Cooper (1983) and Wirth (1983) are references for Pascal and Modula-2.

Many of the references that follow are more advanced than this chapter; they would be useful and interesting to consult as this book is read.

The ACM's Alan M. Turing Award has been given to several people who have done important work in computational complexity. The Turing Award Lectures by Richard M. Karp (1986), Stephen A. Cook (1983), and Michael O. Rabin (1977) give very nice overviews of questions, techniques, and points of view of computational complexity.

Knuth (1976) discusses the meaning and history of the notations $O(f)$ and $\Theta(f)$. Brassard (1985) presents arguments for the variation of the definitions used in this book.

For more advanced mathematical tools for analysis of algorithms, see Purdom and Brown (1985) and Lueker (1980).

Bentley (1982 and 1986) and his column "Programming Pearls" in the *Communications of the ACM* contain beautifully written discussions of algorithm design and techniques for making programs more efficient in practice.

Gries (1981) is concerned with proving correctness of programs and techniques for writing programs that make them more likely to be correct. Hantler and King (1976) is a survey of both formal and informal techniques for proving program correctness.

The reader who wishes to browse through research articles will find a lot of material in the *Journal of the ACM*, the *Proceedings of the ACM Symposium on Theory of Computing* (annual), *SIGACT News*, the *SIAM Journal on Computing*, *Transactions on Mathematical Software*, and the *Communications of the ACM*, to name a few sources.

Knuth (1984), a paper about the space complexity of songs, is recommended very highly for when the going gets rough.

2

Sorting

2.1
Introduction

In this chapter we will study several algorithms for sorting, that is, for rearranging the elements of a list into order. (We use the word *list* in its general English sense. In the descriptions of most of the algorithms, we will assume the list is stored as an array, though some of the algorithms are useful for sorting files and linked lists.) We assume that each item in the list to be sorted contains an identifier, called a *key*, which is an element of some linearly ordered set, and that two keys can be compared to determine which is larger or whether they are equal. We will always sort keys into nondecreasing order. Each entry in the list may contain other information aside from the key. When keys are rearranged during the sorting process, the associated information will also be rearranged as appropriate, but most of the time we will refer only to the keys and will make no explicit mention of the rest of the entry.

There are several good reasons for studying sorting algorithms. First, they are of practical use because sorting is done often. Just as having the entries in telephone books and dictionaries in alphabetical order makes them easy to use, working with large sets of data in computers is facilitated when the data are sorted. Second, quite a lot of sorting algorithms have been devised (more than will be covered here), and studying a number of them should impress upon the reader the fact that one can take many different points of view toward the same problem. The discussion of the algorithms in this chapter should provide some insights on the questions of how to improve a given algorithm and how to choose among several. Third, sorting is one of the few problems for which we can easily derive good lower bounds for worst-case and average behavior. The bounds are good in the sense that there are algorithms that do approximately the minimum amount of work specified.

The algorithms considered in Sections 2.2 through 2.6 are all from the class of sorting algorithms that can compare keys (and copy them) but cannot do other operations on the keys. We call these "algorithms that sort by comparison of keys." The measure of work used for analyzing algorithms in this class is the number of comparisons of keys. In Section 2.4 lower bounds on the number of comparisons performed by such algorithms are established. Section 2.7 discusses sorting algorithms for which different measures of work are appropriate. The algorithms in Sections 2.2 through 2.7 are called *internal sorts* because the data are assumed to be in the computer's high-speed, random-access memory. In Section 2.8 we study an algorithm for sorting large sets of data stored on external, slower storage devices with restrictions on the way data are accessed. Such algorithms are called *external sorts*. When analyzing the sorting algorithms we will consider how much extra space they use (in addition to the input). If the amount of extra space is constant with respect to the input size, the algorithm is said to work *in place*.

To help make the algorithms as clear as possible, we will use *Array*, *Index*, and *Key* as type identifiers.

The reader should do the first few exercises at the end of this chapter as a warm-up before proceeding.

2.2
Insertion Sort

2.2.1   The Strategy

Insertion Sort is a good sorting algorithm to begin with because the idea behind it is a natural and general one, and its worst case and average behavior analysis are easy. It is also used as part of a faster sorting algorithm, Shellsort, which we will describe later (see Section 2.6).

We begin with a list L of n keys in random order, as illustrated by Fig. 2.1. (Insertion Sort may be used on keys from any linearly ordered set, but for the stick figure illustrations, think of the keys as the heights of the sticks.) Suppose we have sorted some initial segment of the list. The general step is to increase the length of the sorted segment by inserting the next key in its proper place. Figure 2.2 shows a snapshot of the list after it has been partially sorted. Let x be the next key to be inserted in the sorted segment. We insert it by moving each key in the sorted segment to the "right" one place until we find a key smaller than or equal to x (or until we run out of keys). Then x is inserted in the gap as shown in Fig. 2.3. To get the algorithm started, we need only observe that the first key alone may be considered a sorted segment.

Figure 2.1   Unsorted keys.

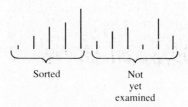

Sorted   Not
                      yet
                      examined

Figure 2.2   Partially sorted.

$x = 1$

Figure 2.3   Insert x.

Algorithm 2.1 Insertion Sort

Input: L, an array of keys, and $n \geq 0$, the number of keys.

Output: L, with keys in nondecreasing order.

```
procedure InsertionSort (var L: Array; n: integer);
var
    x: Key;
    xindex, j : Index;
begin
    for xindex := 2 to n do
        x := L[xindex];
        j := xindex-1;
        while j > 0 and L[j] > x do¹
            L[j+1] := L[j];
            j := j-1
        end { while };
        L[j+1] := x
    end { for }
end { InsertionSort }
```

The correctness of the algorithm can be established by an induction proof.

2.2.2 Worst Case

For the analysis, we use i for *xindex*. For each value of i, the maximum number of key comparisons possible (in the **while** loop) is $i-1$. Thus the total is

$$W(n) \leq \sum_{i=2}^{n}(i-1) = \frac{n(n-1)}{2}.$$

Note that we have established an upper bound on the worst-case behavior; it takes a moment of thought to verify that there are indeed inputs for which $n(n-1)/2$ comparisons are done. One such worst case is when the keys are in reverse (i.e., decreasing) order. So

$$W(n) = \frac{n(n-1)}{2} \in \Theta(n^2).$$

2.2.3 Average Behavior

To simplify the analysis, we assume that the keys are distinct and that all permutations of the keys are equally likely as input. We will first determine how many key comparisons are done on the average to insert one new key into the sorted segment,

---

[1] Reminder to Pascal programmers: By the semantics of our Boolean expressions, no reference to $L[j]$ is made if $j = 0$.

Figure 2.4   Number of comparisons needed to determine the position for x.

i.e., how many iterations of the **while** loop are done, on the average, for each value of i ($=xindex$). (The analysis is very similar to that done for the Sequential Search algorithm in Chapter 1.)

There are i positions where x may go. Figure 2.4 shows how many comparisons are done depending on the position. The probability that x belongs in any one specific position is $1/i$. (This depends on the fact that x has not been examined earlier by the algorithm. If the algorithm had made any earlier decisions based on the value of x, we could not necessarily assume that x is random with respect to the first $i-1$ keys.) Thus the average number of comparisons to insert x is

$$\sum_{j=1}^{i-1} \frac{1}{i} j + \frac{1}{i}(i-1) = \frac{1}{i}\sum_{j=1}^{i-1} j + 1 - \frac{1}{i} = \frac{i+1}{2} - \frac{1}{i}.$$

Now, adding for all insertions,

$$A(n) = \sum_{i=2}^{n}\left(\frac{i+1}{2} - \frac{1}{i}\right) = \frac{n^2}{4} + \frac{3n}{4} - 1 - \sum_{i=2}^{n}\frac{1}{i}.$$

We have seen (Eq. 1.7) that $\sum_{i=2}^{n}\frac{1}{i} \approx \ln n$, so, ignoring lower-order terms, we have

$$A(n) \approx \frac{n^2}{4} \in \Theta(n^2).$$

2.2.4   Space

Clearly, Insertion Sort is an in-place sort.

2.2.5   Lower Bounds on the Behavior of Certain Sorting Algorithms

Think of the key x as occupying the "empty" position in the array while Insertion Sort compares x to the key to its left. After each comparison, Insertion Sort either moves no keys or simply interchanges two adjacent keys. We will show that all sorting algorithms that do such limited, "local" moving of keys after each comparison must do about the same amount of work as Insertion Sort.

A permutation on n items can be described by a one-to-one function from the set $N = \{1, 2, \ldots, n\}$ onto itself. There are $n!$ distinct permutations on n items. Let the keys in the unsorted list L be $x_1, x_2, \ldots, x_n$. There is a permutation π such that, for $1 \le i \le n$, $\pi(i)$ is the correct position of x_i when the list is sorted. Without loss of generality, we can assume that the keys are the integers 1, 2, ..., n since we can substitute 1 for the smallest key, 2 for the next smallest, and so on, without causing any changes in the instructions carried out by the algorithm. Then the unsorted input is $\pi(1)$, $\pi(2)$, ..., $\pi(n)$. For example, consider the input list 2, 4, 1, 5, 3. $\pi(1) = 2$ means that the first key, 2, belongs in the second position, which it clearly does. $\pi(2) = 4$ because the second key, 4, belongs in the fourth position, and so on. We will identify the permutation π with the list $\pi(1)$, $\pi(2)$, ..., $\pi(n)$.

An *inversion* of the permutation π is a pair $(\pi(i), \pi(j))$ such that $i < j$ and $\pi(i) > \pi(j)$. If $(\pi(i), \pi(j))$ is an inversion, the ith and jth keys in the list are out of order relative to each other. For example, the permutation 2, 4, 1, 5, 3 has four inversions (2, 1), (4, 1), (4, 3), and (5, 3). If a sorting algorithm removes at most one inversion after each key comparison (say, by interchanging adjacent keys, as Insertion Sort does), then the number of comparisons performed on the input $\pi(1)$, $\pi(2)$, ..., $\pi(n)$ is at least the number of inversions of π. So we investigate inversions.

It is easy to show that there is a permutation that has $n(n-1)/2$ inversions. (Which permutation?) Thus the worst-case behavior of any sorting algorithm that removes at most one inversion per key comparison must be in $\Omega(n^2)$.

To get a lower bound on the average number of comparisons done by such sorting algorithms, we compute the average number of inversions in permutations. Each permutation π can be paired off with its *transpose permutation* $\pi(n)$, $\pi(n-1)$, ..., $\pi(1)$. For example, the transpose of 2, 4, 1, 5, 3 is 3, 5, 1, 4, 2. Each permutation has a unique transpose and is distinct from its transpose (for $n > 1$). Let i and j be integers between 1 and n, and suppose $j < i$. Then (i, j) is an inversion in exactly one of the permutations π and transpose of π. There are $n(n-1)/2$ such pairs of integers. Hence each pair of permutations has $n(n-1)/2$ inversions between them, and therefore an average of $n(n-1)/4$. Thus, overall, the average number of inversions in a permutation is $n(n-1)/4$, and we have proved the following theorem.

Theorem 2.1   Any algorithm that sorts by comparison of keys and removes at most one inversion after each comparison must do at least $n(n-1)/2$ comparisons in the worst case and at least $n(n-1)/4$ comparisons on the average (for n keys).

Since Insertion Sort does $n(n-1)/2$ key comparisons in the worst case and approximately $n^2/4$ on the average, it is about the best we can do with any algorithm that works "locally," e.g., interchanging only adjacent keys. It is, of course, not obvious at this point that any other strategy can do better, but if there are significantly faster algorithms they must move keys more than one position at a time.

2.3
Quicksort and Mergesort: Divide and Conquer

2.3.1   Introduction to Divide and Conquer

Both of the algorithms in this section use the Divide and Conquer technique. They divide the problem into smaller instances of the same problem (in this case into smaller lists to be sorted), then solve the smaller instances recursively (i.e., by the same method), and finally combine the solutions to obtain the solution for the original input. Quicksort and Mergesort differ in the ways they divide the problem and later combine the solutions, or sorted sublists.

2.3.2   The Quicksort Strategy

Quicksort's strategy is to rearrange the keys to be sorted so that all the "small" keys precede the "large" keys. Then Quicksort sorts the two sublists of "small" and "large" keys recursively, with the result that the entire list is sorted.

Let L be the array of keys and let *first* and *last* be the indexes of the first and last entries, respectively, in the sublist Quicksort is currently sorting. (Initially *first* $=1$ and *last* $=n$, the number of keys.) The *Split* algorithm chooses a key x from the sublist and rearranges the entries, finding an index *splitPoint* such that, for *first* $\leq i <$ *splitPoint*, $L[i] \leq x$; $L[splitPoint] = x$; and for *splitPoint* $< i \leq$ *last*, $L[i] \geq x$. Then x is in its correct position and is ignored in the subsequent sorting. (See Fig. 2.5.)

Algorithm 2.2   Quicksort

procedure *Quicksort (first, last : Index)*;
var
    splitPoint: Index;

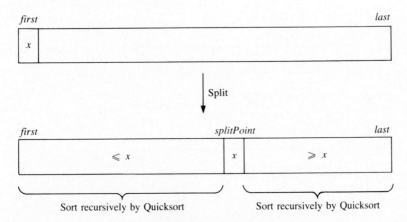

Figure 2.5   Quicksort.

```
begin
    if first < last then
        Split (first, last, splitPoint);
        Quicksort (first, splitPoint−1);
        Quicksort (splitPoint+1, last)
    end { if }
end { Quicksort }
```

All the work of comparing and moving keys is done in the *Split* procedure. There are several different strategies that may be used by *Split*; they yield algorithms with different advantages and disadvantages. We will present one here and consider another in the exercises.

Split may choose as *x* any key in the list between *L*[*first*] and *L*[*last*]; for simplicity, let *x*=*L*[*first*]. The other keys in the list being split are divided into three contiguous sequences, as illustrated in Fig. 2.6: those less than *x*, those greater than or equal to *x*, and those whose relation to *x* is unknown. Initially, all the keys are in the unknown group. At each iteration of its loop *Split* compares the next unknown

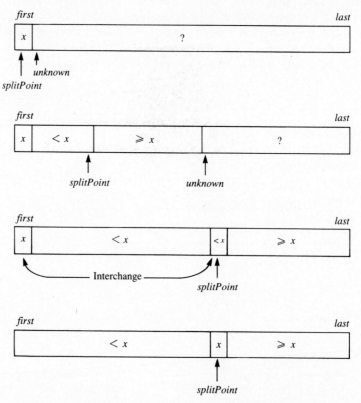

Figure 2.6 How *Split* works: initial, intermediate, and final views.

key, say y, to x. If $y \geq x$, then y can stay where it is. If $y < x$, it is swapped with the first key in the section of larger keys. Finally, after all the keys have been compared to x, x is put in its proper place by swapping with the last of the smaller keys.

Algorithm 2.3   Splitting the List for Quicksort

```
procedure Split (first, last: Index; var splitPoint: Index);
var
    x: Key;
    unknown: Index;
begin
    x := L[first];
    splitPoint := first;
    for unknown := first+1 to last do
        if L[unknown] < x then
            splitPoint := splitPoint+1;
            Interchange(L[splitPoint], L[unknown])
        end { if }
    end { for };
    Interchange(L[first], L[splitPoint])
end { Split }
```

A small example is shown in Fig. 2.7. The detailed operation of *Split* is shown only the first time it is called.

2.3.3   Analysis of Quicksort

Worst Case

Split compares each key with x, so if there are k keys in the section of the array it is working on, it does $k-1$ key comparisons. If $L[first]$ is the smallest key in the section being split, then $splitPoint = first$, and all that has been accomplished is splitting the list into an empty section (keys smaller than x) and a section with $k-1$ keys. Thus, if each time *Split* is called, x is the smallest key, the total number of key comparisons is

$$\sum_{k=2}^{n} (k-1) = \frac{n(n-1)}{2}.$$

This is as bad as Insertion Sort and Maxsort (Exercise 2.1). And, strangely enough, the worst case occurs when the keys are already sorted in ascending order! Is the name Quicksort false advertising?

*Average Behavior

In Section 2.2 we showed that if a sorting algorithm removes at most one inversion from the permutation of the keys after each comparison, then it must do at least $(n^2 - n)/4$ comparisons on the average (Theorem 2.1). Quicksort, however, does not

The keys

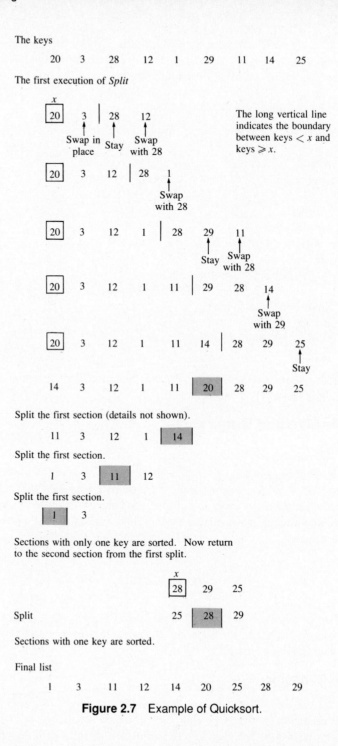

Figure 2.7 Example of Quicksort.

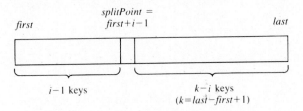

Figure 2.8   Average behavior of Quicksort.

have this restriction. The *Split* algorithm can move a key across a large section of the list, eliminating up to $2n-2$ inversions with one interchange. Quicksort deserves its name because of its average behavior.

We will assume that the keys are distinct and that all permutations of the keys are equally likely. Let k be the number of keys in the section of the list being sorted, and let $A(k)$ be the average number of key comparisons done for lists of this size. Suppose the next time *Split* is executed x gets put in the ith position in this sublist (Fig. 2.8). *Split* does $k-1$ key comparisions, and the sublists to be sorted next have $i-1$ keys and $k-i$ keys, respectively. Each possible position for the split point i is equally likely (has probability $1/k$) so, letting $k = n$, we have the recurrence relation

$$A(n) = n-1+\sum_{i=1}^{n}\frac{1}{n}(A(i-1)+A(n-i)) \quad \text{for } n \geq 2$$

$$A(1) = A(0) = 0.$$

Inspection of the terms in the sum lets us simplify the recurrence relation to

$$A(n) = n-1+\frac{1}{n}\sum_{i=2}^{n-1}2A(i). \tag{2.1}$$

This is a more complicated recurrence relation than the ones we saw before because the value of $A(n)$ depends on all earlier values. We can try to use some ingenuity to solve the recurrence, or we can make a guess at the solution and prove it by induction. The latter technique is especially suitable for recursive algorithms. It is instructive to see both methods, so we will do both.

To form a guess for $A(n)$, let us consider a case in which Quicksort works quite well. Suppose that each time *Split* is executed, it partitions the list into two equal sublists. Since we are just making an estimate to help guess how fast Quicksort is on the average, we will estimate the size of the two sublists at $n/2$ and not worry about whether this is an integer. The number of comparisons done is described by the recurrence relation

$$Q(n) \approx n-1+2Q(n/2).$$

$$Q(1) = 0.$$

Expand the recurrence relation to get

$$Q(n) \approx n-1+2Q(n/2) \approx n-1+2(n/2-1)+4Q(n/4)$$

$$\approx n-1+n-2+n-4+8Q(n/8).$$

Thus

$$Q(n) \approx n \lg n - n \in \Theta(n \lg n).$$

Thus if $L[first]$ were close to the median each time the list is split, the number of comparisons done by Quicksort would be in $\Theta(n\lg n)$. This is significantly better than $\Theta(n^2)$. But if all permutations of the keys are equally likely, are there enough "good" cases to affect the average? We prove that there are.

Theorem 2.2 For $n \geq 1$, $A(n) \leq cn\ln n$ for some constant c. (Note: We have switched to the natural logarithm to simplify some of the computation in the proof.)

Proof by induction on n. For $n=1$, we have $A(1)=0$ and $c1\ln1=0$.

Now for $n>1$, we can use the recurrence relation (Eq. 2.1) and the induction hypothesis to get

$$A(n) = n-1+\frac{2}{n}\sum_{i=2}^{n-1}A(i) \leq n-1+\frac{2}{n}\sum_{i=2}^{n-1}ci\ln i$$

for some constant c. We can bound the sum by integrating (see Eq. 1.7):

$$\sum_{i=2}^{n-1}ci\ln i \leq c\int_2^n x\ln x dx.$$

Integrating by parts gives

$$\int_2^n x\ln x dx = \left[\frac{x^2\ln x}{2} - \frac{x^2}{4}\right]\Big|_2^n = \frac{n^2\ln n}{2} - \frac{n^2}{4} - 2\ln2 + 1.$$

So

$$A(n) \leq n-1+\frac{2c}{n}\left[\frac{n^2\ln n}{2} - \frac{n^2}{4} - 2\ln2 + 1\right]$$

$$= cn\ln n - 1 + n\left[1 - \frac{c}{2}\right] + \frac{2c}{n}(1-2\ln2).$$

To show that $A(n) \leq cn\ln n$, it suffices to show that the third and fourth terms are negative or zero. The third term is less than or equal to zero for $c \geq 2$. The fourth term is always negative since $\ln2 \approx 0.7$. So we may let $c=2$ and conclude that $A(n) \leq cn\ln n$. $\quad\square$

Since $\ln n \approx 0.7\lg n$, we can conclude that $A(n) \leq 1.4n\lg n$.

Although we have established the average behavior of Quicksort, it is still instructive to return to the recurrence relation (Eq. 2.1) and try to solve it directly. We have

$$A(n) = n-1+\frac{2}{n}\sum_{i=2}^{n-1}A(i) \tag{2.2}$$

$$A(n-1) = n-2+\frac{2}{n-1}\sum_{i=2}^{n-2}A(i) \tag{2.3}$$

If we subtract the summation in Eq. 2.3 from the summation in Eq. 2.2, most of the terms drop out. Since the summations are multiplied by different factors, we need a slightly more complicated bit of algebra. Informally, we compute

$$n \times \text{Eq.2.2}-(n-1) \times \text{Eq.2.3}.$$

So

$$nA(n)-(n-1)A(n-1) = n(n-1)+2\sum_{i=2}^{n-1}A(i)-(n-1)(n-2)-2\sum_{i=2}^{n-2}A(i)$$

$$= 2A(n-1)+2n-2.$$

So

$$\frac{A(n)}{n+1} = \frac{A(n-1)}{n}+\frac{2n-2}{n(n+1)}.$$

Now let

$$B(n) = \frac{A(n)}{n+1}$$

and

$$B(1) = 0.$$

The recurrence relation for B is

$$B(n) = B(n-1)+\frac{2n-2}{n(n+1)}.$$

We leave it to the reader to verify that

$$B(n) = \sum_{i=2}^{n}\frac{2i-2}{i(i+1)} \approx 2\ln n$$

and therefore

$$A(n) \approx 1.4(n+1)\lg n.$$

Space Usage

At first glance it may seem that Quicksort is an in-place sort. It is not. While the algorithm is working on one sublist, the beginning and ending indexes (call them the borders) of all the other sublists yet to be sorted must be saved on a stack, and the size of the stack depends on the number of sublists into which the list will be split. This, of course, depends on n. In the worst case, *Split* may split off one entry at a time in such a way that $n-1$ pairs of borders are stored on the stack. Thus the worst-case amount of space used by the stack is in $\Theta(n)$. One of the modifications to the algorithm described next can significantly reduce the maximum stack size.

2.3.4   Improvements on the Basic Quicksort Algorithm

1.   We have seen that Quicksort works well if the key x used by *Split* to partition a sublist is close to the median entry.   Choosing $L[first]$ as x causes Quicksort to do poorly in cases where sorting should be easy (for example, when the list is already sorted).   There are several other strategies for choosing x.   One is to choose a random integer q between *first* and *last* and let $x=L[q]$.   Another is to let x be the median of the entries $L[first]$, $L[(first+last)/2]$, and $L[last]$.   (In either case, the key in $L[first]$ would be swapped with x before proceeding with the *Split* algorithm.)   Both of these strategies require some extra work to choose x, but they pay off by improving the average running time of a Quicksort program.

2.   The version of *Split* presented in the text is easy to understand and program. There is an alternative version in the exercises that is more complicated and prone to bugs. However, the version in the exercises is written so that its inner loops are very fast.   It may be preferred for a program that is used often.   (Also, the time needed for procedure call overhead can be eliminated by writing the code for *Split* directly in the *Quicksort* procedure.)

3.   Every time *Split* partitions a list, some data must be put on the stack.   The manipulation of the stack takes time but is worthwhile when n is large because the algorithm is fast.   Quicksort is not particularly good for small lists.   But, by the nature of the algorithm, for large n Quicksort will break the list up into small sublists and recursively sort them.   Thus whenever the size of a sublist is small, the algorithm becomes inefficient.   This problem can be remedied by choosing a small *smallSize* and sorting sublists of size $\leq smallSize$ by some simple, nonrecursive sort, called *OtherSort* in the modified algorithm.   (Insertion Sort is a good choice.)

Algorithm 2.4   Quicksort

```
procedure Quicksort (first, last: Index);
var
    splitPoint: Index;
begin
    if last − first ≥ smallSize
        then
                Split (first, last, splitPoint);
                Quicksort (first, splitPoint − 1);
                Quicksort (splitPoint + 1, last)
        else OtherSort (first, last)
    end { if }
end { Quicksort }
```

What value should *smallSize* have? The best choice depends on the particular implementation of the algorithm (that is, the computer being used and the details of the program), since we are making some trade-offs between overhead and key comparisons. A value close to 10 may do reasonably well.

4. In a recursive implementation of Quicksort, the recursive calls to the *Quicksort* procedure, i.e., *Quicksort(first, splitPoint*−1) and *Quicksort(splitPoint*+1, *last)* will each cause *first, last,* and *splitPoint* to be stacked. Much of the pushing and popping that will be done is unnecessary. After *Split,* the program will start sorting the sublist *L[first], . . . , L[splitPoint*−1]; later it must sort the sublist *L[splitPoint*+1],..., L[last].* Therefore only *splitPoint*+1 and *last* need be saved on the stack before sorting the first sublist, and nothing need be saved before sorting the second one. Extra stack overhead is avoided by manipulating the stack in the program instead of leaving it to the compiler.

Algorithm 2.5   Quicksort with Explicit Stacking

Input: *L*, the array of keys, and $n \geq 0$, the number of keys.

Output: *L*, with keys in nondecreasing order.

```
procedure QuicksortWithStack (var L: Array; n: integer);
var
    first, last, splitPoint : Index;
begin
    Push([1, n]);
    while stack is not empty do
        Pop ([first, last]);
        while first < last do
            Split (first, last, splitPoint);
            Push ([splitPoint+1, last]);
            last := splitPoint−1
        end { while first < last }
    end { while stack not empty }
end { QuicksortWithStack }
```

5. Let *b* be the amount of space needed on the stack to store the borders of one sublist. *Split* may split a list into one very large section and one very small one. If this happens repeatedly and the smaller section is the one whose borders are stacked for later processing, the stack will need roughly *bn* locations. On the other hand, if we stack the borders of the larger section and immediately work on the smaller one, the stack will never fill more than *b*lg*n* cells because the smaller sublist, the one we will continue to split, is no more than half the size of the sublist from which it was obtained. Thus to keep the stack small, *last*−*splitPoint* and *splitPoint*−*first* should be compared, and the borders of the larger sublist stacked while the smaller is sorted. Note that this change saves space at the expense of time (comparing the size of sublists). If very large lists are being sorted, this space savings may be necessary.

6.   The five modifications just described have been discussed independently, but they are compatible and can be combined in one program.

Remarks

In practice, Quicksort programs run quite fast on the average for large n, and they are widely used. In the worst case, though, Quicksort behaves poorly. Like Insertion Sort, Maxsort, and Bubblesort (Section 2.2 and Exercises 2.1 and 2.2), Quicksort's worst-case time is in $\Theta(n^2)$, but unlike the others, Quicksort's average behavior is in $\Theta(n\lg n)$. Are there sorting algorithms whose worst-case time is in $\Theta(n\lg n)$, or can we establish a worst-case lower bound of $\Theta(n^2)$? The Divide and Conquer approach gave us the improvement in average behavior. Let us examine the general technique again and see how to use it to improve on the worst-case behavior.

2.3.5   Divide and Conquer

The principle behind the Divide and Conquer approach is that it is (often) easier to solve several small instances of a problem than one large one. Thus we decompose the input into smaller inputs, solve the smaller problems recursively, and combine the solutions to obtain a solution for the original input. To escape from the recursion, we solve some small instances of the problem directly. Thus, we can describe Divide and Conquer by the following pseudo-function.

```
function Solution (I);
begin
    if size(I) ≤ smallSize
        then Solution := DirectSolution(I)
        else
            Decompose(I, I₁, . . . , Iₖ);
            for i := 1 to k do
                Sᵢ := Solution (Iᵢ)
            end { for };
            Solution := Combine(S₁, . . . , Sₖ)
    end { if }
end { Solution }
```

To design a specific Divide and Conquer algorithm, we must specify the subalgorithms *DirectSolution*, *Decompose*, and *Combine*. The number of smaller instances into which the input is divided is k. Suppose we let $S(n)$ be the number of steps done by *DirectSolution* for an input of size n, $D(n)$ be the number of steps done by *Decompose* for an input of size n, and $C(n)$ be the number of steps done by *Combine*. Then the general form of the recurrence relation describing the amount of work done by the algorithm is

$$T(n) = \begin{cases} S(n) & \text{for } n \leq smallSize \\ D(n) + \sum_{i=1}^{k} T(size(I_i)) + C(n) & \text{for } n > smallSize \end{cases}$$

For Quicksort (Algorithm 2.2) $k=2$, *smallSize* $= 1$ (and *DirectSolution* does nothing), *Decompose* is *Split* (with $D(n) = n-1$), and *Combine* does nothing ($C(n) = 0$) because the two sorted sublists are already in their appropriate places in the whole array.

The special case of Divide and Conquer where the input is split into two equal-sized pieces has the simplified recurrence relation

$$T(n) = \begin{cases} S(n) & \text{for } n \leq smallSize \\ D(n) + 2T\left(\dfrac{n}{2}\right) + C(n) & \text{for } n > smallSize \end{cases}$$

Our analysis of the best case for Quicksort showed that if the input is split in two equal pieces, and D and C are small enough, then $T(n)$ is in $\Theta(n\lg n)$. Mergesort, described in the next two sections, forces an equal split.

2.3.6 Merging Sorted Lists

In this section we review a straightforward solution to the following problem: Given two lists A and B sorted in nondecreasing order, merge them to create one sorted list C. Merging sorted sublists is essential to the strategy of Mergesort. The measure of work done by a merge algorithm will be the number of comparisons of keys performed by the algorithm.

Let n and m be the number of items in A and B, respectively. Beginning with the complete lists A and B, compare the first remaining keys in A and B and move the smaller one to the next vacant position in C. When A or B is empty, move the items remaining in the other list to C.

Algorithm 2.6   Merge

Input: A and B, lists with keys in nondecreasing order; and n and m, the number of keys in each.

Output: C, a list containing all the keys from A and B in nondecreasing order.

```
indexA := 1; indexB := 1; indexC := 1;
{ indexA indexes A; indexB indexes B; indexC indexes C }
while indexA ≤ n and indexB ≤ m do
    if A[indexA] < B[indexB]
        then C[indexC] := A[indexA]; indexA := indexA+1
        else C[indexC] := B[indexB]; indexB := indexB+1
    end { if };
    indexC := indexC+1
end { while };
if indexA > n
    then move B[indexB], ..., B[m] to C[indexC], ..., C[n+m]
    else move A[indexA], ..., A[n] to C[indexC], ..., C[n+m]
end { if }
```

Worst Case

Whenever a comparison of keys from A and B is done, at least one key is moved to C and never examined again. After the last comparison, at least two keys — the smaller of the two compared and all that remain in the other list — are moved to C. So at most $n+m-1$ comparisons are done. The worst case, using all $n+m-1$ comparisons, occurs when $A[n]$ and $B[m]$ belong in the last two positions in C. If $n=m$, then $2n-1$ comparisons are done in the worst case. We show next that for this special case (i.e., $n=m$) the algorithm is optimal.

Optimality When $n=m$

Theorem 2.3   Any algorithm to merge two sorted lists, each containing n entries, by comparison of keys, does at least $2n-1$ such comparisons in the worst case.

Proof. Suppose we are given an arbitrary merge algorithm. Let a_i and b_i be the ith entries of A and B, respectively. We show that keys can be chosen so that the algorithm must compare a_i with b_i, for $1 \le i \le n$, and a_i with b_{i+1}, for $1 \le i \le n-1$. Specifically, choose keys so that, whenever the algorithm compares a_i and b_j, if $i<j$, the result is that $a_i < b_j$, and if $i \ge j$, the result is that $a_i > b_j$. Choosing the keys so that $b_1 < a_1 < b_2 < a_2 < \cdots < b_n < a_n$ will satisfy these conditions. However, if for some j, the algorithm never compares a_j and b_j, then choosing keys so that $b_1 < a_1 < b_2 < \cdots < a_{j-1} < a_j < b_j < b_{j+1} < \cdots < b_n < a_n$ will also satisfy these conditions, and the algorithm would not be able to determine the correct ordering. Similarly, if for some j, it never compares a_j and b_{j+1}, the arrangement $b_1 < a_1 < \cdots < b_j < b_{j+1} < a_j < \cdots < a_n$ would be consistent with the results of the comparisons done, and again the algorithm could not determine the correct ordering.   □

Space Usage

It might appear from the way in which Algorithm 2.6 is written that merging lists with a total of N entries requires enough memory locations for $2N$ entries, since all entries are copied to C. In some cases, however, the amount of extra space needed can be decreased. Suppose $n \ge m$. If the sorted lists to be merged are in arrays, and A has enough room for $n+m$ keys, then only the extra m locations in A are needed. Simply identify C with A, and do the merging from the ends (larger keys) of A and B, as indicated in Fig. 2.9. The first m entries moved to "C" will fill the extra locations of A. From then on the vacated locations in A are used. There will always be a gap (i.e., some empty locations) between the end of the merged portion of the list and the remaining entries of A until all the entries have been merged. Observe that if this space-saving storage layout is used, the last line in the Merge algorithm (**else** move $A[indexA], \ldots, A[n]$ to $C[indexC], \ldots, C[n+m]$) can be eliminated because, if B empties before A, the remaining items in A are in their correct position and do not have to be moved.

Whether or not C overlaps one of the input lists, the extra space used by the Merge algorithm when $m=n$ is in $\Theta(n)$.

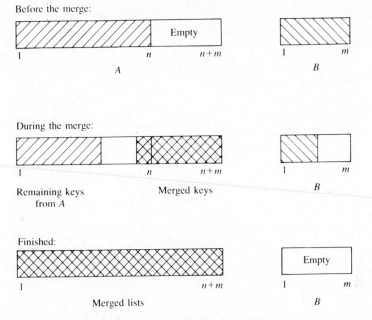

Figure 2.9 Overlapping lists for Merge.

2.3.7 Mergesort

The problem with Quicksort is that *Split* does not always decompose the list into two equal sublists. Mergesort just slices the list in two equal halves and sorts the halves separately (and, of course, recursively). Then it merges the sorted halves. (See Fig. 2.10.) Thus, using the Divide and Conquer terminology, *Decompose* merely computes the middle index of the sublist and does no key comparisons; *Combine* does the merging. We assume that Merge is modified to merge adjacent subsections of an array, putting the resulting merged list back into the cells originally occupied by the keys being merged. Its parameters are the first and last indexes of each of the subsections of the array it is to merge.

Algorithm 2.7   Mergesort

```
procedure Mergesort (first, last : Index);
begin
    if first < last then
        Mergesort (first, ⌊(first+last)/2⌋);
        Mergesort (⌊(first+last)/2⌋+1, last);
        Merge (first, ⌊(first+last)/2⌋, ⌊(first+last)/2⌋+1, last);
    end { if }
end { Mergesort }
```

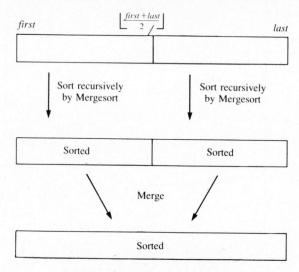

Figure 2.10 Mergesort strategy.

The recurrence relation for the worst-case behavior of Mergesort is

$$W(n) = W(\lfloor n/2 \rfloor) + W(\lceil n/2 \rceil) + n - 1$$

$$W(1) = 0.$$

This is very similar to the informal analysis we did for the best case for Quicksort. Thus $W(n) \approx n\lg n - n$. So we finally have a sorting algorithm whose worst-case behavior is in $\Theta(n\lg n)$, but because of the extra space used for the merging, Mergesort is not an in-place sort.

2.4
Lower Bounds for Sorting by Comparison of Keys

In this section we derive lower bounds for the number of comparisons that must be done in the worst case and on the average by any algorithm that sorts by comparison of keys. To derive the lower bounds we will assume that the keys in the list to be sorted are distinct.

2.4.1 Decision Trees for Sorting Algorithms

Let n be fixed and suppose that the keys are $x_1, x_2, \ldots, x_n$. We will associate with each algorithm and positive integer n a (binary) decision tree that describes the sequence of comparisons carried out by the algorithm on any input of size n. Let Sort be any algorithm that sorts by comparison of keys. Each comparison has a two-way branch (since the keys are distinct), and we assume that Sort has an output instruction that outputs the rearranged list of keys. The decision tree for Sort is defined inductively by associating a tree with each compare and output instruction as follows. The tree associated with an output instruction consists of one node labeled with the rearrangement of the keys. The tree associated with an instruction that compares keys x_i and x_j consists of a root labeled $(i:j)$, a left subtree that is the tree associated with the instruction executed next if $x_i < x_j$, and a right subtree that is the tree associated with the instruction executed next if $x_i > x_j$. The decision tree for Sort is the tree associated with the first compare instruction it executes. An example of a decision tree for $n = 3$ is shown in Fig. 2.11.

The action of Sort on a particular input corresponds to following one path in its decision tree from the root to a leaf. The tree must have at least $n!$ leaves because there are $n!$ ways in which the keys may be permuted. Since the unique path followed for each input depends only on the ordering of the keys and not on their particular values, exactly $n!$ leaves can be reached from the root by actually executing Sort. We will assume that any paths in the tree that are never followed are removed. We also assume that comparison nodes with only one child are removed and replaced by the child, and that this "pruning" is repeated until all internal nodes have degree 2. The pruned tree represents an algorithm that is at least as efficient as the original one, so the lower bounds we derive using trees with exactly $n!$ leaves and all internal nodes of degree 2 will be valid lower bounds for all algorithms that sort by comparison of keys. From now on we assume Sort is described by such a tree.

The number of comparisons done by Sort on a particular input is the number of internal nodes on the path followed for that input. Thus the number of comparisons done in the worst case is the number of internal nodes on the longest path, and that is the depth of the tree. The average number of comparisons done is the average of the lengths of all paths from the root to a leaf. (For example, for $n = 3$,

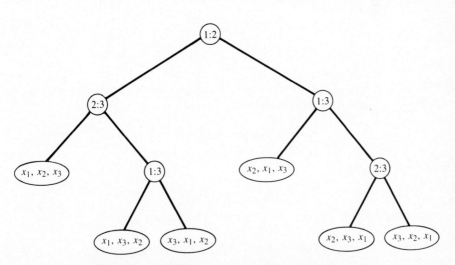

Figure 2.11   Decision tree for a sorting algorithm, $n = 3$.

the algorithm whose decision tree is shown in Fig. 2.11 does three comparisons in the worst case and two and two-thirds on the average.)

2.4.2 Lower Bound for Worst Case

To get a worst-case lower bound for sorting by comparison, we derive a lower bound for the depth of a binary tree in terms of the number of leaves, since the only quantitative information we have about the decision trees is the number of leaves.

Lemma 2.4 Let l be the number of leaves in a binary tree and let d be its depth. Then $l \leq 2^d$.

Proof. A straightforward induction on d. □

Lemma 2.5 Let l and d be as in Lemma 2.4. Then $d \geq \lceil \lg l \rceil$.

Proof. Taking logs of both sides of the inequality in Lemma 2.4 gives $\lg l \leq d$. Since d is an integer, $d \geq \lceil \lg l \rceil$. □

Lemma 2.6 For a given n, the decision tree for any algorithm that sorts by comparison of keys has depth at least $\lceil \lg n! \rceil$.

Proof. Let $l = n!$ in Lemma 2.5. □

So the number of comparisons needed to sort in the worst case is at least $\lceil \lg n! \rceil$. Our best sort so far is Mergesort, but how close is $\lceil \lg n! \rceil$ to $n \lg n$? There are several ways to estimate or get a lower bound for $\lg n!$. Perhaps the simplest is to observe that

$$n! \geq n(n-1) \cdots (\lceil n/2 \rceil) \geq \left[\frac{n}{2}\right]^{n/2}.$$

So

$$\lg n! \geq \frac{n}{2} \lg \frac{n}{2},$$

which is in $\Theta(n \lg n)$. Thus we see already that Mergesort is of optimal order. To get a closer lower bound, we use the fact that

$$\lg n! = \sum_{j=1}^{n} \lg j.$$

Using Eq. 1.9 we get

$$\lg n! \geq n \lg n - 1.5n.$$

Thus the depth of the decision tree is at least $\lceil n \lg n - 1.5n \rceil$.

Theorem 2.7 Any algorithm to sort n items by comparisons of keys must do at least $\lceil \lg n! \rceil$, or approximately $\lceil n \lg n - 1.5n \rceil$, key comparisons in the worst case.

So Mergesort is very close to optimal. There is some difference between the exact behavior of Mergesort and the lower bound. Consider the case where $n = 5$. Insertion Sort does 10 comparisons in the worst case, and Mergesort does 8, but the lower bound is $\lceil \lg 5! \rceil = \lceil \lg 120 \rceil = 7$. Is the lower bound simply not good enough, or can we do better than Mergesort? The reader is encouraged to try to find a way to sort five keys with only seven comparisons in the worst case.

2.4.3 Lower Bound for Average Behavior

We need a lower bound on the average of the lengths of all paths from the root to a leaf. The *external path length* of a tree is the sum of the lengths of all paths from the root to a leaf; it will be denoted by *epl*.

A binary tree in which every node has degree 0 or 2 is called a *2-tree*. Our decision trees are 2-trees, so the next two lemmas give us a lower bound on their epl.

Lemma 2.8   Among 2-trees with *l* leaves, the epl is minimized only if all the leaves are on at most two adjacent levels.

Proof. Suppose we have a 2-tree with depth *d* that has a leaf *X* at level *k*, where $k \le d-2$. We will exhibit a 2-tree with the same number of leaves and lower epl. Choose a node *Y* at level $d-1$ that is not a leaf, remove its children, and attach two children to *X*. (See Fig. 2.12 for an illustration.) The total number of leaves has not

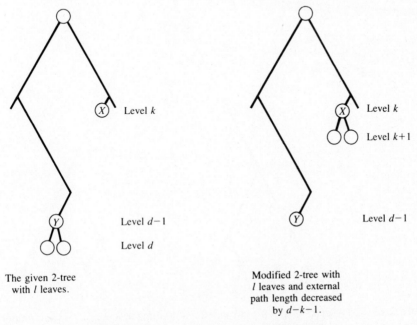

The given 2-tree with *l* leaves.

Modified 2-tree with *l* leaves and external path length decreased by $d-k-1$.

Figure 2.12 Decreasing external path length.

changed. The epl has been decreased by $2d+k$, because the paths to the children of Y and the path to X are no longer counted, and increased by $2(k+1)+d-1 = 2k+d+1$, the sum of the lengths of the paths to Y and the new children of X. There is a net decrease in the epl of $2d+k-(2k+d+1) = d-k-1 > 0$ since $k \leq d-2$.   □

Lemma 2.9   The minimum epl for 2-trees with l leaves is $l\lfloor \lg l \rfloor + 2(l - 2^{\lfloor \lg l \rfloor})$.

Proof. If l is a power of 2, all the leaves are at level $\lg l$. (This statement depends on both the facts that the tree is a 2-tree and that all leaves are on at most two levels. The reader should verify it.) The epl is $l\lg l$, which is the value of the expression in the lemma in this case.

If l is not a power of 2, the depth of the tree is $d = \lceil \lg l \rceil$ and all the leaves are at levels $d-1$ and d. The sum of the path lengths (for all leaves) down to level $d-1$ is $l(d-1)$. For each leaf at level d, 1 must be added to get the total epl. The number of leaves at level d is $2(l - 2^{d-1})$, since for each node at level $d-1$ that is not a leaf, there are two leaves at level d. (See Fig. 2.13.) Thus the sum is $l(d-1) + 2(l - 2^{d-1})$ $= l\lfloor \lg l \rfloor + 2(l - 2^{\lfloor \lg l \rfloor})$.   □

Lemma 2.10   The average path length in a 2-tree with l leaves is at least $\lfloor \lg l \rfloor$.

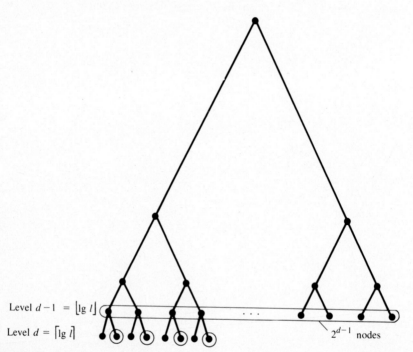

Figure 2.13   Computing external path length for Lemma 2.9. $l = 2^{d-1} +$ the number of nodes at level $d-1$ that are not leaves.

Proof. The minimum average path length is

$$\frac{l\lfloor \lg l \rfloor + 2(l - 2^{\lfloor \lg l \rfloor})}{l} = \lfloor \lg l \rfloor + \varepsilon,$$

where $0 \leq \varepsilon < 1$ since $l - 2^{\lfloor \lg l \rfloor}$ is always less than $l/2$. $\square$

Theorem 2.11   The average number of comparisons done by an algorithm to sort n items by comparison of keys is at least $\lfloor \lg n! \rfloor \approx \lfloor n \lg n - 1.5n \rfloor$.

Thus no algorithm can do substantially better than Quicksort and Mergesort on the average.

2.5
Heapsort

2.5.1 Heaps

The Heapsort algorithm uses a data structure called a *heap*, which is a binary tree with some special properties. The definition of a heap includes a description of the structure and a condition on the data in the nodes. Informally, a heap structure is a complete binary tree with some of the rightmost leaves removed. (See Fig. 2.14 for illustrations.) Let S be a set of keys with a linear ordering and let T be a binary tree with depth d whose nodes contain elements of S. T is a heap if and only if it satisfies the following conditions:[2]

1.   There are 2^l nodes at level l, for $1 \leq l \leq d-1$. At level $d-1$ the leaves are all to the right of the internal nodes. The rightmost internal node at level $d-1$ may have degree 1 (with no right child); the others all have degree 2.

2.   The key at any node is greater than or equal to the keys at each of its children (if it has any).

We will use the term *heap structure* to describe a binary tree that satisfies condition (1). Observe that a complete binary tree is a heap structure. When new nodes are added to a heap, they must be added left to right at the bottom level, and if a node is removed, it must be the rightmost node at the bottom level if the resulting structure is still to be a heap. Note that the root must contain the largest key in the heap.

2.5.2 The Heapsort Strategy

If the keys to be sorted are arranged in a heap, then we can build a sorted list in reverse order by repeatedly removing the key from the root (the largest remaining key), filling the output array from the end, and rearranging the keys still in the heap to reestablish the heap property, thus bringing the next largest key to the root. Since

---

[2] Warning: Some texts use slightly different definitions. Frequently heaps are defined so that the key at a node is less than or equal to the keys at its children.

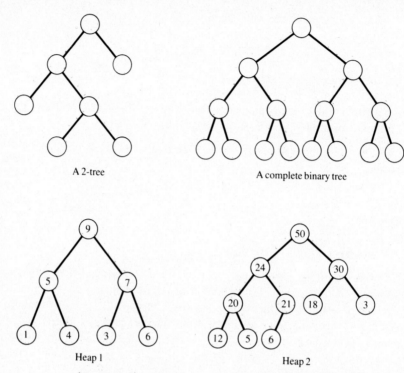

Figure 2.14   2-trees, complete binary trees, and heaps.

this approach requires constructing a heap in the first place, and then repeatedly doing some rearranging of the keys in the heap, it does not look like a promising strategy for getting an efficient sorting algorithm. However, it turns out to do quite well. Thus we outline the strategy here, and then proceed to work out the details. As usual, we assume the array L contains n keys.

> Construct a heap from the list of keys to be sorted;
> for $i := 1$ to n do
>   Copy the key from the root to the array;
>   Remove the rightmost leaf from the last level of the heap structure,
>   saving its key;
>   Rearrange the keys (including the one from the deleted node,
>   but not the one from the root) to reestablish the heap property
> end { for }

The first and last pictures in Fig. 2.15 show an example before and after one iteration of the **for** loop.

We now need an algorithm to construct a heap and an algorithm to rearrange the keys to reestablish the heap property. The latter, which we will call *FixHeap*, can be used to solve the construction problem as well, so we consider it next.

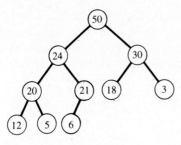

The heap.

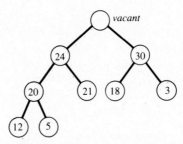

The key at the root has been removed; the rightmost leaf at the bottom level has been removed. $K = 6$ must be reinserted.

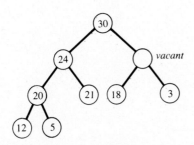

The larger child of *vacant*, 30, is greater than K so it moves up and *vacant* moves down.

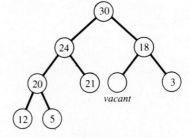

The larger child of *vacant*, 18, is greater than K so it moves up and *vacant* moves down.

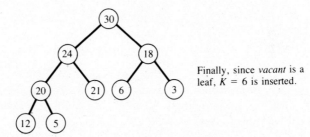

Finally, since *vacant* is a leaf, $K = 6$ is inserted.

Figure 2.15 Deleting the key at the root and reestablishing the heap property.

2.5.3 FixHeap

We have a heap structure with a vacant root. The two subtrees are heaps, and we have an extra key, say K, to be inserted. Since the root is vacant, we begin there and let K filter down to its correct position. At its final position, K must be greater than or equal to each of its children, so at each step K is compared to the larger of the children of the currently vacant node. If K is larger (or equal) it can be inserted; otherwise, the larger child is moved up to the vacant node and the process is repeated.

FixHeap assumes that there is at least one node in the heap. Its action is illustrated in the last four pictures of Fig. 2.16. Although the tree structure of the heap is essential to motivating and understanding Heapsort, we will see later that we will not use pointers to represent the heap. Thus, in the *FixHeap* algorithm, we will not use pointer notation.

Algorithm 2.8 FixHeap

Input: The *root* of a heap and a key K to be inserted.

Output: The heap with keys properly rearranged.

```
procedure FixHeap (root: Node; K: Key);
var
    vacant, largerChild : Node;
begin
    vacant := root;
    while vacant is not a leaf do
        largerChild := the child of vacant with the larger key;
        if K < largerChild's key
            then
                copy largerChild's key to vacant;
                vacant := largerChild
            else exitloop
        end { if }
    end { while };
    put K in vacant
end { FixHeap }
```

Lemma 2.12 *FixHeap* does $2d$ comparisons of keys in the worst case on a heap with depth d.

Proof. Two comparisons of keys are done in each iteration of the **while** loop (except when the vacant node is the sole node that may have only one child). The maximum number of iterations is the depth of the heap. □

2.5.4 Heap Construction

Suppose we start by putting all the keys in a heap structure in arbitrary order. The *FixHeap* algorithm suggests a Divide and Conquer approach to establishing the heap-ordering property. The two subtrees can be turned into heaps recursively, then *FixHeap* can be used to "insert" the key at the root into its proper place, thus combining the two smaller heaps and the root into one large heap. The boundary case is a tree consisting of one node (i.e., a leaf); it is already a heap. So we can describe the heap construction by:

```
procedure ConstructHeap (root: Node);
begin
    if root is not a leaf then
        ConstructHeap (left child of root);
        ConstructHeap (right child of root);
        FixHeap (root, key in root)
    end { if }
end { ConstructHeap }
```

If we unravel the work done by this Divide and Conquer algorithm, we see that it really starts rearranging keys near the leaves first, and works its way up the tree. The same strategy can be described non-recursively in the following algorithm. (The work is done in a different order, but that does not matter.) See Fig. 2.16 for an illustration.

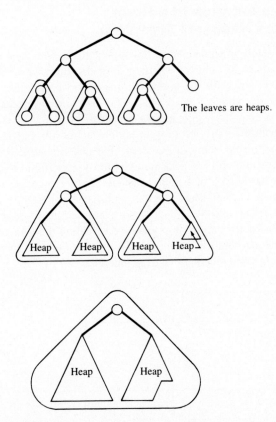

The leaves are heaps.

Figure 2.16   Constructing the heap. (*FixHeap* is called for each circled subtree.)

Algorithm 2.9   Heap Construction

Input: A heap structure (Property (1)) with keys in arbitrary nodes.

Output: The same structure satisfying the heap-ordering property (Property (2)).

> **for** *level* := *depth*−1 **to** 0 **by** −1 **do**
>   **for** each non-leaf *node* at level *level* **do**
>   *K* := the key at *node*;
>   *FixHeap(node, K)*
>   **end** { for }
> **end** { for }

It is not yet clear that Heapsort is a good algorithm; it seems to require extra space, and the implementation of some steps in the algorithms (e.g., controlling the inner **for** loop in the Heap Construction algorithm) may be complicated. It is time to consider the implementation of a heap.

2.5.5   Implementation of a Heap and the Heapsort Algorithm

Binary trees are usually implemented as linked structures, with each node containing pointers to the roots of its subtrees. Setting up and using such a structure requires extra time and extra space for the pointers. However, we can store and use a heap efficiently without any pointers at all. In a heap there are no nodes at, say, level *l* unless level *l*−1 is completely filled, so a heap can be stored in an array level by level (beginning with the root), left to right within each level. Figure 2.17 shows the storage arrangement for the heaps in Fig. 2.14. For this scheme to be useful we must be able to find the children of a node quickly and to determine if a node is a leaf quickly. Suppose the index *i* of a node is given. Then a counting argument can be used to show that its left child has index $2i$ and that its right child has index $2i+1$. (The proof is left for an exercise.) For Heap Construction we must move across a level efficiently, but that is done simply by incrementing the array index. (For the final version of the algorithm, we will simplify the loop control by moving right to left across the levels, decrementing the array index.)

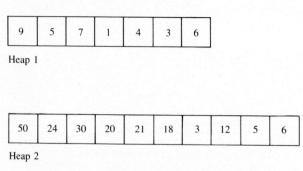

Heap 1

Heap 2

Figure 2.17   Storage of the heaps in Fig. 2.14.

The startling feature of Heapsort is that the whole sorting procedure can be carried out in place; the small heaps built during the construction phase, and later the heap and the deleted keys, can occupy the array *L* that originally contained the unsorted list of keys. During the deletion phase, when the heap contains, say, *k* keys, it will occupy the first *k* locations in the array. Thus, just one variable is needed to mark the end of the heap. Figure 2.18 illustrates the sharing of the array between the heap and the sorted keys.

Algorithm 2.10   Heapsort

Input: *L*, an unsorted array, and $n \geq 1$, the number of keys.

Output: *L*, with keys in nondecreasing order.

```
procedure Heapsort (var L: Array; n: integer);
var
    i, heapsize : Index;
    max : Key;
begin

    { Heap Construction }
    for i := ⌊n/2⌋ to 1 by −1 do
        FixHeap (i, L[i], n)
    end { for };

    { Repeatedly remove the key at the root and rearrange the heap. }
    for heapsize := n to 2 by −1 do
        max := L[1];
        FixHeap (1, L[heapsize], heapsize−1);
        L[heapsize] := max
    end { for }
end { Heapsort }
```

The *FixHeap* algorithm, revised for the array implementation, is:

```
procedure FixHeap (root: Index; K: Key; bound: Index);
var
    vacant, largerChild : Index;
begin
    vacant := root;
    while 2*vacant ≤ bound do
        largerChild := 2*vacant;
        if 2*vacant < bound and L[2*vacant+1] > L[2*vacant]
            then largerChild := 2*vacant+1;
        if K < L[largerChild]
            then
                L[vacant] := L[largerChild];
                vacant := largerChild
```

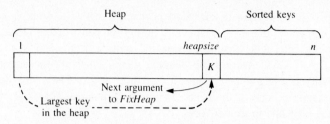

Figure 2.18 The heap and sorted keys in the array.

```
        else exitloop
      end { if }
    end { while };
    L[vacant] := K
  end { FixHeap }
```

We can now see clearly that Heapsort is an in-place sort.

2.5.6 Worst-Case Analysis

Consider first the heap construction loop. Let $d = \lfloor \lg n \rfloor$, the depth of the heap with all n nodes. By Lemma 2.12 the number of comparisons done by *FixHeap* in the worst case is twice the depth of the subtree it works on, which is d minus the level of the subtree's root. Thus the number of comparisons done by the heap construction is at most

$$\sum_{l=0}^{d-1} 2(d-l)(\text{the number of nodes at level } l) = 2\sum_{l=0}^{d-1}(d-l)2^l.$$

Using Eqs. 1.3 and 1.5 from Section 1.2 and some algebraic manipulation, it can be shown that

$$2\sum_{l=0}^{d-1}(d-l)2^l = 2^{d+2} - 2d - 4 \in \Theta(n).$$

Thus the heap is constructed in linear time! (An alternative counting argument is given in the exercises.)

Now consider the second loop. By Lemma 2.12 again, the number of comparisons done by *FixHeap* on a heap with k nodes is at most $2\lfloor \lg k \rfloor$, so the total for all the deletions is at most $2\sum_{k=1}^{n-1}\lfloor \lg k \rfloor$. To evaluate this sum we use Fig. 2.19, which illustrates the case where $n = 10$. The summation we need equals the sum of the areas of the rectangles shown in the figure. Again letting $d = \lfloor \lg n \rfloor$, the sum is

$$\sum_{j=1}^{d-1} j2^j + d(n - 2^d),$$

where the summation term includes the areas of all the complete rectangles (height j

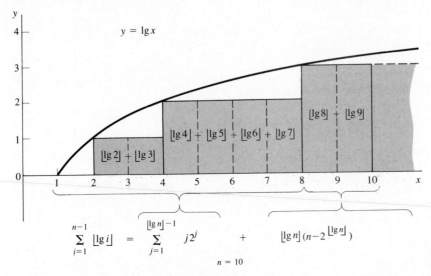

$$\sum_{i=1}^{n-1} \lfloor \lg i \rfloor \;=\; \sum_{j=1}^{\lfloor \lg n \rfloor -1} j2^j \;+\; \lfloor \lg n \rfloor (n-2^{\lfloor \lg n \rfloor})$$

$$n = 10$$

Figure 2.19

and width 2^j) and the second term is the area of the last, incomplete rectangle with height d and width $n-2^d$. The total area, using Eq. 1.5, is

$$2(d\,2^{d-1}-2^d+1)+d\,(n-2^d) = nd-2^{d+1}+2 \in \Theta(n\lg n).$$

The following theorem sums up our results.

Theorem 2.13   Heapsort does at most $2(n-1)\lfloor \lg n \rfloor$ comparisons of keys in the worst case. It is an $\Theta(n\lg n)$ sorting algorithm.

Proof.   The heap construction phase does at most $2^{d+2}-2d-4$ comparisons, and the deletions do at most $2(nd-2^{d+1}+2)$. The total is $2nd-2d=2(n-1)\lfloor \lg n \rfloor$.   □

2.5.7 Remarks

Heapsort does $\Theta(n\lg n)$ comparisons on the average as well as in the worst case. Table 2.1 sums up the results of the analysis of the behavior of the four sorting algorithms that have been discussed so far. Although Mergesort is close to optimal in the worst case, there are algorithms that do fewer comparisons. The lower bound obtained in Section 2.4 is quite good. It is known to be exact for some values of n; that is, $\lceil \lg n! \rceil$ comparisons are sufficient to sort, for some values of n. It is also known that $\lceil \lg n! \rceil$ comparisons are not sufficient for all n. For example, $\lceil \lg 12! \rceil = 29$, but it has been proved that 30 comparisons are necessary (and sufficient) to sort 12 items in the worst case. See the notes and references at the end of the chapter for references on sorting algorithms whose worst-case behavior is close to the lower bound.

Table 2.1
Results of analysis of four sorting algorithms. (Entries are approximate.)

Algorithm	Worst case	Average	Space usage
Insertion Sort	$\dfrac{n^2}{2}$	$\dfrac{n^2}{4}$	In place
Quicksort	$\dfrac{n^2}{2}$	$\Theta(n\lg n)$	Extra space proportional to $\lg n$
Mergesort	$n\lg n$	$\Theta(n\lg n)$	Extra space proportional to n for merging
Heapsort	$2(n-1)\lg n$	$\Theta(n\lg n)$	In place

2.6
Shellsort

2.6.1   The Algorithm

The technique used by Shellsort (named for its inventor, Donald Shell) is interesting, and the algorithm is easy to program and runs fairly quickly. Its analysis, however, is very difficult and incomplete. Shellsort sorts a list L with n keys by successively sorting sublists whose entries are intermingled in the whole list. The sublists to be sorted are determined by a sequence, $h_t, h_{t-1}, \ldots, h_1$, of parameters called increments. Suppose, for example, that the first increment, h_t, is 6. Then the list is divided into six sublists, each beginning with one of the first six keys and consisting of every sixth key from there on. After these sublists are sorted, the next increment, h_{t-1}, is used to separate the list again into sublists, this time with entries h_{t-1} elements apart, and again sublists are sorted. The process is repeated for each increment. The final increment, h_1, is always 1, so at the end, the entire list will be sorted. Figure 2.20 illustrates the action of this method on a small list.

The informal description of Shellsort should prompt a number of questions. What algorithm should be used to sort the sublists? Considering that the last increment is 1 and that the entire list is sorted on the last pass, is Shellsort any more efficient than the algorithm used to sort the sublists? Can the algorithm be written to minimize all the bookkeeping that seems to be needed to control the sorting of all the sublists? What increments should be used?

We tackle the first two questions first. As the example in Fig. 2.20 shows, when the last few passes are made using small increments, few keys will be out of order because of all the work that was done in earlier passes. So Shellsort may be efficient if, and indeed would be efficient only if, the method used to sort sublists is one that does very little work if the list is already sorted or nearly sorted. Insertion Sort (Section 2.2) has this property. It does only $n-1$ comparisons if the list is completely sorted, it is simple to program, and it has very little overhead.

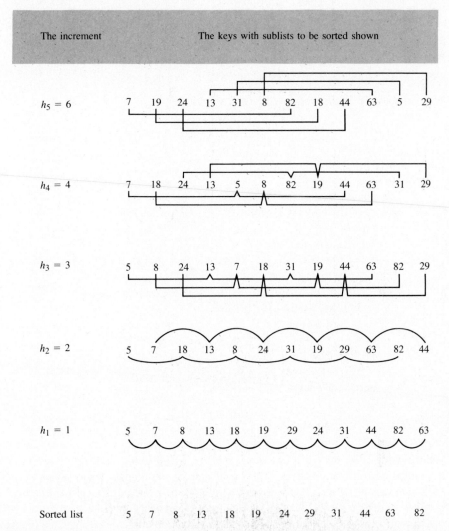

The increment   The keys with sublists to be sorted shown

$h_5 = 6$ 7   19   24   13   31   8   82   18   44   63   5   29

$h_4 = 4$ 7   18   24   13   5   8   82   19   44   63   31   29

$h_3 = 3$ 5   8   24   13   7   18   31   19   44   63   82   29

$h_2 = 2$ 5   7   18   13   8   24   31   19   29   63   82   44

$h_1 = 1$ 5   7   8   13   18   19   29   24   31   44   82   63

Sorted list   5   7   8   13   18   19   24   29   31   44   63   82

Figure 2.20   Shellsort. (Note that only two pairs of keys are interchanged on the final pass.)

Now suppose Shellsort is using an increment h and is to sort h sublists, each containing approximately n/h entries. If each sublist is to be completely sorted before any work is begun on the next, the algorithm would need to keep track of which sublists have been sorted and which remain to be done. This bookkeeping is avoided by having the algorithm make one pass through the entire list (for each increment), intermingling its work on all the sublists. Since consecutive keys of a sublist are h cells apart rather than one apart, "1" is replaced by "h" in Insertion Sort. (Compare Algorithm 2.1 with the inner **for** loop that follows.)

Algorithm 2.11   Shellsort

Input: L, an unsorted array; n, the number of keys; and increments $h_t, h_{t-1}, \ldots, h_1$, where $h_1 = 1$.

Output: L, with keys in nondecreasing order.

```
procedure Shellsort (var L: Array; n: integer);
var
    x: Key;
    i, j : Index;
    h, s: integer;
begin
    Compute (or initialize computation of) increments;
    for s := t to 1 by −1 { s indexes the increments } do
        h := hₛ;
        for i := h+1 to n do
        { i begins at the second key of the first sublist. }
            x := L[i];
            j := i−h;   { the preceding key in the current sublist }
            while j > 0 and L[j] > x do³
                L[j+h] := L[j];
                j := j−h
            end { while };
            L[j+h] := x
        end { for i }
    end { for s }
end { Shellsort }
```

Observe that, although after each comparison Insertion Sort removes at most one inversion from the list it is sorting, the way it is used in Shellsort causes keys to be moved across many others with the chance of eliminating many inversions for each comparison. Thus there is a possibility that the average behavior of Shellsort is better than $\Theta(n^2)$. The efficiency of Shellsort stems from the fact that sorting with one increment, say k, will not undo any of the work done previously when a different increment, say h, was used. More precisely, we say that a list is h-ordered if $L[i] \leq L[i+h]$ for $1 \leq i \leq n-h$, in other words, if all the sublists consisting of every hth key are sorted. To k-sort a list means to sort sublists using increment k.

Theorem 2.14   If an h-ordered list is k-sorted, the list will still be h-ordered.

Proof.   See Knuth (1973) for the proof. The reader should examine Fig. 2.20 to see that the theorem is true for the example given there.   □

---

[3] Reminder to Pascal programmers: By the semantics of our Boolean expressions, no reference to $L[j]$ is made if $j = 0$.

2.6.2   Analysis and Remarks

The number of comparisons done by Shellsort is a function of the sequence of increments used. A complete analysis is extremely difficult and requires answers to some mathematical problems that have not yet been solved. Therefore, the best possible sequence of increments has not been determined, but some specific cases have been thoroughly studied. One of these is the case where $t=2$, that is, where exactly two increments, h and 1, are used. It has been shown that the best choice for h is approximately $1.72\sqrt[3]{n}$, and with this choice the average running time is proportional to $n^{5/3}$. This may seem surprising since using the increment 1 is the same as doing Insertion Sort, which has $\Theta(n^2)$ average behavior; just doing one preliminary pass through the list with increment h lowers the order of the running time. By using more than two increments, the running time can be improved even more.

It is known that, if the increments $h_k = 2^k - 1$ for $1 \le k \le \lfloor \lg n \rfloor$ are used, the number of comparisons done in the worst case is in $O(n^{3/2})$. These increments are very easy to compute in assembly language; each one (after the first) can be obtained from the previous one by shifting one bit to the right. Empirical studies (with values of n as high as 250,000) have shown that another set of increments gives rise to very fast-running programs. These are defined by $h_i = (3^i - 1)/2$ for $1 \le i \le t$, where t is chosen as the smallest integer such that $h_{t+2} \ge n$. These increments are easy to compute iteratively. We can find h_t at the beginning of the sort by using the relation $h_{s+1} = 3h_s + 1$ and comparing the results to n. Instead of storing all the increments, they can be recomputed in reverse order during the sort using the formula $h_s = (h_{s+1} - 1)/3$.

It has been proved that, if the increments are all integers of the form $2^i 3^j$ that are less than n (used in decreasing order), then the number of comparisons done is in $O(n(\lg n)^2)$. The worst-case running times for the other sets of increments are known or expected to be of higher order. However, because of the large number of integers of the form $2^i 3^j$, there will be more passes through the list, hence more overhead, with these increments than with others. Therefore, they are not particularly useful unless n is fairly large.

Shellsort is clearly an in-place sort. Although the analysis of the algorithm is far from complete, and it is not known which increments are best, its speed and simplicity make it a good choice in practice.

2.7
Radix Sorting

2.7.1   Using Properties of the Keys

For the sorting algorithms in Sections 2.2 through 2.6, only one assumption was made about the keys: that they are elements of a linearly ordered set. The basic operation of the algorithms is a comparison of two keys. If we make more assumptions about the keys, we can consider algorithms that perform other operations on them. Suppose the keys are names and are printed on cards, one name per card. To alphabetize the cards by hand we might first separate them into 26 different piles according to the first letter of the name, or fewer piles with several letters in each; alphabetize the cards in each pile by some other method (perhaps similar to Insertion Sort); and finally combine the sorted piles. If the keys are all five-digit decimal integers, we might separate them into 10 piles according to the first digit. If they are integers between 1 and m, for some m, we might make a pile for each of the k intervals $[1, m/k], [m/k+1, 2m/k]$, and so on. In each of these examples the keys are distributed into different piles as a result of examining individual letters or digits in a key or by comparing keys to predetermined values. Then the piles are sorted individually and recombined. Algorithms that sort by such methods are called "bucket sorts," "radix sorts," or "algorithms that sort by distribution." These algorithms are not in the class of algorithms previously considered because to use them we must know something about either the structure or the range of the keys.

We will present one radix sort algorithm in detail later. To distinguish the specific algorithm from others of the same type, we will use the term "bucket sorts" for the general class of algorithms.

How fast are bucket sorts? A bucket sort has three phases, which we may call distribution, sorting buckets, and combining buckets. The type of work done in each phase is different, so our usual approach of choosing one basic operation to count will not work well here. Suppose there are k buckets. During the distribution phase, the algorithm examines each key once (either examining a particular field of bits or comparing the key to at most k preset values). Then it does some work to indicate in which bucket the key belongs. This might involve copying the key or setting some indexes or pointers. The number of operations performed by a reasonable implementation of the first phase should be in $\Theta(n)$. Suppose that to sort the buckets we use an algorithm that sorts by comparison of keys doing, say, $S(m)$ comparisons for a bucket with m keys. Let n_i be the number of keys in the ith bucket. The algorithm does $\sum_{i=1}^{k} S(n_i)$ comparisons during the second phase. The third phase may require, at worst, that all of the keys be copied from the buckets into one list; the amount of work done is in $O(n)$. Thus, most of the work is done while sorting buckets. Suppose $S(m)$ is in $\Theta(m\lg m)$. Then if the keys are uniformly distributed among the buckets, the algorithm does roughly $ck(n/k)\lg(n/k) = cn\lg(n/k)$ comparisons of keys in the second phase, where c is a constant that depends on the sorting algorithm used in the buckets. Increasing k, the number of buckets, decreases the number of comparisons done. If we choose $k = n/10$, then $n\lg 10$ comparisons would be done and the running time of the bucket sort would be linear in n, assuming that the keys are evenly distributed and that the running time for the first phase does not depend on k. However, in the worst case, all of the keys will go into one bucket and the entire list will be sorted in the second phase, turning all of the work of the first and last phases into wasteful overhead. Thus in the worst case a bucket sort would be very inefficient. If the distribution of the keys is known in advance, the range of keys to go into each bucket can be adjusted so that all buckets receive an approximately equal number of keys.

The amount of space needed by a bucket sort depends on how the buckets are stored. If every bucket is to consist of a set of sequential locations (e.g., an array), then each must be allocated enough space to hold the maximum number of keys that might belong in one bucket, and that is n. Thus kn locations would be used to sort n keys. As the number of buckets increases, the speed of the algorithm would increase but so would the amount of space used. Linked lists would be better; only $\Theta(n+k)$ space (for n keys plus links and a listhead for each bucket) would be used. Distributing keys among the buckets would require adjusting pointers. But then how would the keys in each bucket be sorted? Quicksort and Heapsort, for example, two of the faster algorithms discussed, cannot easily sort linked lists. If the number of buckets is large, the number of keys in each will generally be small and a slower algorithm could be used. Insertion sort can be easily modified to sort keys in a simply linked list. With approximately n/k keys per bucket, Insertion Sort will do approximately $n^2/4k^2$ comparisons on the average for each bucket, or $n^2/4k^2$ comparisons in all. Here again, as k increases, so does the speed, but so also does the amount of space used.

The reader might wonder why we do not use a bucket sort algorithm recursively to create smaller and smaller buckets. There are several reasons. The bookkeeping would quickly get out of hand; pointers indicating where the various buckets begin and information needed to recombine the keys into one list would have to be stacked and unstacked often. Due to the amount of bookkeeping necessary for each recursive call, the algorithm should not count on ultimately having only one key per bucket, so another sorting algorithm will be used anyway to sort small buckets. Thus if a fairly large number of buckets is used in the first place, there is little to gain and a lot to lose by bucket sorting recursively. However, although recursively distributing keys into buckets is not efficient, something quite useful can be salvaged from this idea.

2.7.2 Radix Sort

Suppose that the keys are five-digit numbers. A recursive algorithm, as just suggested, could first distribute the keys among 10 buckets according to the leftmost, or most significant, digit and then distribute the keys in each bucket among 10 more buckets according to the next most significant digit, and so on. The buckets could not be coalesced until they are completely sorted, hence the large amount of messy bookkeeping. It is startling that if the keys are distributed into buckets first according to their *least significant* digits (or bits, letters, or fields), then the buckets can be coalesced in order before distributing on the next digit. The problem of sorting the buckets has been completely eliminated. If there are, say, five digits in a key, then the algorithm distributes keys into buckets and then coalesces the buckets five times, distributing on each digit position in turn, right to left, as illustrated in Fig. 2.21. Does this always work? On the final pass when two keys are put into the same bucket because they both start with, say, 9, what ensures that they are in the proper order relative to each other? In Fig. 2.21, the keys 90283 and 90583 differ in

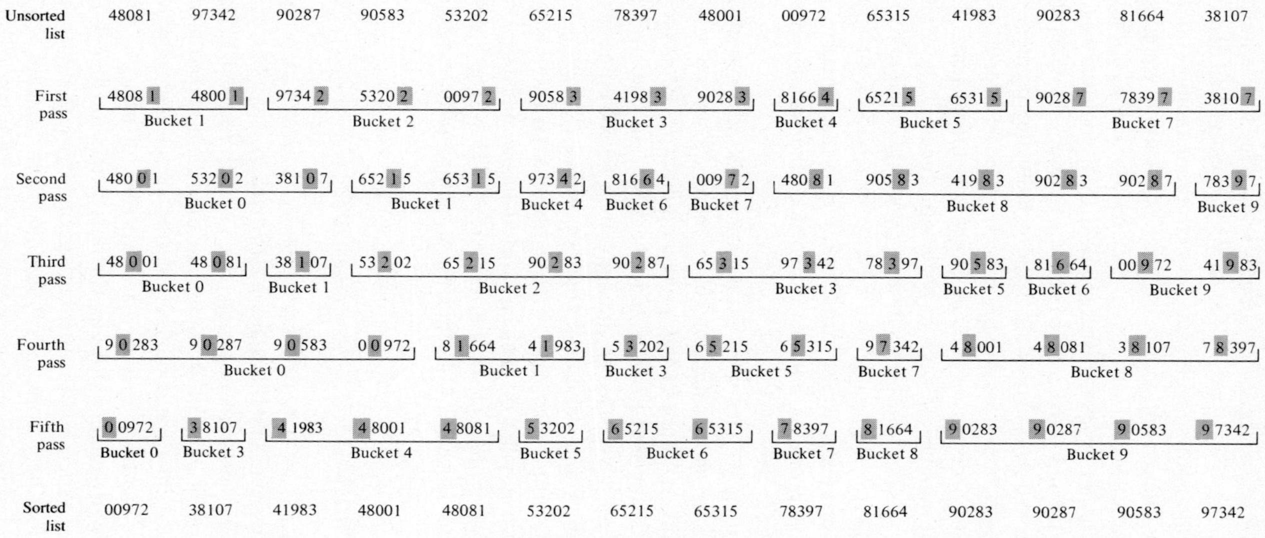

Unsorted list: 48081 97342 90287 90583 53202 65215 00972 48001 78397 65215 90283 81664 38107

First pass:
48081, 48001 — Bucket 1
53202, 97342, 00972 — Bucket 2
90583, 41983, 90283 — Bucket 3
81664 — Bucket 4
65315, 65215 — Bucket 5
78397, 38107 — Bucket 7

Second pass:
48001, 53202, 38107 — Bucket 0
65315, 65215 — Bucket 1
90283 — Bucket 2
97342 — Bucket 4
90583 — Bucket 5
81664 — Bucket 6
00972 — Bucket 7
41983, 90287 — Bucket 8
78397 — Bucket 9

Third pass:
48001, 38107 — Bucket 0
41983 — Bucket 1
90283 — Bucket 2
97342 — Bucket 3
48081 — Bucket 0
90287 — Bucket 2
65315, 65215 — Bucket 3
81664, 90583 — Bucket 5, 6
78397 — Bucket 3
00972 — Bucket 9

Fourth pass:
00972 — Bucket 0
81664 — Bucket 1
48081, 48001 — Bucket 1
53202, 65215, 65315 — Bucket 5
41983 — Bucket 1
90283, 90287, 90583 — Bucket 8
97342 — Bucket 7
38107 — Bucket 8
78397 — Bucket 8

Fifth pass:
00972 — Bucket 0
38107 — Bucket 3
48081, 48001 — Bucket 4
53202, 65215, 65315 — Bucket 6
78397 — Bucket 7
81664 — Bucket 8
90283, 90287, 90583 — Bucket 9
97342 — Bucket 9

Sorted list: 00972 38107 41983 48001 48081 53202 65215 65315 78397 81664 90283 90287 90583 97342

Figure 2.21 Radix Sort.

the third digit only and are put in the same bucket in each pass except the third. After the third pass, so long as the buckets are coalesced in order and the relative order of two keys placed in the same bucket is not changed, these keys remain in proper order relative to each other. In general, if the leftmost digit position in which two keys differ is the ith position (from the right), they will be in the proper order relative to each other after the ith pass.

This sorting method is used by card-sorting machines. On old machines, the machine did the distribution step; the operator collected the piles after each pass and combined them into one for the next pass.

The distribution into piles, or buckets, may be controlled by a column on a card, a digit position, or a bit field in the key. The algorithm is called *Radix Sort* because it treats the keys as numbers in a particular radix, or base. In the example in Fig. 2.21, the radix is 10. If the keys are 32-bit positive integers, the algorithm could use, say, four-bit fields, implicitly treating the keys as numbers in radix 16. It would distribute them among 16 buckets. Thus the radix is also the number of buckets. In

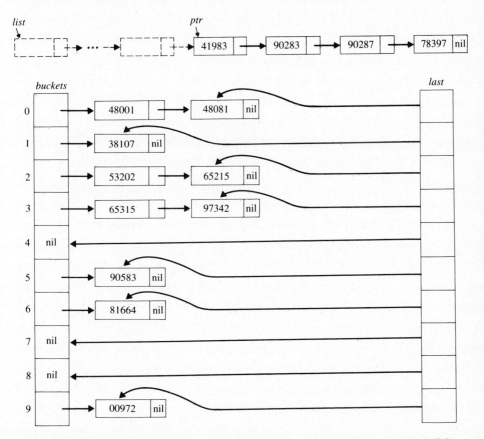

Figure 2.22   The data structure for Radix Sort during distribution on the third digit.

the Radix Sort algorithm that follows, we will assume that distribution is done on bit fields. The fields are extracted from the keys beginning with the low-order bits. The radix (number of buckets) and the number of fields are constants of the algorithm; they do not depend on the number of keys in the input.

The data structure is illustrated in Fig. 2.22 for the third pass of the example in Fig. 2.21.

Algorithm 2.12   Radix Sort

Input: *list, numBuckets,* and *fields* as described below.

Output: The sorted list.

Note: Some of the detail on linked list manipulation is omitted.

```
procedure RadixSort
    (var list: pointer;
        { It is assumed that the unsorted list is given as a linked list with
        list pointing to the first node. Each node has a key field and
        a link field. }
    numBuckets: integer;
        { The number of buckets. }
    numFields: integer
        { The number of fields in a key on which the distribution is done. } );
var
    buckets: ArrayOfPointers;
        { These are listheads for the buckets. Each bucket is a linked list.
        Using listheads will simplify some of the list manipulation detail. }
    last: ArrayOfPointers;
        { last[i] will point to the last key in the ith bucket to facilitate
        insertion of keys at the end of a bucket to maintain the relative order. }
    i, j : integer;   { field number and bucket number, respectively }

procedure Distribute;
    { Distribute keys into buckets. }
var
    j: integer;   { bucket number }
    ptr: pointer;
begin
    ptr := list;
    while ptr ≠ nil do
        Extract the ith field from ptr↑.key and assign to j
            the appropriate bucket number;
        Add ptr↑ to the end of the jth bucket list;
        ptr := ptr↑.link
    end { while }
end { Distribute };
```

```
procedure Coalesce;
var
    j: integer;   { bucket number }
begin
    for j := 1 to numBuckets do
        if the jth bucket is not empty then
            Link the last node in the list constructed so far
                to the first node of bucket j
        end {if}
    end {for};
    Store nil in the link field of the last node.
end { Coalesce };

begin { RadixSort }
    for i := 1 to numFields do
        { Initialize empty buckets. }
        for j := 1 to numBuckets do
            buckets[j] := nil;
            last[j] := nil
        end { for j };
        Set a mask for extracting the ith field (from the right,
            or least significant, end) of a key;
        Distribute;
        Coalesce
    end { for }
end { RadixSort }
```

Analysis and Remarks

Distributing one key requires extracting a field and doing a few link operations; the number of steps is bounded by a constant. So, for all keys, *Distribute* does $\Theta(n)$ steps. *Coalesce* does a few steps for each bucket, hence $\Theta(numBuckets)$ steps overall, but we are assuming *numBuckets* is constant with respect to n. The number of distribution and coalescing passes is *numFields*, the number of fields used for distribution. This is a small constant. Thus the total number of steps done by *Radix-Sort* is linear in n.

Our implementation of *RadixSort* used $\Theta(n)$ extra space for the link fields. Other implementations that do not use links also use extra space in $\Theta(n)$.

2.8
External Sorting

2.8.1   The Problem

We have seen that bucket-sorting algorithms may do few or no comparisons of keys. It is therefore appropriate to separate them from the class of algorithms studied at the beginning of this chapter. There is a class of algorithms that does sort by comparison of keys but that is also distinguished from those studied earlier because the number of comparisons done is not the most appropriate measure of the efficiency of the algorithms. These are the *external sorts*. Here we assume that the number of keys is so large that they cannot all fit in the computer's high-speed memory at one time; some must be stored on a slower external storage device. The time required to transfer data back and forth between the high-speed memory and the external storage device usually outweighs the time required to perform comparisons on data in the high-speed memory.

Most of the algorithms discussed in this section assume that the lists to be sorted are on tape or disk. We will consider some approaches specifically for sorting on tape with a small number of tape drives. With such storage devices it is most efficient to process the keys sequentially because tape rewinds, disk seeks, and other operations needed to locate a particular datum are very expensive in terms of time. The emphasis in devising good algorithms is on decreasing the number of times the data are accessed.

2.8.2 Polyphase Merge Sorting

In Section 2.3.6 we studied a straightforward algorithm to merge sorted lists. It made one pass over the data, processing the keys of each of the two files to be merged in sequential order. The algorithm is easily adapted to merge two files. The external sorting methods that we will study in this section rely heavily on the Merge algorithm. Because merging must be done repeatedly, these algorithms are called Polyphase Merge Sorting algorithms.

We define a *run* as a sequence of keys in nondecreasing order. The general scheme of the external sorts has two phases:

I.   Run construction and distribution: arrange the keys in runs in two (or more) files; and

II. Polyphase merge: repeatedly merge the runs until there is only one.

Let m be the number of keys that can fit in the high-speed memory at one time along with whatever programs are necessary. (To some extent, m will depend on the algorithm being used. In describing sorting algorithms throughout this chapter, we have usually ignored the fact that each key is associated with a record, perhaps a very large one. This point is relevant here since, if the records are large, m does not vary very much from one algorithm to another.) As usual, let n be the total number of keys.

The first, and simplest, external sorting algorithm uses four files: T_0, T_1, T_2, and T_3. The keys to be sorted are initially in T_0. The following outline describes the algorithm.

Algorithm 2.13   External Sort with Four Files (Outline)

{Phase I: Run construction and distribution}
while there are more records in T_0 **do**
    1. Read in m records (perhaps fewer the last time).
    2. Sort them using an internal sorting algorithm.
    3. If the previous run went in T_2, add this one to T_3; else to T_2.
end;

Rewind tapes, or reset disk files.

{Phase II: Merging the runs}
$i_1 := 2; i_2 := 3$ {indexes of input files}
$j_1 := 0; j_2 := 1$ {indexes of output files}
while there is more than one run **do**
    1. Merge the first run in T_{i_1} with the first run in T_{i_2}
        and put the resulting run in T_{j_1}.
    2. Merge the next run from T_{i_1} and T_{i_2} and
        put the result in T_{j_2}.
    3. Repeat steps 1 and 2 (putting the output alternately
        in T_{j_1} and T_{j_2}) until the end of T_{i_1} and T_{i_2}.
    4. Rewind tapes, or reset files.
        Add 2 (modulo 4) to i_1, i_2, j_1, and j_2
        to reverse the roles of input and output files.
end

At various times there may be one more run in T_{i_1} than in T_{i_2}. The odd run
would simply be copied to the appropriate output file. Figure 2.23 illustrates the
algorithm.

To analyze this and similar algorithms, we will count several operations,
including comparisons of keys and complete passes over the data. Rewinding tapes
is a slow operation and will take significant time if tapes are used. Thus we will
count tape rewinds too; this can be ignored for disks.

In one execution of the loop in the merge phase (lines 1 through 4), each
record is transferred from an input file to the high-speed memory where some operations (comparisons of keys) are performed, and then is transferred to an output file.
Then the tapes are rewound (simultaneously). Thus the number of rewinds and
passes over the data in the merge phase is the number of passes through the loop.
Let k be the number of keys in each run formed during the run construction phase.
(To simplify the analysis we will assume that all of the initial runs have the same
length; in fact, the last one may be smaller.) Then the total number of runs in the
files when the merge phase begins is $r = n/k$.

After the first merge pass there is a total of $\lceil r/2 \rceil$ runs in T_0 and T_1. After the
next merge pass there are $\lceil r/4 \rceil$ runs, and after the ith merge pass, there are $\lceil r/2^i \rceil$.
The algorithm terminates when there is only one run so the number of merge passes

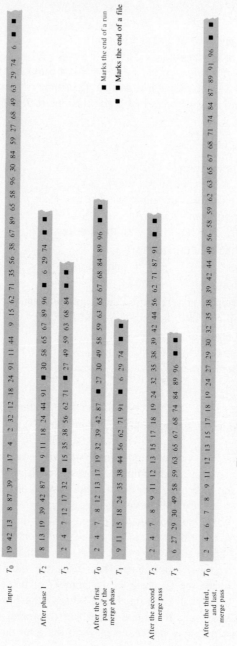

Figure 2.23 Polyphase merge with four files, $m = 6$.

is $\lceil \lg r \rceil$. Since the run construction phase does just one pass over the data followed by one rewind (of three tapes simultaneously), the total number of passes is $\lceil \lg r \rceil + 1$. Recall that $r = n/k$ and that in Algorithm 2.13 $k = m$, the number of keys that can be sorted in the high-speed memory. If we are working with a specific computer, m cannot be changed, but we see that the number of passes would be smaller if the merge phase started with fewer, larger runs, that is, if $k > m$. This observation motivates an important improvement in external sorting that will be described after we complete the analysis of Algorithm 2.13.

The number of comparisons done by Algorithm 2.13 depends on the sorting algorithm used in the run construction phase. Suppose an $\Theta(m \lg m)$ sort is used. Then the number of comparisons done in the sorting phase is in $\Theta((n/m)m \lg m) = \Theta(n \lg m)$. In the first merge pass $r/2$ pairs of runs, each of size k, are merged, requiring at most $r(2k-1)/2$ comparisons. The ith merge pass requires at most $r(2^i k-1)/2^i = n - (r/2^i)$ comparisons, so the total number of comparisons done in the merge phase is

$$\sum_{i=1}^{\lceil \lg r \rceil} \left(n - \frac{r}{2^i} \right) = n \lceil \lg r \rceil - \beta r,$$

where $1/2 < \beta \le 1$. Observe that the number of comparisons done in the merge phase would also be reduced if r were reduced, i.e., if k could be increased. For Algorithm 2.13, however, with $k = m$, the total number of comparisons done is in $\Theta(n \lg m + n \lg(n/m)) = \Theta(n \lg n)$. The total number of passes over the n keys is $\lceil \lg(n/m) \rceil + 1$.

2.8.3 Run Construction by Replacement Selection

The analysis of the merging in the Polyphase Merge Sort has shown that the number of passes over the keys would decrease if the size of the runs constructed in the first phase could be increased. Our approach to the run construction was very simple and straightforward. Assuming m is the number of keys that fit in the high-speed memory at one time, we read in m keys, sort them using a fast internal sort, and output a run of size m. All of the runs, except perhaps the last, will have exactly m keys. Now we will consider an alternative method of constructing runs that is expected to produce longer ones. This method is called *replacement selection*. The main idea behind it is that, as soon as one record has been written out to one of the output files, another may be read in from the input file. If its key is larger than the key just written, the new record may be made part of the current run. If its key is smaller, it will be held for the next run. Thus records may be added to the current run until all of memory is filled with records whose keys are smaller than the last key put in the run. At that point the current run is ended and the smallest key in memory is the first key of the next run. While a run is being constructed, we may think of the keys in memory as being divided into two classes: the active keys, those that are at least as large as the last key written out and therefore will be part of the current run, and the inactive keys, those that are smaller than the last key written and

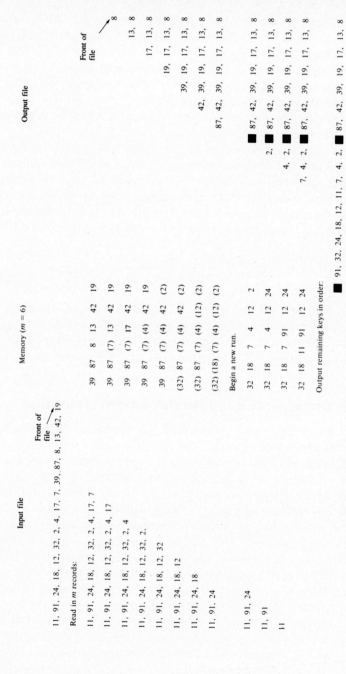

Figure 2.24 Run construction using replacement selection. (■ marks the end of a run).

therefore will be part of the next run. Figure 2.24 illustrates the run construction. Inactive keys are shown in parentheses.

How can we easily distinguish between active and inactive keys in the run construction algorithm? What methods and data structure should be used?

Constructing a run requires that we first find the smallest key in a set of keys and output it, then find the next smallest, and so on. The straightforward algorithm for finding the smallest key in a list (similar to Algorithm 1.3) does not provide us with any useful information for finding the next smallest; we would have to run the whole algorithm repeatedly. If, instead, we run a "tournament" on the keys to find the smallest, it is not hard to save some information that enables us to keep finding the next smallest key quickly. In a tournament, keys are paired up and compared, the "winners" of those comparisons (i.e., the smaller keys) are compared, and so on until the champion (the smallest key) is found. Figure 2.25 illustrates a tournament on the first six keys in the example from Fig. 2.24. Each internal node shows the winner of the match between the two keys at the previous level. Now suppose we know the champion. The next smallest key must be one of the keys that was compared directly with the champion and lost. (Any key that lost to some other key can be at best in third place.) The internal nodes in Fig. 2.25 do not show enough information for us to find the next smallest key. The data structure for the run construction algorithm is a tree in which the leaves contain the records and keys as they are read in, and the internal nodes contain pointers to the loser, not the winner, of each comparison in the tournament. See Fig. 2.26.

Each time a record is written to a file, a new one is read in and the structure must be updated. We will assume that a pointer *winner* points to the leaf with the smallest key. To describe the core of the algorithm, finding the next winner and updating the loser pointers, we will temporarily ignore the problem of initializing the tree (i.e., finding the champion in the first place) and the case when the new key

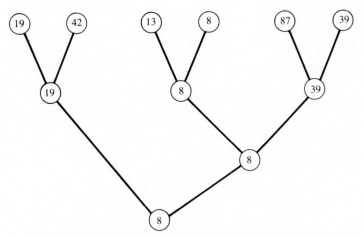

Figure 2.25   A tournament for finding the smallest key.

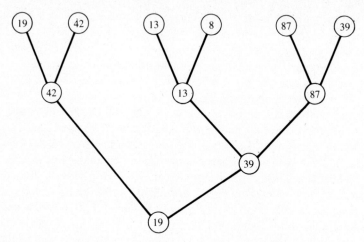

Figure 2.26 The tournament tree with internal nodes showing the loser, not the winner.

must wait for the next run because it is too small to fit in the current one. We also assume that *m* is even.

Algorithm 2.14 Update the Tournament Structure

Input: The tree structure described in the previous paragraph, just after the smallest key (at *winner*) has been written and a new key has been read into the node at *winner*. Each internal node has a *loser* field that contains a pointer to the loser of the comparison represented at that node. Each node has a *parent* field that contains a pointer to the node's parent.

Output: The tree with *winner* pointing to the leaf with the smallest key, and the *loser* pointers updated.

```
procedure UpdatePointers (var winner: pointer);
var
    ptr: pointer;
begin
    ptr := winner↑.parent;
    while ptr ≠ nil do
        if ptr↑.loser↑.key < winner↑.key then
            Interchange (ptr↑.loser, winner);
        end { if };
        ptr := ptr↑.parent
    end { while }
end { UpdatePointers }
```

What shall we do with the inactive records, i.e., those that are not part of the current run? A simplistic answer is to mark them somehow, skip over them during the updating procedure just described, and, finally, when all the records in memory are so marked, set up a new tournament tree and begin working on a new run. This solution is inefficient and certainly inelegant. Fortunately we can do better. Suppose we associate a run number with each record in memory. The number can easily be assigned when the record is read in; if its key is larger than the key just written out, its run number is the same as the latter's; otherwise, it is one higher. Each key may be considered as a pair $(runNumber, key)$ where $runNumber$ is the run number and key is the original key. When the *loser* pointers are updated, run numbers, as well as keys, are compared, and the pair $(runNumber_1, key_1)$ is considered smaller than $(runNumber_2, key_2)$ if $runNumber_1 < runNumber_2$ or if $runNumber_1 = runNumber_2$ and $key_1 < key_2$. Now we can correctly manipulate keys from different runs in the same tree at one time. When the tree is filled with keys that must go in a new run, the *loser* pointers are already properly set and no initialization need be done to the data structure for the new run. The expanded algorithm that handles the case when the new record does not belong in the current run is presented later. Here we assume that there is a $runNumber$ field in each leaf, that run is the number of the run currently being constructed, and that $winner$ points to the next record to be written out.

{ Output a key }
if $winner\uparrow.runNumber \neq run$ **then**
    do whatever is necessary for beginning a new run in an output file,
        including determining which file the next run goes in;
    $run := winner\uparrow.runNumber$
end { if };
$lastKeyOut := winner\uparrow.key$;
write out the record at $winner$;

{ Input a key }
read the next record from the input file into the node at $winner$;
if $winner\uparrow.key < lastKeyOut$ **then**
    $winner\uparrow.runNumber := run+1$
    { else $winner\uparrow.runNumber$ is still run as before }
end { if };

{ Update *loser* pointers }
UpdatePointers ($winner$)
    { The key comparison in *UpdatePointers* must be modified
    to consider run numbers. }

In Fig. 2.27 the tree is shown at several stages of the run construction for the example in Fig. 2.24. The initial arrangement of the records in Fig. 2.27(a) may seem arbitrary; it is not. We will see how the initial configuration is achieved later. At this point, it is strongly suggested that the reader begin with the tree in

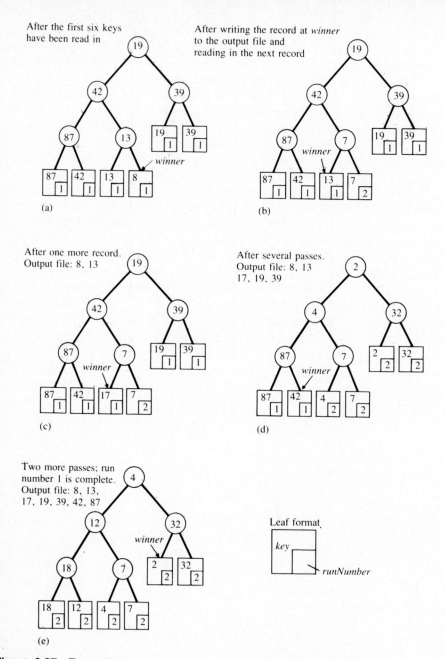

Figure 2.27 Trees for run construction. (The internal nodes contain pointers to the keys shown in the node.)

Fig. 2.27(a), with *run* = 1, and follow the steps of the algorithm to see how the structure is modified to produce the remaining trees in Fig. 2.27.

Now how do we get the tree in Fig. 2.27(a) set up in the first place? The trick to the initialization is to start with a tree with "empty" nodes and assume the leaves contain keys with value ∞ (or *maxint*) and run number 0, a run that will not be written to any output file. The *loser* pointers may be initialized in any way such that each of the leaves except one is pointed to by exactly one *loser* pointer, and *winner* is initialized to point to the one leaf that is not a loser. (For the example in Fig. 2.27, *winner* was initially assigned the location of the first leaf, and the *loser* pointers were assigned level by level, left to right, as shown in Fig. 2.28.) Then we use the run construction algorithm as we have developed it so far, but not writing out keys from run number 0. Each time a new record is read in, exactly one *loser* pointer that had pointed to a leaf with *runNumber* = 0 will be modified to point to a leaf containing a record. Until the tree is full, *winner* will point to leaves with *runNumber* = 0, thus indicating an "empty" leaf where the next record may be put. The tree in Fig. 2.27(a) was obtained in this way from the first six keys in Fig. 2.24.

Terminating the run construction algorithm smoothly requires the use of another "dummy" run; when there are no more records in the input file, the run number of the leaf to be filled is incremented just as if there were a new key that did not belong to the current run. A counter is used to keep track of the number of real runs. Thus, whenever a record is to be written out, its run number must be tested to see if it is 0 or if it is higher than the maximum run number (the value of which is not known in advance, since it depends on the number and arrangement of the keys in the input). With the details of initialization and termination, we now have the complete Replacement Selection algorithm.

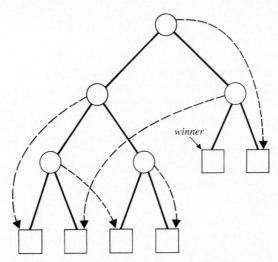

Figure 2.28 Initialization of *winner* and *loser* pointers; *loser* pointers are shown as dashed lines.

Algorithm 2.15   Replacement Selection

Input: A file of records.

Output: Files of sorted runs.

```
procedure ReplacementSelection;
var
    winner: pointer;
    trueRuns: integer; { the number of real runs }
    run: integer;        { the current run number }
    lastKeyOut: Key;

begin
{ Initialization }
    trueRuns := 0; run := 0; lastKeyOut := ∞;
    Initialize tree structure with parent and loser pointers;
    Initialize winner;
    Assign 0 to the runNumber field in each leaf;

    while winner↑.runNumber ≤ trueRuns do

        { Output a key }
        if winner↑.runNumber ≠ run then
            do whatever is necessary to begin a new run in an output file,
                including determining which file the next run goes in;
            run := winner↑.runNumber
        end { if };
        if winner↑.runNumber ≠ 0 then
            lastKeyOut := winner↑.key;
            write out the record at winner;
        end { if };

        { Input a key }
        if there is more input then
            read the next record from the input file into winner;
            if winner↑.key < lastKeyOut then
                winner↑.runNumber := run+1;
                trueRuns := winner↑.runNumber
            end { if new key < last key out }
        else winner↑.runNumber := trueRuns+1 { dummy run }
        end { if more input };

        { Update loser pointers }
        UpdatePointers (winner)
            { The key comparison in UpdatePointers must be modified to consider
            run numbers. }

    end { while }
end { ReplacementSelection }
```

Many details about I/O have been omitted from the algorithm. There are also details to be worked out for the representation of the tree. Paths in the tree are traversed only in an upward direction — i.e., from leaf to root — so from each node we need a way of finding the location of its parent. Since the tree doesn't have "missing" nodes, we can use a storage scheme similar to the one used for heaps, that is, store the nodes sequentially so that the parent of the ith node is the $\lfloor i/2 \rfloor$th node. Allocating space for the tree structure and determining the exact location of the parent of a node is complicated by the facts that the leaves, which contain the records, require much more space than the internal nodes, and the size of the records is probably not known when the program is written. A clever storage layout solves these problems: Each leaf is stored with one internal node so that there are, say, m_0 large nodes. (Since there are m_0 leaves and only $m_0 - 1$ internal nodes, a dummy internal node number 0 is used.) Thus if the tree nodes are numbered consecutively from the root (1) to the last leaf ($2m_0 - 1$), then for each i such that $0 \le i < m_0$, node i, an internal node, is stored with node $m_0 + i$, a leaf. The location of the parent of the internal node and the location of the parent of the leaf may be computed when the structure is initially set up and stored with the data. If this storage arrangement is used, *winner* would be initialized to point to node 0, the dummy internal node and first leaf. Each *loser* pointer may be initialized to point to the double node that contains it. This is how the *loser* pointers in Fig. 2.28 were initialized. Possible type declarations for the nodes would be:

> *InternalNodeType* = **record**
>   *internalParent, loser* : *pointer*
> **end** { InternalNodeType };
> *LeafNodeType* = **record**
>   *runNumber*: *integer*;
>   *key*: *Key*;
>   *leafParent*: *pointer*;
>   { other fields in the records to be sorted }
> **end** { LeafNodeType };
> *NodeType* = **record**
>   *internalNode*: *InternalNodeType*;
>   *leafNode*: *LeafNodeType*
> **end** { NodeType };

2.8.4   Analysis of Algorithm 2.15

How much time does the run construction algorithm take? The amount of work done by the parts of the algorithm that output and input keys is bounded by a constant for each key. Thus the total number of steps is dominated by the number of comparisons of keys done in *UpdatePointers*. Each time the structure is updated, a comparison is done for each *loser* pointer on the path from a leaf to the root. Let m_0 be the number of records that can fit in memory, along with the tree structure and program, at one time. Then m_0 is the number of leaves in the tree and the number of comparisons done for each update is $\lceil \lg m_0 \rceil$ or $\lfloor \lg m_0 \rfloor$. (The leaves may be on two levels as in Fig. 2.27.) Suppose that there are n records in the input file. There are m_0 dummy keys for run number 0. The structure is updated after the output step for each of these (even though no writing is done for run 0). The algorithm terminates when it encounters a dummy key from run $trueRuns+1$ at the output step, so the run construction algorithm uses $\Theta((n+m_0)\lg m_0)$ time. The time used by the first method we considered for run construction, i.e., reading in as many records as will fit, sorting them, and writing out a run, is in $O(n\lg m)$, where m is the number of records that fit in memory along with the sorting program. If m and m_0 are approximately equal (and we will say more about that later), the two methods take similar amounts of time. (For neither method did we take into account the time required for input and output since it would be highly dependent on the sophistication of the system on which the algorithms are implemented.)

The point of studying the replacement selection method was not to decrease the time required for run construction, but rather to increase the size of the runs generated. That, we saw, would decrease the amount of time required for the merge phase of an external sort. Have we succeeded? In the example used in this section (Figs. 2.24 and 2.27) $m_0=6$ but the runs constructed had lengths 7 and 9. It is possible to prove that with random keys the average run length produced by replacement selection is $2m_0$. (See the notes and references at the end of the chapter.) In practice, the records to be sorted are often already partly in order and even longer runs may be produced, but for the purpose of comparing the run construction methods we will use the length $2m_0$ for replacement selection. The exact values of m and $2m_0$ depend in part on the programs used, so our comments will not be quantitatively precise. However, we can draw one important conclusion: When the size of the records is very large relative to the size of the keys, m and m_0 are fairly close and replacement selection produces runs about twice as long as run construction by sorting. With large records m and m_0 are fairly close because the space needed for the *loser* pointers will not significantly affect the total number of records that can fit in memory at once. The sorting algorithm may use extra pointers also if the records are large. (On the other hand, if the records contained only the keys, then m_0 could be less than $m/2$ because of the extra fields used in the tree structure for replacement selection. In this case the two methods would produce runs of comparable sizes.) The advantage of using replacement selection increases with the size of the records.

An Application: Priority Queues

A priority queue is a queue in which each item has an assigned priority number and the usual first in, first out rule is modified so that an item is not removed if there are any items with higher priority still in the queue. Priority queues are often used for programming jobs submitted to a computer.

The data structure and updating algorithm for replacement selection can be used for a priority queue with priority numbers playing the role of run numbers and sequentially assigned item (or job) numbers playing the role of keys.

*2.8.5 Polyphase Merging with Three Tapes

The polyphase merge with four files (discussed in Section 2.8.2) would need four tape drives if the files are on tapes. Many systems do not have four tape drives, so we will now consider the problem of doing the merge phase with three tapes.

We assume that all records were initially on T_0 and that in the first phase r runs of size k were constructed and written on the other two tapes, T_1 and T_2. The straightforward, though inefficient, approach we consider first would put an equal number of runs on the two tapes. We can merge runs from the input tapes as before, but since there are only three tapes, all of the resulting runs will go on the one output tape. Each merge pass must begin with the runs on two "input" tapes, so before beginning the next merge, half of the runs would be copied onto another tape. Thus Phase II of the algorithm would be rewritten as follows:

Phase II′   Merging the runs using three tapes

> $i_1 := 1; i_2 := 2;$ { indexes of "input" tapes }
> $j := 0;$ { index of "output" tape }
> **while** there is more than one run **do**
> > 1′   Merge runs from T_{i_1} and T_{i_2} and put the resulting runs on T_j.
> > 2′   Rewind all three tapes. Copy half the runs from T_j onto T_{i_2}. Rewind T_{i_2}.
> > 3′ $i_1 := i_1+1; i_2 := i_2+1; j := j+1$ (all modulo 3)
> **end** { while }

The number of comparisons done when three tapes are used is the same as the number of comparisons done with four tapes.

We define the number of passes over the data to be the average number of times each key is moved from one tape to another, i.e., the total number of keys moved divided by n. (For the four-file sort every key was moved the same number of times, so we did not need to state this definition earlier.) The number of passes in the three-tape merge will be larger than for a four-tape merge because of the extra partial passes for copying. There is one pass for the run construction phase, then $1.5\lceil \lg r\rceil - 1/2$ passes in the merge phase. (The $-1/2$ reflects the fact that no keys must be copied after the final merge pass. That is, in practice the algorithm would terminate in line 2′.) Thus, using three tapes as described, the number of passes is $1.5\lceil \lg r\rceil + 1/2$.

Copying keys from one tape to another seems like a very nonproductive effort. Can it be avoided if we are restricted to using only three tapes? Clearly if, in the run construction phase, the runs are distributed equally on the two output tapes, then after the first merge pass all of the keys will be on one tape and some must be copied. Suppose the runs are not distributed equally. For example, suppose eight runs are on tape T_1 and five are on tape T_2, as in Fig. 2.29(a), where runs are shown as rectangles labeled with their size (number of keys). After merging five pairs of runs onto T_0, T_2 is empty and T_1 still contains three runs. Now, T_2 and T_0 may be rewound and three pairs of runs from T_1 and T_0 may be merged and placed on T_2.

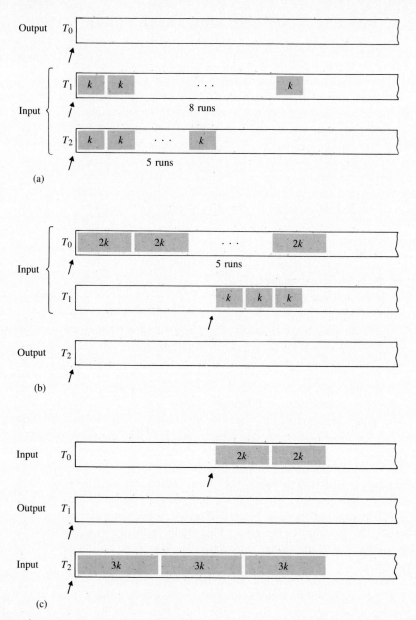

Figure 2.29 Merging with three tapes (no copying). Arrows indicate the location of the tape head.

(See Figs. 2.29(b) and 2.29(c). Note that runs of different sizes are being merged.) After rewinding T_1 and T_2, two pairs of runs from T_2 and T_0 are merged and placed on T_1. After two more merge passes, the details of which may be worked out by the reader, there will be one run of size $13k$ on T_2, and the sort will be completed.

How many passes over the data were made in this example? To get from the situation in Fig. 2.29(a) to that in Fig. 2.29(b), only 10/13 of the keys are transferred from T_1 and T_2 to T_0; three runs remain untouched on T_1. To get to the situation in Fig. 2.29(c), 9/13 of the keys are moved. In the remaining steps not shown 10/13, then 8/13, and finally, on the last pass, 13/13 of the keys are moved. The total number of passes during the merge phase is 50/13, or approximately 3.8. If the 13 runs had been arranged with six on one tape and seven on another, and the Phase II' method with copying were used, it would do four complete passes over the data while merging runs and (at most) 22/13 copying passes for a total of 74/13, or approximately 5.7. At no point in the example in Fig. 2.29 did we get "stuck" and have to copy keys; that is, at no point were all the keys on one tape before they were completely sorted. Were we just lucky this time? Or can we arrange the runs constructed in the first phase of the algorithm so that copying can always be avoided? Clearly, copying can be avoided by putting one run on T_2 and all the other $r-1$ runs on T_1, but this arrangement would require rewinding two tapes each time a pair of runs is merged. The total number of rewinds (of two or three tapes simultaneously) would be r. As an exercise, the reader may compute the total number of passes over the keys; it is not in $O(\lg r)$. So it is not difficult to avoid copying keys. Our goal, however, is to eliminate copying *and* to keep the number of passes over the keys relatively small.

Look at Table 2.2, which provides a summary of the example illustrated in Fig. 2.29. The table lists the number and size of the runs on the tapes after each rewind. The columns of numbers in the table are all sequences of Fibonacci

Table 2.2
Data on the example in Fig. 2.29. The total number of runs at the beginning of the merge phase is $13 = F_7$.

Runs on the two input tapes		Runs merged	
Number	Size (As a multiple of k)	Number of pairs	Total number of size k runs transferred from one tape to another
8, 5	1, 1	5	10
5, 3	2, 1	3	9
3, 2	3, 2	2	10
2, 1	5, 3	1	8
1, 1	8, 5	1	13
1	13		

numbers.[4] In general, if the number of runs constructed in the first phase of the sort is a Fibonacci number F_s, for some s, then distributing the runs so that F_{s-1} runs are on one tape and F_{s-2} are on another will ensure that there will always be runs on two tapes ready for merging without copying until the keys are sorted. Assuming that the run construction phase has put F_{s-2} runs on T_1 and F_{s-1} on T_2, lines 1' and 2' of the Phase II' algorithm for merging using three tapes would be changed to the following:

> 1'' Merge runs from T_{i_1} and T_{i_2}, putting the resulting runs on T_j, until T_{i_1} is empty.
>
> 2'' Rewind T_{i_1} and T_j.

With these changes, we will refer to the algorithm as Phase II'': merging with three tapes using a Fibonacci distribution. If initially there are F_s runs, the Phase II'' loop will be executed $s-1$ times. To aid in computing the number of passes (that is, the average number of times each key is transferred from one tape to another), we use Table 2.3, a generalized version of Table 2.2.

The number of passes is the sum of the righthand column divided by n (i.e., $F_s k$, where k is the initial run size). Thus we must compute

$$\frac{\sum_{i=1}^{s-2} F_i F_{s-i+1} k}{F_s k} = \frac{\sum_{i=1}^{s-2} F_i F_{s-i+1}}{F_s}. \tag{2.4}$$

Lemma 2.15

$$\sum_{i=1}^{s-2} F_i F_{s-i+1} = \frac{s-5}{5} F_{s+1} + \frac{2s+2}{5} F_s, \text{ for } s \geq 2.$$

Proof. The proof is by induction on s. In the induction step we will need the assumption that $s \geq 4$, so for the basis of the induction the equation in the lemma must be checked for $s=2$ and $s=3$. The reader may easily verify that, for $s=2$, both sides have value 0, and for $s=3$ both sides have value 2.

Now, assuming $s \geq 4$, and using the recurrence relation for the Fibonacci numbers and the induction assumption, we have

$$\sum_{i=1}^{s-2} F_i F_{s-i+1} = \sum_{i=1}^{s-2} F_i (F_{s-i} + F_{s-i-1})$$

$$= \sum_{i=1}^{s-2} F_i F_{s-i} + \sum_{i=1}^{s-2} F_i F_{s-i-1}$$

$$= \sum_{i=1}^{s-2} F_i F_{(s-1)-i+1} + \sum_{i=1}^{s-2} F_i F_{(s-2)-i+1}$$

---

[4] The Fibonacci numbers are a sequence of integers defined by the recurrence $F_i = F_{i-1} + F_{i-2}$ for $i > 1$, with boundary conditions $F_0 = 0$ and $F_1 = 1$.

Table 2.3

Runs on the input tapes (k = initial run size)					Merged runs
Number_1	Size_1 (as a multiple of k)	Number_2	Size_2 (as a multiple of k)	Number of pairs	Number of keys transferred from one tape to another = Number of pairs × (Size_1 + Size_2), as a multiple of k
F_{s-1}	$1 = F_2$	F_{s-2}	$1 = F_1$	F_{s-2}	$F_{s-2}(F_2 + F_1) = F_{s-2}F_3$
F_{s-2}	$2 = F_3$	F_{s-3}	$1 = F_2$	F_{s-3}	$F_{s-3}(F_3 + F_2) = F_{s-3}F_4$
F_{s-3}	$3 = F_4$	F_{s-4}	$2 = F_3$	F_{s-4}	$F_{s-4}(F_4 + F_3) = F_{s-4}F_5$
.	.	.	.	.	.
F_{s-i}	F_{i+1}	$F_{s-(i+1)}$	F_i	$F_{s-(i+1)}$	$F_{s-(i+1)}(F_{i+1} + F_i) = F_{s-(i+1)}F_{i+2}$
.	.	.	.	.	.
$F_2 = 1$	F_{s-1}	$F_1 = 1$	F_{s-2}	F_1	$F_1(F_{s-1} + F_{s-2}) = F_1F_s$
$F_1 = 1$	F_s				

$$= \left[\sum_{i=1}^{s-3} F_i F_{(s-1)-i+1} + F_{s-2} \right] + \left[\sum_{i=1}^{s-4} F_i F_{(s-2)-i+1} + F_{s-3} + F_{s-2} \right]$$

$$= \left[\frac{(s-1)-5}{5} F_s + \frac{2(s-1)+2}{5} F_{s-1} + F_{s-2} \right]$$

$$+ \left[\frac{(s-2)-5}{5} F_{s-1} + \frac{2(s-2)+2}{5} F_{s-2} + F_{s-3} + F_{s-2} \right]$$

$$= \frac{s-6}{5} F_s + \frac{3s-2}{5} F_{s-1} + \frac{2s+3}{5} F_{s-2}$$

$$= \frac{3s-3}{5} F_s + \frac{s-5}{5} F_{s-1} = \frac{s-5}{5} F_{s+1} + \frac{2s+2}{5} F_s. \qquad \square$$

(The formula in Lemma 2.15 was not pulled out of a hat. The reader interested in its derivation may consult the notes and references at the end of this chapter.) Using Eq. 2.4 we see that

$$\text{the number of passes} = \frac{s-5}{5} \frac{F_{s+1}}{F_s} + \frac{2s+2}{5}. \qquad (2.5)$$

We want to relate this expression to the number of runs, $r = F_s$, so that the number of passes done using the Fibonacci distribution can be compared with the number of passes done by the other methods discussed. We will use the following lemma, the proof of which is easy and is left as an exercise.

Lemma 2.16 Let F_j be the jth Fibonacci number. Then for $j \geq 5$,

$$\frac{F_{j+1}}{F_j} \leq \frac{13}{8}.$$

We also use the fact that

$$F_j \approx \frac{[^1\!/_2(1+\sqrt{5})]^j}{\sqrt{5}}. \qquad (2.6)$$

(See the references. In fact, F_j is $\dfrac{[^1\!/_2(1+\sqrt{5})]^j}{\sqrt{5}}$ rounded to the nearest integer.)

Theorem 2.17 The number of passes done in Phase II″ (merging with three tapes using a Fibonacci distribution) is approximately $1.04 \lg r$, where r is the number of runs produced by the sort phase.

Proof. Using Eq. 2.6, with $j = s$, we have

$$\sqrt{5} F_s \approx [^1\!/_2(1+\sqrt{5})]^s$$

and, taking logs,

$$\lg(\sqrt{5} F_s) = s [\lg(1+\sqrt{5}) - 1].$$

So

$$s \approx \frac{\lg\sqrt{5}+\lg F_s}{\lg(1+\sqrt{5})-1} \approx 1.67+1.43\lg F_s.$$

Therefore, using Eq. 2.5 and Lemma 2.16, the number of passes is approximately

$$\frac{1.67+1.43\lg F_s-5}{5} \frac{13}{8} + \frac{2(1.67+1.43\lg F_s)+2}{5} \approx 1.04\lg F_s$$

$$= 1.04\lg r . \qquad \square$$

The result in Theorem 2.17 compares quite favorably with the number of passes done by Phase II', where runs had to be copied from one file to another. In fact, using the Fibonacci distribution is almost as good as using four tapes. The Fibonacci distribution is also very useful if *more* tape drives are available; it can be generalized to produce very fast algorithms. With six tapes, for example, the number of passes would be roughly $0.55\lg r+0.86$.

*2.8.6   Arranging the Runs for Merging

We have seen that, if the number of sorted runs is a Fibonacci number, say F_s for some s, the number of merge passes will be relatively low if the runs are distributed so that F_{s-1} are on one tape and F_{s-2} on another. But the total number of runs is not known in advance, and it may not be a Fibonacci number. (The size of each run, and hence the total number, depends on the particular sequence of keys to be sorted if replacement selection is used to construct the runs.) Thus our fast merge phase algorithm is not of much use without an efficient way of arranging the runs produced by the run construction phase to ensure that their arrangement on the tapes approximates the Fibonacci arrangement. This is accomplished by writing the runs so that first there is one run on one tape and none on the other, then two on one tape and one on the other, then three and two, then five and three, and so on, so that after each round (except perhaps the last) the numbers of runs on the tapes are two consecutive Fibonacci numbers. The writing out of runs ends when there are no more keys. At that time the number of runs on one tape is between F_{s-2} and F_{s-1}, and the number of runs on the other is between F_{s-3} and F_{s-2}. The merge phase can then act as though there are enough "dummy" runs on each tape to give F_{s-1} and F_{s-2} runs, respectively. By keeping track of the number of dummy runs and treating them appropriately, the merge phase can be made to mimic a merge with F_s runs.

To write the algorithm for arranging the runs, we must determine how many runs are to be added to each tape during each round of writing. Let $runs_{t,j}$ be the number of runs on T_t after j rounds. Then

$$runs_{1,1} = 1 = F_1, \quad runs_{2,1} = 0 = F_0,$$

and

$$runs_{1,j} = F_j, \quad runs_{2,j} = F_{j-1}, \quad \text{for } j > 1.$$

Let $add_{t,j}$ be the number of runs to be added to T_t during round j. Then

$$add_{1,j} = runs_{1,j} - runs_{1,j-1} = F_j - F_{j-1} = F_{j-2} = runs_{2,j-1}$$

and

$$add_{2,j} = runs_{2,j} - runs_{2,j-1} = runs_{1,j-1} - runs_{2,j-1}.$$

Thus $add_{t,j}$, for $t = 1,2$, can easily be computed from $runs_{t,j-1}$, $t = 1,2$. Also, of course,

$$runs_{1,j} = runs_{1,j-1} + runs_{2,j-1} \quad \text{and} \quad runs_{2,j} = runs_{1,j-1}.$$

In the algorithm the second subscripts are not needed; the values of add_1, add_2, $runs_1$, and $runs_2$ are updated after each round of writing. A straightforward algorithm would write out add_1 runs on T_1, then add_2 runs on T_2, then update the values of the add's and $runs$'s and begin the next round of run construction and writing. It would halt when there are no more runs to be written on the tapes. If the runs are written on the tapes in this manner, it is possible that when the algorithm terminates all of the dummy runs will be on T_2. "Merging" a run with a dummy means just copying the run from an input tape to the output tape. "Merging" two dummy runs requires only that the number of dummies on each input tape be decremented and the count of dummies on the output tape be incremented. Hence, when possible, we would like the number of dummies on each tape to be equal. At the beginning of each round of writing, $add_{1,j} = F_{j-2}$ and $add_{2,j} = F_{j-3}$. Since $add_{1,j}$ is much larger than $add_{2,j}$ the scheme is to write F_{j-4} runs on T_1, that is, write only on T_1 until $add_{1,j} = F_{j-3}$, and then alternate between the two tapes. Thus we have the following algorithm.

Algorithm 2.16   Fibonacci Run Distribution

```
runs₁ := 0; runs₂ := 0;   { runs on tape after round 0 }
add₁ := 1; add₂ := 0;     { runs to be added in round 1 }
while true do
    while add₁ ≠ add₂ do
        if end-of-input then return end;
        Write a run on T₁;
        add₁ := add₁ - 1
    end { while add₁ ≠ add₂ };
    while add₂ > 0 do
        for t := 1 to 2 do
            if end-of-input then return end;
            Write a run on Tₜ;
            addₜ := addₜ - 1
        end { for }
    end { while add₂ > 0 };
    { Compute number of runs after current round. }
```

$temp := runs_1; \ runs_1 := runs_1 + runs_2; \ runs_2 := temp;$
{ Compute number of runs to be added on next round. }
$add_1 := runs_2; \ add_2 := runs_1 - runs_2$
end { while true }

Exercises

Section 2.1: Introduction

2.1. One of the easiest sorting algorithms to understand is one that we call Maxsort. It works as follows: Find the largest key, say *max*, in the unsorted section of the list (initially the whole list) and then interchange *max* with the key in the last position in the unsorted section. Now *max* is considered part of the sorted section consisting of larger keys at the end of the list; it is no longer in the unsorted section. Repeat this procedure until the whole list is sorted.

a)   Write out an algorithm for Maxsort assuming an array L that contains n keys to be sorted.

b)   How many comparisons of keys does Maxsort do in the worst case? On the average?

2.2. The next few exercises are about a sorting method called Bubblesort. It sorts by making several passes through the list, comparing pairs of keys in adjacent locations and interchanging them if they are out of order. That is, the first and second keys are compared and interchanged if the first is larger than the second; then the (new) second and the third keys are compared and interchanged if necessary, and so on. It is easy to see that the largest key will filter down to the end of the list; on subsequent passes it will be ignored. If on any pass no entries are interchanged, the list is completely sorted and the algorithm can halt. The following algorithm makes this informal description of the method precise.

Algorithm 2.17 Bubblesort

Input: L, an array of keys, and $n \geq 0$, the number of keys.

Output: L with keys in nondecreasing order.

```
procedure Bubblesort (var L: Array; n: integer);
    { Sorts keys into nondecreasing order. }
var
    numPairs: integer; { the number of pairs to be compared }
    didSwitch: boolean; { true if an interchange is done }
    j: integer;
begin
    numPairs := n;
    didSwitch := true;
    while didSwitch do
        numPairs := numPairs-1;
        didSwitch := false;
```

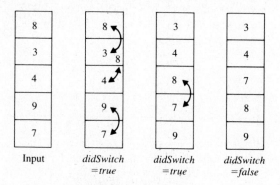

| Input | didSwitch
=true | didSwitch
=true | didSwitch
=false |

Figure 2.30 Bubblesort.

```
for j := 1 to numPairs do
  if L[j] > L[j+1] then
      Interchange (L[j], L[j+1]);
      didSwitch := true
  end { if }
 end { for }
end { while }
end { Bubblesort }
```

The example in Fig. 2.30 illustrates how Bubblesort works.

a) How many key comparisons does Bubblesort do in the worst case? What arrangement of keys is a worst case?

b) What arrangement of keys is a "best case" for Bubblesort, i.e., for what input does it do the fewest comparisons? How many comparisons does it do in the best case?

2.3. The correctness of Bubblesort (Exercise 2.2) depends on several facts. These are easy to verify but worth doing in order to consciously recognize the mathematical properties involved.

a) Prove that, after one pass through the list, the largest entry will be at the end.

b) Prove that, if there is no pair of consecutive entries out of order, then the entire list is sorted.

2.4. We can modify Bubblesort (Exercise 2.2) to avoid unnecessary comparisons in the tail of the list by keeping track of where the last interchange occurred in the **for** loop.

a) Prove that if the last exchange made in some pass occurs at the jth and $(j+1)$st positions, then all entries from the $(j+1)$st to the nth are in their correct position. (Note that this is stronger than saying simply that these items are in order.)

b) Modify the algorithm so that if the last exchange made in a pass occurs at the jth and $(j+1)$st positions, the next pass will not examine any entries from the $(j+1)$st position to the end of the list.

c) Does this change affect the worst-case behavior of the algorithm? If so, how?

2.5. Can something similar to the improvement in the preceding exercise be done to avoid unnecessary comparisons when the keys at the beginning of the list are already in order? If so, write out the modifications to the algorithm. If not, explain why not.

Section 2.2: Insertion Sort

2.6. We observed that a worst case for Insertion Sort occurs when the keys are initially in decreasing order. Describe some other initial arrangements of the keys that are also worst cases.

2.7. Consider the following variation of Insertion Sort: For $2 \leq i \leq n$, to insert the key $L[i]$ among $L[1] \leq L[2] \leq \cdots \leq L[i-1]$, do a binary search to find the correct position for $L[i]$.

   a)   How many key comparisons would be done in the worst case?

   b)   What is the total number of times keys are moved in the worst case?

   c)   What is the order of the worst-case running time?

   d)   Can the number of moves be reduced by putting the keys in a linked list instead of an array? Explain.

2.8. In the average analysis of Insertion Sort we assumed that the keys were distinct. Would the average for all possible inputs, including cases with duplicate keys, be higher or lower? Why?

2.9. Show that a permutation on n items has at most $n(n-1)/2$ inversions. Which permutation(s) have exactly $n(n-1)/2$ inversions?

Section 2.3: Quicksort and Mergesort: Divide and Conquer

2.10. How many key comparisons does Quicksort (Algorithms 2.2 and 2.3) do if the list is already sorted? How many interchanges of keys does it do?

2.11. Prove that if after each call to *Split* (Algorithm 2.3), the borders of the larger sublist are stacked and the other sublist is sorted first, then the maximum stack size is in $O(\lg n)$.

2.12. How would the action of Quicksort differ if a queue were used instead of a stack to store the borders of the sublists yet to be sorted? Consider correctness, time, and space usage.

2.13. Suppose that, instead of choosing $L[first]$ as x, *Split* lets x be the median of $L[first]$, $L[\lfloor (first+last)/2 \rfloor]$, and $L[last]$. How many key comparisons will Quicksort do in the worst case to sort n keys? (Remember to count the comparisons done in choosing x.)

*2.14. This problem examines an alternative algorithm for *Split*. Its inner loops are extremely fast. Let $x = L[first]$. The strategy (illustrated in Fig. 2.31) is to use two index variables i and j that move inward from opposite ends of the list looking for keys that are "out of place" with respect to x. That is, i moves along from the beginning of the list until it finds a key larger than or equal to x, and j moves back from the end of the list until it finds a key smaller than or equal to x. These two keys are interchanged, and then i and j continue scanning for "out of place" keys until they meet (or cross). The trick that keeps the inner loops so short is to avoid explicit testing for when i or j run off the ends of the section of the list we are working on. This is done by using x itself as a "sentinel" at the beginning of our sublist (that's why the test in j's loop is "$L[j] \leq x$" rather than "$L[j] < x$") and by requiring a large sentinel at the end of L. A side effect of keeping the loops fast and simple is that one extra interchange is done before discovering that i and j have crossed; it is undone after exiting the loops.

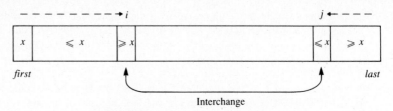

Figure 2.31   Alternative Split Strategy.

Algorithm 2.18   Alternative Split for Quicksort

> **procedure** *Split2* (*first*, *last*: *Index*; **var** *splitPoint*: *Index*);
>   { This procedure requires a sentinel value at the end of
>   the array *L* that is larger than any of the keys. }
> **var**
>   *x*: *Key*;
>   *i*, *j* : *Index*;
> **begin**
>   *x* := *L*[*first*];
>   *i* := *first*; *j* := *last*+1;
>   **repeat**
>   **repeat** *j* := *j*−1 **until** $L[j] \le x$;
>   **repeat** *i* := *i*+1 **until** $L[i] \ge x$;
>   *Interchange* (*L*[*i*], *L*[*j*])
>   **until** $i \ge j$;
>   *Interchange* (*L*[*i*], *L*[*j*]);   { Undo the extra interchange. }
>   *Interchange* (*L*[*first*], *L*[*j*]);   { Put *x* at the splitpoint. }
>   *splitPoint* := *j*
> **end** { Split2 }

a)   If the section of the array being split has *k* keys, how many key comparisons does *Split2* do in the worst case? (It's not *k*−1.) Will this affect the order of the total worst-case number of comparisons done by Quicksort?

b)   For what case (what initial arrangement of keys) is the sentinel at the end of the array actually examined?

c)   Since *Split2* works on small subsections of the array, why isn't a sentinel needed at the end of each subsection?

d)   Suppose the section of the list being split by *Split2* contains many duplicates of the key *x*. Will all the copies of *x* be arranged on the same side of the *split-Point*?

2.15.   Suppose the array *L* contains the keys 10, 9, 8, ..., 2, 1, and is to be sorted using *Quicksort*.

a)   Using the version of *Split* in Algorithm 2.3, show how the keys would be arranged after each of the first two calls to the *Split* procedure, tell how many interchanges are done by each of these two calls to *Split*, and, from this example, estimate the total number of interchanges that would be done to sort *n* keys initially in decreasing order.

*b) Do the same for *Split2*, which is described in the preceding exercise. (Be sure to read the explanation that precedes Algorithm 2.18; do not just read the code.)

c) List some of the relative advantages and disadvantages of the two versions of the *Split* algorithm.

2.16. How many key comparisons are done by Mergesort if the keys are already in order when the sort begins?

2.17. Suppose we have a straightforward algorithm for a problem that does $\Theta(n^2)$ steps for inputs of size n. Suppose we devise a Divide and Conquer algorithm that divides an input into two inputs half as big, and does $D(n) = n \lg n$ steps to divide the problem and $C(n) = n \lg n$ steps to combine the solutions to get a solution for the original input. Is the Divide and Conquer algorithm more or less efficient than the straightforward scheme? Justify your answer.

Section 2.4: Lower Bounds for Sorting by Comparison of Keys

2.18. Draw the decision tree for Quicksort with $n = 3$. (You will have to modify the conventions a bit. Some branches should be labeled "≤" or "≥.")

2.19. a) Give an algorithm to sort four keys using only five key comparisons in the worst case.

*b) Give an algorithm to sort five keys that is optimal in the worst case.

Section 2.5: Heapsort

2.20. Suppose a character array to be sorted (into alphabetical order) by Heapsort initially contains the following sequence of letters:

C O M P L E X I T Y

Show how they would be arranged in the array after the heap construction phase. How many key comparisons are done to construct the heap with these keys?

2.21. The nodes of a heap are stored in an array level by level, beginning with the root and left to right within each level. Prove that the left child of the node in the ith cell is in the $2i$th cell.

2.22. A list of distinct keys in decreasing order is to be sorted (into increasing order) by *Heapsort*.

a) How many comparisons of keys are done in the heap construction phase if there are 10 keys?

b) How many are done if there are n keys? Show how you derive your answer.

c) Is a list in decreasing order a best case, worst case, or intermediate case for the heap construction? Justify your answer.

2.23. This exercise provides an alternative argument for the worst-case analysis of the heap construction phase of *Heapsort*. The *height* of a node in a binary tree is the depth of the subtree rooted at that node. Prove that the sum of the heights of the nodes in a heap of n nodes is at most $n-1$. (Hint: Use a marking strategy, systematically marking off one branch in the tree for each unit of height in the sum.)

2.24. We could eliminate one call to *FixHeap* from the second phase of *Heapsort* in Algorithm 2.10 by changing the **for** loop control to

for *heapsize* := n **to** 3 **by** −1 **do**

What statement, if any, must be added after the **for** loop to take care of the case when two keys remain in the heap? How many comparisons, if any, are eliminated?

2.25. Suppose we have a heap with n keys stored in an array H, and we want to add a new key, K. The following algorithm adds a new leaf at $H[n+1]$ and lets small keys on the path from this leaf to the root filter down until the proper place for the new key is found. (For an index i, $parent(i) = i$ **div** 2.)

```
procedure Insert (K: Key; var H: IntegerArray; var n: integer:);
begin
    n := n+1;
    vacant := n;
    while vacant > 1 { not the root } and K > H[parent(vacant)] do
        H[vacant] := H[parent(vacant)];
        vacant := parent(vacant)
    end { while };
    H[vacant] := K
end { Insert }
```

a)   How many comparisons of keys does *Insert* do in the worst case on a heap with n keys after the insertion?

b)   An earlier version of *Heapsort* used *Insert* to construct a heap from the keys to be sorted by inserting the keys, one at a time, into a heap that was initially empty. How many comparisons are done by this method in the worst case to construct a heap of n keys?

c)   How many comparisons would be done in the worst case by *Heapsort* if *Insert* were used, as described in (b), to construct the heap?

Section 2.7: Radix Sorting

2.26. Suppose Radix Sort does m distribution passes on keys with w bits (where m is a divisor of w) and there is one bucket for each pattern of w/m bits, hence *numBuckets* $= 2^{w/m}$. Since mn key distributions are done, it may seem advantageous to decrease m. How large must the new *numBuckets* be if m is halved?

Section 2.8: External Sorting

2.27. Suppose three tapes are to be used for an external sort and the r runs constructed during the run construction phase are distributed so that one run is on one tape and all the others are on another tape. How many passes over the keys will be done in the merge phase?

2.28. Prove Lemma 2.16.

2.29. Suppose that you have a large unsorted file on a tape and only two tape drives. Outline a method for sorting the file.

Additional Problems

2.30. To sort or not to sort: Outline a reasonable method of solving each of the following problems. Give the order of the worst-case complexity of your methods.

a)   You are given a pile of thousands of telephone bills and thousands of checks sent in to pay the bills. (Assume telephone numbers are on the checks.) Find out who did not pay.

b)   You are given a list containing title, author, call number, and publisher of all the books in a school library and another list of 30 publishers. Find out how many of the books were published by each of those 30 companies.

c)   You are given all of the book checkout cards used in the campus library during the past year. Determine how many distinct people checked out at least one book.

2.31. Solve the following recurrence relations. You may assume $T(1) = 1$, the recurrence is for $n > 1$, and c is some positive constant.

a) $T(n) = T(n/2) + c \lg n$.

b) $T(n) = T(n/2) + cn$.

c) $T(n) = 2T(n/2) + cn$.

d) $T(n) = 2T(n/2) + cn \lg n$.

e) $T(n) = 2T(n/2) + cn^2$.

*2.32. Solve the following recurrence relation.

$$T(n) = \sqrt{n} \, T(\sqrt{n}) + cn \text{ for } n > 2 \text{ and some positive constant } c,$$
$$T(2) = 1.$$

2.33. Give an efficient in-place algorithm to rearrange an array of n keys so that all the negative keys precede all the nonnegative keys. How fast is your algorithm?

2.34. A sorting method is *stable* if equal keys remain in the same relative order in the sorted list as they were in the original list. (That is, a sort is stable if for any $i < j$ such that initially $L[i] = L[j]$, the sort moves $L[i]$ to $L[k]$ and moves $L[j]$ to $L[l]$ for some k and l such that $k < l$.) Which of the following algorithms are stable? For each that is not, give an example in which the relative order of two equal keys is changed.

a)   Insertion Sort.

b)   Maxsort (Exercise 2.1).

c)   Bubblesort (Exercise 2.2).

d)   Quicksort.

e)   Heapsort.

f)   Shellsort.

g)   Radix Sort.

h)   The external sort in Section 2.8 using Replacement Selection.

2.35. Suppose you have a list of 1000 records in which only a few are out of order, not very far from their correct position. Which sorting algorithm would you use to put the whole list in order? Justify your choice.

2.36. Briefly describe how to adapt each of the following sorting algorithms to sort keys stored in a linked list (without changing the order of the worst-case complexity). Tell which, if any, cannot be easily adapted.

a)   Insertion Sort.

b)   Maxsort (Exercise 2.1).

c)   Bubblesort (Exercise 2.2).

d)   Quicksort.

e)   Mergesort.

f)   Heapsort.

g)   Shellsort.

2.37. Throughout most of this chapter we have assumed that the keys in the list to be sorted were distinct. Often, there are duplicate keys. Such duplication could make sorting easier, but algorithms that were designed for distinct (or mostly distinct) keys may not take advantage of the duplication. Let us consider the extreme case where there are only two possible key values, say 0 and 1.

a)   What is the order of the number of key comparisons done by Insertion Sort in the worst case? (Describe a worst-case input.)

b)   What is the order of the number of key comparisons done by Quicksort in the worst case? (Describe a worst-case input.)

c)   Give an efficient algorithm for sorting a list of n keys that may each be either 0 or 1. What is the order of the worst-case running time of your algorithm?

*2.38. Each of n keys in an array may have one of the values *red*, *white*, or *blue*. Give an efficient algorithm for rearranging the keys so that all the *red*s come before all the *white*s, and all the *white*s come before all the *blue*s. (It may happen that there are no keys of one or two of the colors.) The only operations permitted on the keys are examination of a key to find out what color it is, and a swap, or interchange, of two keys (specified by their indexes). What is the order of the worst-case running time of your algorithm? (There is a linear solution.)

2.39. Suppose that you have a computer with n memory locations, numbered 1 through n, and one instruction CEX, called "compare-exchange." For $1 \le i,j \le n$, CEX i, j compares the keys in memory cells i and j and interchanges them if necessary so that the smaller key is in the cell with the smaller index. The CEX instruction can be used to sort. For example, the following program sorts for $n = 3$:

```
CEX 1,2
CEX 2,3
CEX 1,2
```

a)   Write an efficient program using only CEX instructions to sort six keys. (Suggestion: Write programs for $n = 4$ and $n = 5$ first. It is easy to write programs for $n = 4$, 5, and 6 using 6, 10, and 15 instructions, respectively. However, none of these is optimal.)

b)   Write a CEX program to sort n keys in n cells for a fixed but arbitrary n. Use as few instructions as you can. Describe the strategy your program uses and include comments where appropriate. Since there are no loop and test instructions, you may use ellipses to indicate repetition of instructions of a certain form; for example:

CEX 1,2

CEX 2,3

.

.

.

CEX $n-1,n$

c) How many CEX instructions does your program for (b) have?

d) Give a lower bound on the number of CEX instructions needed to sort n keys.

2.40. a) Suppose CEX instructions (described in the preceding exercise) can be carried out simultaneously if they are working on keys in different memory cells. For example, CEX 1,2, CEX 3,4, CEX 5,6, etc. can all be carried out at the same time. Give an algorithm to sort four keys in only three time units. (Recall that sorting four keys requires five comparisons.)

*b) Give an algorithm using (simultaneous) CEX instructions to sort n keys in $o(\lceil \lg n! \rceil)$ time units.

*2.41. M is an $n \times n$ integer matrix in which the entries of each row are in increasing order (reading left to right) and the entries in each column are in increasing order (reading top to bottom). Give an efficient algorithm to find the position of an integer x in M, or to determine that x is not there. Tell how many comparisons of x with matrix entries your algorithm does in the worst case. You may use 3-way comparisons; that is, a comparison of x with $M[i,j]$ tells if $x < M[i,j]$, $x = M[i,j]$, or $x > M[i,j]$.

*2.42. L is an array containing n integers, and we want to find the maximum sum for a contiguous subsequence of elements of L. (If all elements of a sequence are negative, we define the maximum subsequence to be the empty sequence with sum equal to zero.) For example, consider the array with elements

$$38 \quad -62 \quad 47 \quad -33 \quad 28 \quad 13 \quad -18 \quad -46 \quad 8 \quad 21 \quad 12 \quad -53 \quad 25$$

The maximum subsequence sum for this array is 55. The maximum subsequence occurs in positions 3 through 6 (inclusive).

a) Give an algorithm that finds the maximum subsequence sum in an array. What is the order of the running time of your algorithm? (The data in Tables 1.1 and 1.2 come from various algorithms for this problem. As those tables indicate, there are many solutions of varying complexity, including a linear one.)

b) Show that any algorithm for this problem must examine all elements in the array in the worst case. (So any algorithm does $\Omega(n)$ steps in the worst case.)

2.43. *a) Write an algorithm that, when given an array L of records with keys $a_1, a_2, \ldots, a_n$ and a permutation π of the numbers 1, 2, ..., n, rearranges the records in the order $a_{\pi(1)}, a_{\pi(2)}, \ldots, a_{\pi(n)}$. You may assume that the values of π are given in another array. Assume that the records in L are large; in particular, they will not fit in the π array. Your algorithm may destroy π. If you use extra space, state how much.

b) What is the total number of times records are moved by your algorithm in the worst case? Is the running time of your algorithm proportional to the number of moves? If so, explain why. If not, what is the order of the running time?

*2.44. Let f be a function defined on the interval from 0 to n, i.e., for $0 \le x \le n$, which achieves its maximum value at x_M, a point in the interval, and is strictly increasing for $0 \le x < x_M$ and strictly decreasing for $x_M < x \le n$. (Note: x_M may be 0 or n.)

a) The only operation you can do with f is evaluate it at points in the interval. Write an algorithm to find x_M with an error of at most ε, where $0 < \varepsilon < 1$; i.e., if your algorithm's answer is $\bar{x}$, then $|\bar{x} - x_M| < \varepsilon$. Use instructions of the form $y := f(x)$ to indicate evaluation of f at the point x.

b) How many evaluations does your algorithm do in the worst case? (You should be able to devise an algorithm that is in $o(n)$.)

2.45. What sorting method would you use for each of the following problems? Explain your choice.

a) A university in Southern California has about 30,000 full-time students and about 10,000 part-time students. (There is a cap of 50,000 students allowed to enroll in the university at one time, due to parking limitations.) Each student record contains the student's name, nine-digit identification number, address, grades, etc. A name is stored as a string of 41 characters, 20 characters each for the first and last name, and one character for the middle initial.

The problem is to produce an alphabetized class list for each of approximately 5000 courses at the beginning of each semester. These lists are given to the instructors before the first day of class. The maximum class size is about 200. Most of the classes have about 30 students. The input for each class is an array with at most 200 records. These records contain a student's name, identification number, and university standing (freshman, sophomore, junior, senior, grad), and a pointer to the student's full record on disk.

b) The problem is to sort 500 exam papers alphabetically by student's last name. The sorting will be done by one person in an office with two desktops temporarily cleared of all other papers, books, and coffee cups. It is 1:00 AM, and the person would like to go home as soon as possible.

Programs

For each program include a counter that counts comparisons of keys. Include among your test data lists in which the keys are in decreasing order, increasing order, and random order. Use lists of several sizes. Output should include the number of keys and the number of comparisons done.

1. Quicksort: Use the improvements described in Section 2.3.4.

2. Heapsort: Show the full heap after all the keys have been inserted.

3. Radix Sort

4. External sort: Do run construction by replacement selection; the run construction algorithm may be implemented and tested with the "memory size" set to, say, 100. Determine the maximum, minimum, and average run size.

Notes and References

Much of the material in this chapter is based on Knuth (1973), without a doubt the major reference on sorting and related problems. The interested reader is strongly encouraged to consult this book for more algorithms, analyses, exercises, and references. Some of the original sources of the algorithms are: Hoare (1962) for Quicksort, including variations and applications; Williams (1964) for Heapsort (with an early improvement by Floyd (1964)); and Shell (1959) for Shellsort. The version of the *Split* procedure used with Quicksort in the text appears in Bentley (1986) where it is attributed to N. Lomuta.

The concise argument given in Section 2.2 for the average number of inversions in a permutation was pointed out by Sampath Kannan. At the end of Section 2.5 we commented that there are algorithms that do fewer comparisons than Mergesort in the worst case. The Ford-Johnson algorithm, called Merge Insertion, is one such algorithm. It is known to be optimal for small values of n. Binary Insertion is another algorithm that does approximately $n\lg n$ comparisons in the worst case. See Knuth (1973) for descriptions of these algorithms, a discussion of various choices of increments for Shellsort, a proof of Theorem 2.14, and more discussion of external sorting. The formula in Lemma 2.15 of Section 2.8 is derived with the help of generating functions. See Knuth (1968) for this and other properties of the Fibonacci numbers that are useful in the analysis of algorithms.

Many sorting algorithms are known by more than one name, and it will aid the reader who consults other references to be aware of the alternative names. Bubblesort is sometimes called Exchange Sort; Quicksort is sometimes called Partition-Exchange Sort. Shellsort is referred to as "sorting by diminishing increments."

The sorting problem in Exercise 2.38 is solved in Dijkstra (1976) where it is called "The Dutch National Flag Problem." Bentley (1986) gives some history and several solutions for the maximum subsequence problem (Exercise 2.42). The data in Table 1.2 and all but the exponential column of Table 1.1 come from solutions to this problem. Exercise 2.45 was contributed by Roger Whitney.

3

Selection and Adversary Arguments

3.1
Introduction

3.1.1 The Selection Problem

Suppose L is an array containing n keys from some linearly ordered set, and let k be an integer such that $1 \leq k \leq n$. The *selection* problem is the problem of finding the kth smallest key in L. As with most of the sorting algorithms we studied, we will assume that the only operations that may be performed on the keys are comparisons of pairs of keys (as well as copying or moving keys).

In Chapter 1 we solved the selection problem for the case $k = n$, for that problem is simply to find the largest key. We considered a straightforward algorithm that did $n-1$ key comparisons, and we proved that no algorithm could do fewer. The dual case for $k = 1$, that is, finding the smallest key, can be solved similarly. Another very common instance of the selection problem is the case where $k = \lceil n/2 \rceil$, that is, where we want to find the middle, or *median*, element.

Of course the selection problem can be solved in general by sorting L; then $L[k]$ would be the answer. Sorting requires $\Theta(n \lg n)$ key comparisons, and we have just observed that for some values of k, the selection problem can be solved in linear time. Finding the median seems, intuitively, to be the hardest instance of the selection problem. Can we find the median in linear time? Or can we establish a lower bound for median finding that is more than linear, maybe $\Theta(n \lg n)$? We will answer these questions in this chapter.

3.1.2 Lower Bounds

So far we have used the decision tree as our main technique to establish lower bounds. Recall that the internal nodes of the decision tree for an algorithm represent the comparisons the algorithm performs, and that the leaves represent the outputs. (For the search problem in Section 1.5, the internal nodes also represented outputs.) The number of comparisons done in the worst case is the depth of the tree; the depth is at least $\lceil \lg l \rceil$, where l is the number of leaves.

In Chapter 1 we used decision trees to get the (worst-case) lower bound of $\lfloor \lg n \rfloor + 1$ for the search problem. That is exactly the number of comparisons done by Binary Search, so a decision tree argument gave us the best possible lower bound. In Chapter 2 we used decision trees to get a lower bound of $\lceil \lg n! \rceil$, or roughly $\lceil n \lg n - 1.5n \rceil$, for sorting. There are algorithms whose performance is very close to this lower bound, so once again a decision tree argument gave a very strong result. However, decision tree arguments do not work very well for the selection problem.

A decision tree for the selection problem must have at least n leaves because any one of the n keys in the list may be the output, i.e., the kth smallest. Thus we can conclude that the depth of the tree (and the number of comparisons done in the worst case) is at least $\lceil \lg n \rceil$. But this is not a good lower bound; we already know that even the easy case of finding the largest key requires at least $n-1$ comparisons. What is wrong with the decision tree argument? In a decision tree for an algorithm that finds the largest key, some outputs appear at more than one leaf, and there will in fact be more than n leaves. To see this, draw the decision tree for *FindMax* (Algorithm 1.3) with $n = 4$. The decision tree argument fails to give a good lower bound because we do not have an easy way to determine how many leaves will contain duplicates of a particular outcome. Instead of a decision tree, we will use a technique called an *adversary argument* to establish better lower bounds for the selection problem.

Suppose you are playing a guessing game with a friend. You are to pick a date (a month and day), and the friend will try to guess the date by asking yes/no questions. You want to force your friend to ask as many questions as possible. If the first question is "Is it in the winter?" and you are a good adversary, you will answer "No," because there are more dates in the three other seasons. To the question "Is the first letter of the month's name in the first half of the alphabet?" you should answer "Yes." But is this cheating? You did not really pick a date at all. In fact, you will not pick a specific month and day until the need for consistency in your answers pins you down. This may not be a friendly way to play a guessing game, but it is just right for finding lower bounds for the behavior of an algorithm.

Suppose we have an algorithm that we think is efficient. Imagine an adversary who wants to prove otherwise. At each point in the algorithm where a decision (a key comparison, for example) is made, the adversary tells us the result of the decision. The adversary chooses its answers to try to force the algorithm to work hard, i.e., to make a lot of decisions. You may think of the adversary as gradually constructing a "bad" input for the algorithm while it answers the questions. The only constraint on the adversary's answers is that they must be internally consistent; there must be *some* input for the problem for which its answers would be correct. If the adversary can force the algorithm to perform $f(n)$ steps, then $f(n)$ is a lower bound for the number of steps in the worst case.

We want to find a lower bound on the complexity of a *problem*, not just a particular algorithm. Therefore, when we use adversary arguments, we will assume that the algorithm is any algorithm whatsoever from the class being studied, just as we did with the decision tree arguments. To get a good lower bound we need to construct a clever adversary that can thwart any algorithm.

In the rest of this chapter we present algorithms for selection problems and adversary arguments for lower bounds for several cases, including the median. In most of the algorithms and arguments, we will use the terminology of contests, or tournaments, to describe the results of comparisons. The comparand that is found to be larger will be called the *winner*; the other will be called the *loser*.

3.2
Finding *max* and *min*

Throughout this section we will use the names *max* and *min* to refer to the largest and smallest keys, respectively, in a list of n keys.

We can find *max* and *min* by using Algorithm 1.3 to find *max*, eliminating *max* from the list, and then using the appropriate variant of the algorithm to find *min* among the remaining $n-1$ keys. Thus *max* and *min* can be found by doing $(n-1)+(n-2)$, or $2n-3$, comparisons. This is not optimal. Although we know (from Chapter 1) that $n-1$ key comparisons are needed to find *max* or *min* independently, when finding both, some of the work can be "shared." Exercise 1.12 asks for an algorithm to find *max* and *min* with only about $3n/2$ key comparisons. The solution (for even n) is to pair up the keys and do $n/2$ comparisons, then find the largest of the winners, and, separately, find the smallest of the losers. (If n is odd, the last key may have to be considered among the winners and the losers.) In this section we give an adversary argument to show that this solution is optimal. Specifically, we will prove

Theorem 3.1   Any algorithm to find *max* and *min* of n keys by comparison of keys must do at least $3n/2-2$ key comparisons in the worst case.

To establish the lower bound we may assume that the keys are distinct. To know that a key x is *max* and that a key y is *min*, an algorithm must know that every key other than x has lost some comparison and that every key other than y has won some comparison. If we count each win as one unit of information and each loss as one unit of information, then an algorithm must have (at least) $2n-2$ units of information to be sure of giving the correct answer. We give a strategy for an adversary to use in responding to the comparisons so that it gives away as few units of new information as possible with each comparison. Imagine the adversary constructing a specific input list as it responds to the algorithm's comparisons.

We denote the status of each key at any time during the course of the algorithm as follows:

Key status	Meaning
W	Has won at least one comparison and never lost
L	Has lost at least one comparison and never won
WL	Has won and lost at least one comparison
N	Has not yet participated in a comparison

The adversary strategy is described in Table 3.1. The main point is that, except in the case where both keys have not yet been in any comparison, the adversary can give a response that provides at most one unit of new information. We need to verify that if the adversary follows these rules, its replies are consistent with some input. Then we need to show that this strategy forces any algorithm to do as many comparisons as the theorem claims.

Observe that in all cases in Table 3.1 except the last, either the key chosen by the adversary as the winner has not yet lost any comparison, or the key chosen as the loser has not yet won any. Consider the first possibility: Suppose that the algorithm compares x and y, that the adversary chooses x as the winner, and that x has not yet

Table 3.1
The adversary strategy for the min and max problem.

Status of keys x and y compared by an algorithm	Adversary response	New status	Units of new information
N, N	$x > y$	W, L	2
W, N or WL, N	$x > y$	W, L or WL, L	1
L, N	$x < y$	L, W	1
W, W	$x > y$	W, WL	1
L, L	$x > y$	WL, L	1
W, L or WL, L or W, WL	$x > y$	No change	0
WL, WL	Consistent with assigned values	No change	0

lost any comparison. Even if the value already assigned by the adversary to x is smaller than the value it has assigned to y, the adversary can change x's value to make it beat y without contradicting any of the responses it gave earlier. The other situation, where the key chosen as the loser has never won, can be handled similarly — by reducing the value of the key if necessary. So the adversary can construct an input consistent with the rules in the table for responding to the algorithm's comparisons. This is illustrated in the following example.

Example 3.1   Constructing an input using the adversary's rules

The first column in Table 3.2 shows a sequence of comparisons that might be carried out by some algorithm. The remaining columns show the status and value assigned to the keys by the adversary. (Keys that have not yet been assigned a value are denoted by asterisks.) Each row after the first contains only the entries relevant to the current comparison. Note that when x_3 and x_1 are compared (in the fifth comparison), the adversary increases the value of x_3 because x_3 is supposed to win. Later, the adversary changes the values of x_6 and x_4 consistent with its rules. After

Table 3.2
An example of the adversary strategy.

	x_1		x_2		x_3		x_4		x_5		x_6	
Comparison	Status	Value	Status	Value	Status	Value	Status	Value	Status	Value	Status	Value
x_1, x_2	W	20	L	10	N	*	N	*	N	*	N	*
x_1, x_5	W	20							L	5		
x_3, x_4					W	15	L	8				
x_3, x_6					W	15					L	12
x_3, x_1	WL	20			W	25						
x_2, x_4			WL	10			L	8				
x_5, x_6									WL	5	L	3
x_6, x_4							L	2			WL	3

the first five comparisons, every key except x_3 has lost at least once, so x_3 is *max*. After the last comparison x_4 is the only key that has never won, so it is *min*. In this example the algorithm did eight comparisons; the worst-case lower bound for six keys (still to be proved) is $3/2 \times 6 - 2 = 7$. ∎

To complete the proof of Theorem 3.1, we need only show that the adversary rules will force any algorithm to do at least $3n/2 - 2$ comparisons to get the $2n - 2$ units of information it needs. The only case where an algorithm can get two units of information from one comparison is the case where the two keys have not been included in any previous comparisons. Suppose for the moment that n is even. An algorithm can do at most $n/2$ comparisons of previously unseen keys, so it can get at most n units of information this way. From each other comparison, it gets at most one unit of information. Thus to get $2n - 2$ units of information, an algorithm must do at least $n/2 + n - 2 = 3n/2 - 2$ comparisons in total. The reader can easily check that for odd n, at least $3n/2 - 3/2$ comparisons are needed. This completes the proof of Theorem 3.1.

3.3
Finding the Second-Largest Key

3.3.1   Introduction

Throughout this section we will use *max* and *secondLargest* to refer to the largest and second-largest keys, respectively. For simplicity in describing the problem and algorithms, we will assume that the keys are distinct.

The second-largest key can be found with $2n - 3$ comparisons by using *FindMax* (Algorithm 1.3) twice, but this is not likely to be optimal. We should expect that some of the information discovered by the algorithm while finding *max* can be used to decrease the number of comparisons performed in finding *secondLargest*. Specifically, any key that loses to a key other than *max* cannot possibly be *secondLargest*. All such keys discovered while finding *max* may be ignored during the second pass through the list. (The problem of keeping track of them will be considered later.)

Using Algorithm 1.3 on a list with five keys, the results might be as follows:

Comparands	Winner
$L[1], L[2]$	$L[1]$
$L[1], L[3]$	$L[1]$
$L[1], L[4]$	$L[4]$
$L[4], L[5]$	$L[4]$

Then $max = L[4]$ and *secondLargest* is either $L[5]$ or $L[1]$ because both $L[2]$ and $L[3]$ lost to $L[1]$. Thus only one more comparison is needed to find *secondLargest* in this example.

It may happen, however, that during the first pass through the list to find *max* we do not obtain any information useful for finding *secondLargest*. If *max* were $L[1]$, then each other key would be compared only with *max*. Does this mean that in the worst case $2n-3$ comparisons must be done to find *secondLargest*? Not necessarily. In the preceding discussion we used a specific algorithm. No algorithm can find *max* by doing fewer than $n-1$ comparisons, but another algorithm may provide more information useful for eliminating some keys during the second pass through the list. The tournament method, described next, provides such information.

3.3.2 The Tournament Method

The tournament method is so named because it performs comparisons in the same way that tournaments are played. Keys are paired off and compared in "rounds." In each round after the first one, the winners from the preceding round are paired off and compared. (If at any round the number of keys is odd, one of them simply waits for the next round.) A tournament can be described by a tree diagram as shown in Fig. 3.1. Each leaf contains a key, and at each subsequent level the parent of each pair contains the winner. The root will contain the largest key. As in Algorithm 1.3, $n-1$ comparisons are done to find *max*.

In the process of finding *max*, every key except *max* loses in one comparison. How many lose directly to *max*? Roughly half the keys in one round will be losers and will not appear in the next round. If n is a power of 2, there are exactly $\lg n$ rounds; in general, the number of rounds is $\lceil \lg n \rceil$. Since *max* is involved in at most

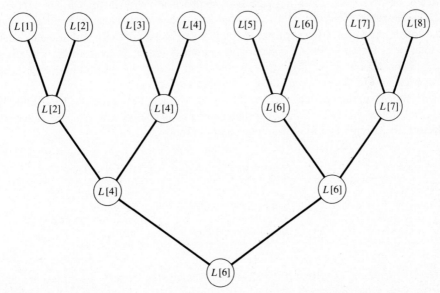

Figure 3.1 An example of a tournament; *max* = $L[6]$; *secondLargest* may be $L[4]$, $L[5]$, or $L[7]$.

one comparison in each round, there are at most $\lceil \lg n \rceil$ keys that lost only to *max*, and thus could possibly be *secondLargest*. The method of Algorithm 1.3 can be used to find the largest of these $\lceil \lg n \rceil$ keys by doing $\lceil \lg n \rceil - 1$ comparisons. Thus the tournament finds *max* and *secondLargest* by doing a total of $n + \lceil \lg n \rceil - 2$ comparisons. This is an improvement over our first result of $2n - 3$. Can we do better?

3.3.3 An Adversary Lower Bound Argument

Both methods we considered for finding the second-largest key first found the largest key. This is not wasted effort. Any algorithm that finds *secondLargest* must also find *max* because, to know that a key is the second largest, one must know that it is not the largest; that is, it must have lost in one comparison. The winner of the comparison in which *secondLargest* loses must, of course, be *max*. This argument gives a lower bound on the number of comparisons needed to find *secondLargest*, namely $n - 1$, because we already know that $n - 1$ comparisons are needed to find *max*. But one would expect that this lower bound could be improved because an algorithm to find *secondLargest* should have to do more work than an algorithm to find *max*. We will prove the following theorem, which has as a corollary that the tournament method is optimal.

Theorem 3.2   Any algorithm (that works by comparing keys) to find the second largest in a list of n keys must do at least $n + \lceil \lg n \rceil - 2$ comparisons in the worst case.

Proof.   For the worst case, we may assume that the keys are distinct. We have already observed that there must be $n - 1$ comparisons with distinct losers. If *max* was a comparand in $\lceil \lg n \rceil$ of these comparisons, then all but one of the $\lceil \lg n \rceil$ keys that lost to *max* must lose again for *secondLargest* to be correctly determined. Then a total of at least $n + \lceil \lg n \rceil - 2$ comparisons would be done. Therefore we will show that there is an adversary strategy that can force any algorithm that finds *secondLargest* to compare *max* to $\lceil \lg n \rceil$ distinct keys.

The adversary assigns a "weight" $w(x)$ to each key x in the list. Initially $w(x) = 1$ for all x. When the algorithm compares two keys x and y, the adversary determines its reply and modifies the weights as follows.

Case	Adversary reply	Updating of weights
$w(x) > w(y)$	$x > y$	$w(x) := w(x) + w(y);\ \ w(y) := 0$
$w(x) = w(y) > 0$	Same as above	Same as above
$w(y) > w(x)$	$y > x$	$w(y) := w(y) + w(x);\ \ w(x) := 0$
$w(x) = w(y) = 0$	Consistent with previous replies	No change

We need to verify that if the adversary follows this strategy, its replies are consistent with some input, and that *max* will be compared to at least $\lceil \lg n \rceil$ distinct keys. These conclusions follow from a sequence of easy observations.

1.   A key has lost a comparison if and only if its weight is zero.

2.   In the first three cases, the key chosen as the winner has nonzero weight, so it has not yet lost. The adversary can give it an arbitrarily high value to make sure it wins without contradicting any of its earlier replies.

3.   The sum of the weights is always n. This is true initially, and the sum is preserved by the updating of the weights.

4.   When the algorithm stops, only one key can have nonzero weight. Otherwise there would be at least two keys that never lost a comparison, and the adversary could choose values to make the algorithm's choice of *secondLargest* incorrect.

Lemma 3.3   Let x be the key that has nonzero weight when the algorithm stops. Then $x = max$, and x has directly won against at least $\lceil \lg n \rceil$ distinct keys.

Proof. By facts 1, 3, and 4, when the algorithm stops, $w(x) = n$. Let $w_k = w(x)$ just after the kth comparison won by x against a previously undefeated key. Then by the adversary's rules,

$$w_k \leq 2w_{k-1}.$$

Now let K be the number of comparisons x wins against previously undefeated keys. Then

$$n = w_K \leq 2^K w_0 = 2^K.$$

Thus $K \geq \lg n$, and since K is an integer, $K \geq \lceil \lg n \rceil$. The K keys counted here are of course distinct, since once beaten by x, a key is no longer "previously undefeated" and will not be counted again (even if an algorithm foolishly compares it to x again).   □

Another way of looking at the adversary's activity is that it builds trees to represent the ordering relations between the keys. If x is the parent of y, then x beat y in a comparison. Figure 3.2 shows an example. The adversary combines two trees only when their roots are compared. If the algorithm compares nonroots, no change is made in the trees. The weight of a key is simply the number of nodes in that key's tree, if it is a root, and zero otherwise.

Example 3.2   The adversary strategy in action

To illustrate the adversary's action and show how its decisions correspond to the step-by-step construction of an input, we show an example for $n = 5$. Keys in the list that have not yet been specified are denoted by asterisks. Thus initially the keys are *, *, *, *, *. Note that values assigned to some keys may be changed at a later time. See Table 3.3, which shows just the first few comparisons (those that find *max*, but not enough to find *secondLargest*). The weights and the values assigned to the keys will not be changed by any subsequent comparisons.   ■

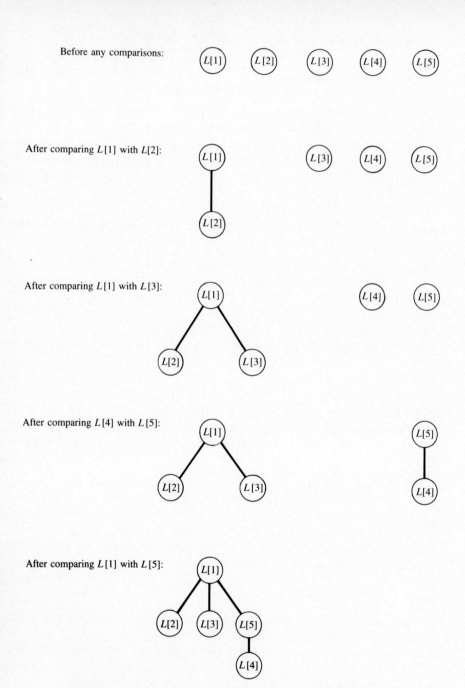

Before any comparisons:

After comparing $L[1]$ with $L[2]$:

After comparing $L[1]$ with $L[3]$:

After comparing $L[4]$ with $L[5]$:

After comparing $L[1]$ with $L[5]$:

Figure 3.2   Trees for the adversary decisions in Example 3.2.

Table 3.3
An example of the adversary strategy.

Comparands	Weights	Winner	New weights	Keys
$L[1], L[2]$	$w(L[1]) = w(L[2])$	$L[1]$	2,0,1,1,1	20,10,*,*,*
$L[1], L[3]$	$w(L[1]) > w(L[3])$	$L[1]$	3,0,0,1,1	20,10,15,*,*
$L[5], L[4]$	$w(L[5]) = w(L[4])$	$L[5]$	3,0,0,0,2	20,10,15,30,40
$L[1], L[5]$	$w(L[1]) > w(L[5])$	$L[1]$	5,0,0,0,0	41,10,15,30,40

3.3.4   Implementation of the Tournament Method for Finding *max* and *secondLargest*

To conduct the tournament to find *max* we need a way to keep track of the winners in each round. This can be done by using an extra array of pointers or by careful indexing if the keys may be moved so that the winner is always placed in, say, the higher-indexed cell of the two being compared. We leave the choice and the details to the reader.

After *max* has been found by the tournament, only those keys that lose to it are to be compared to find *secondLargest*. How can we keep track of the elements that lose to *max* when we do not know in advance which key is *max*? One way is to maintain linked lists of keys that lose to each undefeated key. This can be done by allocating an array for links indexed to correspond to the keys. Initially all links are zero. After each comparison in the tournament, the key that lost would be added to the winner's loser list (at the beginning of the list); see Fig. 3.3 for an example. When the tournament is complete and *max* has been found, it is easy to find *secondLargest* by traversing *max*'s loser list.

Time and Space

The tournament method for finding *max* and *secondLargest* uses $\Theta(n)$ extra space for links. The running time of the algorithm is in $\Theta(n+\lceil \lg n \rceil -2) = \Theta(n)$, since the number of operations for the links is roughly proportional to the number of comparisons done.

We can find the largest and second-largest keys in a list by using *FindMax* twice, doing $2n-3$ comparisons, or we can use the more complicated tournament method, doing $n+\lceil \lg n \rceil -2$ comparisons at most. Which method is better? The results of Exercise 3.8 should be instructive. Both algorithms are in $\Theta(n)$. Since the tournament method does more instructions per comparison while finding *max*, it may well be slower. It is also more complicated. The main point of considering this problem was not to find an algorithm that beats the straightforward one in practice, but to illustrate the adversary argument for the lower bound and, by exhibiting both the adversary argument and the tournament algorithm, to determine the optimal number of comparisons.

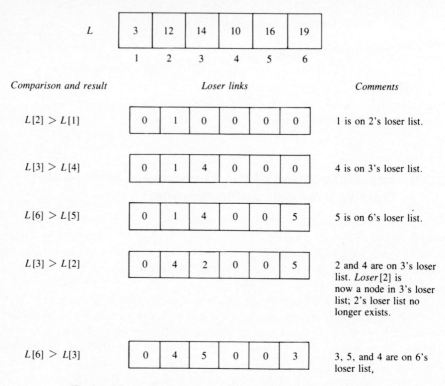

Figure 3.3   Loser link values for a tournament on six keys.

The Selection Problem

3.4.1   The Selection Algorithm

Suppose we want to find the median of n keys. For simplicity we assume that n is odd; the median is the middle, or $(n+1)/2$th-smallest, key. We have already observed that we can find the median by sorting in $\Theta(n\lg n)$ steps. (Some algorithms that do not sort completely but that use ideas from sorting algorithms are investigated in the exercises.)

The algorithm we present here is linear. It is a simplification of the first linear algorithm discovered for solving the selection problem. The simplification makes the general strategy easier to understand (though the details are complicated and tricky to implement), but it is less efficient than the original. The algorithm is important and interesting because it solves the selection problem in general, not just for the median, because it *is* linear, and because it opened the way for improvements.

Figure 3.4   Partitioning smaller and larger keys.

As usual, to simplify the description of the algorithm, we assume that the keys are distinct. However, the algorithm is not hard to modify if there are duplicate keys.

Suppose we can partition the keys into two sets, S_1 and S_2, such that all keys in S_1 are smaller than all keys in S_2. (See Fig. 3.4.) Then the median is in the larger of the two sets (that is, the set with more keys, not necessarily the set with the larger keys). We can throw away all the keys in the smaller set and restrict our search for the median to the larger one. But what do we look for in the larger set? Its median is not the median of the original list of keys. Consider Fig. 3.4. Suppose that $n = 255$, that S_1 has, say, 96 elements, and that S_2 has 159 elements. Then the median is in S_2, and it is the 32nd-smallest key in S_2. Thus this approach to solving the median problem naturally suggests that we solve the general selection problem.

In the following outline of the algorithm we assume that n is an odd multiple of five. Since the algorithm is recursive, even if the original n satisfies this requirement, the subsets operated on recursively may not. The details of patching the algorithm for general n are not hard and are omitted.

Algorithm 3.1   Selection

Input: S, a set of n keys; and k, an integer such that $1 \le k \le n$. (This description of the algorithm assumes that $n = 5(2r+1)$ for some nonnegative integer r.)

Output: the kth smallest key in S.

procedure *Select* (S: *SetOfKeys*; k: *integer*);

1.   Divide the keys into sets of five each, and find the median of each set (directly, not recursively). At this point we can imagine the keys arranged as shown in Fig. 3.5(a). In each set of five keys, the two larger than the median appear above the median, and the two smaller than the median appear below the median.

2.   Find m^*, the median of the medians, by using *Select* recursively. Now imagine the keys as in Fig. 3.5(b) where the sets of five keys have been arranged so that the sets whose medians are larger than m^* appear to the right of m^*'s set, and the sets with smaller medians appear to the left of m^*'s set. Observe that, by transitivity, all keys in the section labeled B are larger than m^*, and all keys in the section labeled C are smaller than m^*.

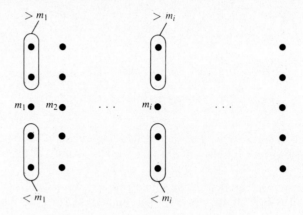

(a) m_i is the median of the ith group of five keys.

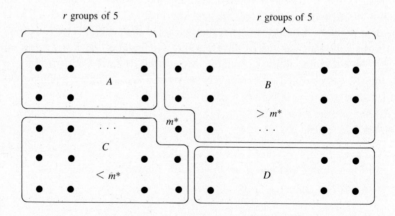

(b) Medians less than m^* are to its left; medians greater than m^* are to its right.

Figure 3.5   The selection algorithm.

3.   Compare each key in the sections labeled A and D to m^*.

4.   Let $S_1 = C \cup \{$ keys from $A \cup D$ that are smaller than $m^* \}$.
Let $S_2 = B \cup \{$ keys from $A \cup D$ that are larger than $m^* \}$.
If $|S_1|+1 = k$, then m^* is the kth-smallest key, so output m^*,
else if $k \leq |S_1|$, then the kth-smallest key is in S_1, so $Select(S_1, k)$,
else the kth-smallest key is in S_2, so $Select(S_2, k-|S_1|-1)$.
end { Select }

3.4.2 Analysis of the Selection Algorithm

We will show that *Select* is a linear algorithm. Since the algorithm was not described in full detail for general n, we will not completely prove this claim, but we will give the structure of the argument.

Let $W(n)$ be the number of key comparisons done by *Select* in the worst case on inputs with n keys. Assuming $n = 5(2r+1)$ for some nonnegative integer r (and ignoring the problem that this might not be true of the sizes of the inputs for the recursive calls), we count the comparisons done by each step of *Select*. Brief explanations of the computation are included after some of the steps.

1.   Find the medians of sets of five keys: $6(n/5)$ comparisons.

     The median of five keys can be found using six comparisons. There are $n/5$ such sets.

2.   Recursively find the median of the medians: $W(n/5)$ comparisons.

3.   Compare all keys in sections A and D to m^*: $4r$ comparisons.

     See Fig. 3.5(b).

4.   Call *Select* recursively: $W(7r+2)$ comparisons.

     In the worst case, all keys in sections A and D will be on the same side of m^* (i.e., all smaller than m^* or all greater than m^*). B and C each have $3r+2$ elements. So the size of the largest possible input for the recursive call to *Select* is $4r+3r+2 = 7r+2$.

Since $n = 5(2r+1)$, r is approximately $.1n$. So

$$W(n) \approx 1.2n + W(.2n) + .4n + W(.7n) = 1.6n + W(.2n) + W(.7n).$$

Now suppose that W is a continuous function satisfying

$$W(n) \leq 6 \quad \text{for } n \leq 5$$

$$W(n) = 1.6n + W(.2n) + W(.7n) \quad \text{for } n > 5.$$

We want to show that for $n \geq 1$, $W(n) \leq 16n$. This is clearly true for $n \leq 5$. We use an induction-like argument for larger n. For $n > 5$, by the recurrence relation for W and the "induction assumption," we have

$$W(n) = 1.6n + W(.2n) + W(.7n) \leq 1.6n + 3.2n + 11.2n = 16n.$$

Thus the selection algorithm is a linear algorithm. The original presentation of the algorithm included improvements to cut down the number of key comparisons to approximately $5.4n$. The best currently known algorithm for finding the median does $3n + o(n)$ comparisons in the worst case. (It, too, is complicated.)

Since *Select* is recursive, it uses space on a stack; it is not an in-place algorithm.

3.5
A Lower Bound for Finding the Median

We are assuming that L is a list of n keys and that n is odd. We will establish a lower bound on the number of key comparisons that must be done by any key-comparison algorithm to find *median*, the $(n+1)/2$th key. Since we are establishing a lower bound, we may, without loss of generality, assume that the keys are distinct.

We claim first that to know *median*, an algorithm must know the relation of every other key to *median*. That is, for each other key, x, the algorithm must know that $x > median$ or $x < median$. In other "words," it must establish relations as illustrated by the tree in Fig. 3.6. Each node represents a key, and each branch represents a comparison. The key at the higher end of the branch is the larger key. Suppose there were some key, say y, whose relation to *median* was not known. (See Fig. 3.7(a) for an example.) An adversary could change the value of y, moving it to the opposite side of *median*, as in Fig. 3.7(b), without contradicting the results of any of the comparisons done. Then *median* would not be the median; the algorithm's answer would be wrong.

Since there are n nodes in the tree in Fig. 3.6, there are $n-1$ branches, so at least $n-1$ comparisons must be done. This is neither a surprising nor an exciting lower bound. We will show that an adversary can force an algorithm to do other "useless" comparisons before it performs the $n-1$ comparisons it needs to establish the tree of Fig. 3.6.

Definition A comparison involving a key x is a *crucial comparison for x* if it is the first comparison where $x > y$, for some $y \geq median$, or $x < y$ for some $y \leq median$. (This is the comparison that establishes the relation of x to *median*. Note that the

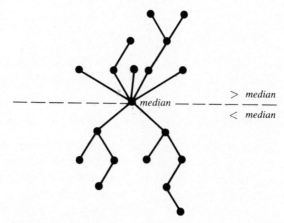

Figure 3.6 Comparisons relating each key to *median*.

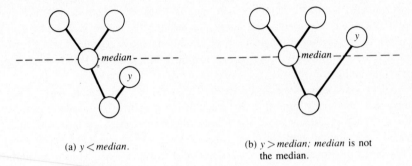

(a) $y < median$.

(b) $y > median$; *median* is not the median.

Figure 3.7   An adversary conquers a bad algorithm.

definition does not require that the relation of y to *median* be already known at the time the crucial comparison for x is done.)

Comparisons of x and y where $x > median$ and $y < median$ are noncrucial. We will exhibit an adversary that forces an algorithm to perform such comparisons. The adversary chooses some value (but not a particular key) to be *median*. It will assign a value to a key when the algorithm first uses that key in a comparison. So long as it can do so, the adversary will assign values to new keys involved in a comparison so as to put the keys on opposite sides of *median*. The adversary may not assign values larger than *median* to more than $(n-1)/2$ keys, nor values smaller than *median* to more than $(n-1)/2$ keys. It keeps track of the assignments it has made to be sure not to violate these restrictions. We indicate the status of a key during the running of the algorithm as follows:

L Has been assigned a value *Larger* than *median*.
S Has been assigned a value *Smaller* than *median*.
N Has not yet been in a comparison.

The adversary's strategy is summed up in Table 3.4. In all cases, if there are already $(n-1)/2$ keys with status S (or L), the adversary ignores the rule in the table and

Table 3.4
The adversary strategy for the median-finding problem.

Comparands	Adversary's action
N, N	Make one key larger than *median*, the other smaller.
L, N or N, L	Assign a value smaller than *median* to the key with status N.
S, N or N, S	Assign a value larger than *median* to the key with status N.

assigns value(s) larger (or smaller) than *median* to the new key(s). When only one key without a value remains, the adversary assigns the value *median* to that key. Whenever the algorithm compares two keys with statuses L and L, S and S, or L and S, the adversary simply gives the correct response based on the values it has already assigned to the keys.

All of the comparisons described in Table 3.4 are noncrucial. How many can the adversary make any algorithm do? Each of these comparisons creates at most one L key, and each creates at most one S key. Since the adversary is free to make the indicated assignments until there are $(n-1)/2$ L keys or $(n-1)/2$ S keys, it can force any algorithm to do at least $(n-1)/2$ noncrucial comparisons. (Since an algorithm could start out by doing $(n-1)/2$ comparisons involving two N keys, this adversary cannot guarantee any more than $(n-1)/2$ noncrucial comparisons.)

We can now conclude that the total number of comparisons must be at least $n-1$ (the crucial comparisons) $+ (n-1)/2$ (the noncrucial comparisons). We sum up the result in the following theorem.

Theorem 3.4 Any algorithm to find the median of n keys (for odd n) by comparison of keys must do at least $3n/2-3/2$ comparisons in the worst case.

Our adversary was not as clever as it could have been in its attempt to force an algorithm to do noncrucial comparisons. In the past several years the lower bound for the median problem has crept up to roughly $1.75n-\log n$, then roughly $1.8n$, then a little higher. The best lower bound currently known is slightly above $2n$ (for large n). There is still a small gap between the best-known lower bound and the best-known algorithm for finding the median.

Exercises

Section 3.1: Introduction

3.1. Draw the decision tree for *FindMax* (Algorithm 1.3) with $n = 4$.

*3.2. Use a decision tree argument to get a lower bound on the number of comparisons needed to merge two sorted lists each containing n keys. How does your result compare with the lower bound derived in Section 2.3.6?

Section 3.2: Finding max *and* min

3.3. We used an adversary argument to establish the lower bound for finding the minimum and maximum of n keys. What lower bound do we get from a decision tree argument?

Section 3.3: Finding the Second-Largest Key

3.4. Write an algorithm in detail for the tournament method to find *max* and *secondLargest*.

3.5. How many comparisons are done by the tournament method to find *secondLargest* on the average if n is a power of 2?

3.6. The following algorithm finds the largest and second largest keys in an array L of n keys by sequentially scanning the array and keeping track of the two largest keys seen so far. (It assumes $n \geq 2$.)

```
if L[1] > L[2] then
    max := L[1];
    second := L[2]
else
    max := L[2];
    second := L[1]
end { if };
for i := 3 to n do
    if L[i] > second then
        if L[i] > max then
            second := max;
            max := L[i]
        else second := L[i]
        end { if L[i] > max }
    end { if L[i] > second }
end { for }
```

a) How many key comparisons does this algorithm do in the worst case? Give a worst-case input for $n = 6$ using integers for keys.

*b) How many key comparisons does this algorithm do on the average for n keys assuming any permutation of the keys (from their proper ordering) is equally likely?

*3.7. Write an efficient algorithm to find the third-largest key from among n keys. How many key comparisons does your algorithm do in the worst case? Is it necessary for such an algorithm to determine which key is *max* and which is *secondLargest*?

 3.8. Write assembly language routines to find *max* and *secondLargest* by the tournament method and by using Algorithm 1.3 twice. How many instructions are executed in the worst case by each routine?

 3.9. The Replacement Selection algorithm of Section 2.8.3 can be adapted to find *max* and *secondLargest*. What are the pros and cons of using Replacement Selection instead of the tournament method?

Section 3.4: The Selection Problem

3.10. Show that the median of five keys can be found with only six comparisons in the worst case. (Recall that at least seven comparisons are needed to sort five keys.)

3.11. Quicksort can be modified to find the kth-smallest key among n keys so that in most cases it does much less work than is needed to sort the list completely. Write a modified Quicksort algorithm called *FindKth* for this purpose. Analyze your algorithm.

3.12. Suppose we use the following algorithm to find the k largest keys in a list of n keys.

```
Build a heap out of the n keys;
for i := 1 to k do
    remove and output the key at the root;
    Use the FixHeap procedure to rearrange the remaining
        keys as needed to reestablish the heap property
end
```

How large can k be (as a function of n) for this algorithm to be linear in n?

*3.13. Generalize the tournament method to find the k largest of n keys (where $1 \leq k \leq n$). Work out any implementation details that affect the order of the running time. How fast is your algorithm as a function of n and k?

Section 3.5: A Lower Bound for Finding the Median

3.14. Suppose that n is even and that we define the median to be the $n/2$th-smallest key. Make the necessary modifications in the lower bound argument and in Theorem 3.4 (where we assumed n was odd).

Additional Problems

*3.15. Suppose that *L1* and *L2* are arrays, each with n keys sorted in ascending order.

   a)   Devise an $O((\lg n)^2)$ algorithm to find the nth smallest of the $2n$ keys. (For simplicity, you may assume the keys are distinct.)

   b)   Give a lower bound for this problem.

3.16. a)   Give an algorithm to determine if the n keys in an array are all distinct. Assume three-way comparisons; that is, the result of a comparison of two keys is <, =, or >. How many key comparisons does your algorithm do?

   *b)   Give a lower bound on the number of (three-way) key comparisons needed. (Try for $\Theta(n\lg n)$.)

3.17. Consider the problem of determining if a bit string of length n contains two consecutive zeros. The basic operation is to examine a position in the string to see if it is a 0 or a 1. For each $n = 2, 3, 4, 5$ either give an adversary strategy to force any algorithm to examine every bit, or give an algorithm that solves the problem by examining fewer than n bits.

*3.18. Suppose that you have a computer with a small memory and that you are given a list of n keys in an external file (on a disk or tape). Keys may be read into memory for processing, but no key may be read more than once.

   a)   What is the minimum number of storage cells needed for keys in memory to find the largest key in the file? Justify your answer.

   b)   What is the minimum number of cells needed for keys in memory to find the median? Justify your answer.

3.19. a)   You are given n keys and an integer k such that $1 \leq k \leq n$. Give an efficient algorithm to find *any one* of the k smallest keys. (For example, if $k = 3$, the algorithm may provide the first-, second-, or third-smallest key. It need not know the exact rank of the key it outputs.) How many key comparisons does your algorithm do? (Hint: Don't look for something complicated. One insight gives a short, simple algorithm.)

   b)   Give a lower bound, as a function of n and k, on the number of comparisons needed to solve this problem.

*3.20. M is an $n \times n$ integer matrix in which the keys in each row are in increasing order (reading left to right) and the keys in each column are in increasing order (reading top to bottom). Consider the problem of finding the position of an integer x in M, or determining that x is not there. Give an adversary argument to establish a lower bound on the number of comparisons of x with matrix entries needed to solve this problem. You may use 3-way comparisons; that is, a comparison of x with $M[i,j]$ tells if

$x < M[i,j]$, $x = M[i,j]$, or $x > M[i,j]$. (Note: Finding an efficient algorithm for the problem was an exercise in Chapter 2. If you did a good job on both your algorithm and your adversary argument, the number of comparisons done by the algorithm should be the same as your lower bound.)

Notes and References

Knuth (1973) is an excellent reference for the material in this chapter. It contains some history of the selection problem, including the attempt by Charles Dodgson (Lewis Carroll), in 1883, to work out a correct algorithm so that second prize in lawn tennis tournaments could be awarded fairly. The tournament algorithm for finding the second largest appeared in a 1932 paper by J. Schreier (in Polish). It was proved optimal in 1964 by S.S. Kislitsin (in Russian). The lower bound argument given here is based on Knuth (1973).

The algorithm and lower bound for finding *min* and *max* and Exercise 3.18 are attributed to I. Pohl by Knuth.

The first linear selection algorithm is in Blum *et al.* (1973). Other selection algorithms and lower bounds appear in Hyafil (1976) and Schonhage, Paterson and Pippenger (1976).

4

Graphs and Digraphs

4.1
Definitions and Representations

4.1.1   Some Elementary Definitions and Examples

Informally, a graph is a finite set of points, some of which are connected by lines, and a digraph (short for "directed graph") is a finite set of points, some of which are connected by arrows. Graphs and digraphs are useful abstractions for numerous problems and structures in operations research, computer science, electrical engineering, economics, mathematics, physics, chemistry, communications, game theory, and many other areas. Consider the following examples:

Example 4.1   A (hypothetical) map of airline routes between several California cities

The points are the cities; a line connects two cities if and only if there is a nonstop flight between them in both directions. See Fig. 4.1.   ■

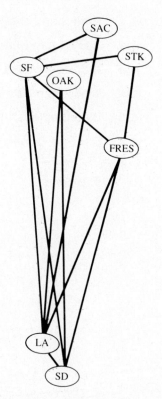

Figure 4.1   A (hypothetical) graph of nonstop airline flights between California cities.

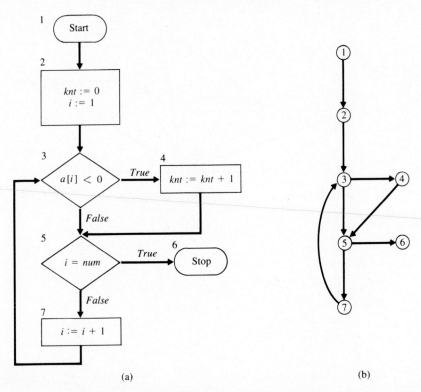

Figure 4.2   (a) A flowchart. (b) A directed graph. Arrows indicate the direction of flow.

Example 4.2   The flow of control in a flowchart

The points are the flowchart boxes; the connecting arrows are the flowchart arrows.
See Fig. 4.2.   ■

Example 4.3   A binary relation

Let S be the set $\{1, 2, \ldots, 9, 10\}$ and let R be the relation on S defined by xRy if and
only if $x \neq y$ and x divides y. In the digraph in Fig. 4.3 the points are the elements of
S and there is an arrow from x to y if and only if xRy.   ■

Example 4.4   Computer networks

In Fig. 4.4 the points are the computers and the edges are the communication links. ■

Example 4.5   An electrical circuit

The points could be diodes, transistors, capacitors, switches, etc. Two points would
be connected by a line if there is a wire connecting them.   ■

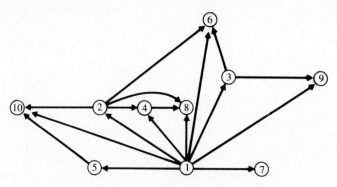

Figure 4.3 The relation *R* in Example 4.3.

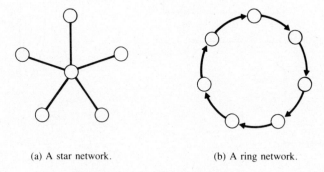

(a) A star network.   (b) A ring network.

Figure 4.4 Computer networks.

Formally, a *graph*, *G*, is a pair (V, E) where V is a finite, nonempty set whose elements are called vertices, and E is a set of subsets of V of order two. Elements of E are called edges.

For the graph in Fig. 4.1, for example, we have $V = \{$SF, OAK, SAC, STK, FRES, LA, SD$\}$ and $E = \{\{$SF, STK$\}$, $\{$SF, SAC$\}$, $\{$SF, LA$\}$, $\{$SF, SD$\}$, $\{$SF, FRES$\}$, $\{$SD, OAK$\}$, $\{$SAC, LA$\}$, $\{$LA, OAK$\}$, $\{$LA, FRES$\}$, $\{$LA, SD$\}$, $\{$FRES, STK$\}$, $\{$SD, FRES$\}\}$.

A *digraph*, *G*, is a pair (V, E) where V is a finite, nonempty set whose elements are called vertices, and E is a set of ordered pairs of distinct elements of V. Elements of E are called edges (or sometimes arcs). For (v,w) in E, v is the *tail* and w the *head* of (v, w). That is, (v, w) is represented in the diagrams as $v{\rightarrow}w$.

In the flowchart example (Fig. 4.2b), using the numbers assigned to the flowchart boxes, $V = \{1, 2, \ldots, 7\}$ and $E = \{(1,2), (2,3), (3,4), (3,5), (4,5), (5,6), (5,7), (7,3)\}$.

The definitions of graph and digraph imply that there cannot be an edge that connects a vertex to itself and that there cannot be two edges between one pair of vertices in a graph, or two edges with the same orientation (direction), between one pair of vertices in a digraph. In some applications these restrictions may be dropped.

For example, let R be a binary relation on a finite set S and let $V = S$. We may wish to have an edge (x, y) for each pair x and y in S such that xRy, even if $x = y$. In applications where the definitions of graph and digraph are modified to permit such edges, care must be taken to see that any theorems and algorithms used for graphs and digraphs are still correct. In this chapter all graphs and digraphs will satisfy the definitions given here.

The preceding five examples should be sufficient to illustrate that graphs and digraphs provide a natural abstraction of relationships of diverse objects, including both physical objects and their arrangement, such as cities connected by airline routes, highways, or railway lines, and abstract objects, such as binary relations and the control structure of a program. These examples should also suggest some of the questions we may wish to ask about the objects represented, questions that will be rephrased in terms of the graph or digraph and that can be answered by algorithms that work on these structures. The question "Is there a nonstop flight between San Diego and Sacramento?" translates into "Is there an edge between the vertices SD and SAC in Fig. 4.1?" Consider the following questions:

What is the cheapest way to fly from Stockton to San Diego?

Which route involves the least flying time?

If one city's airport is closed by bad weather, can you still fly between any other pair of cities?

If one computer in a network goes down, can messages be sent between any other pair of computers in the network?

How much traffic can flow from one specified point to another using certain specified roads?

Is a given binary relation transitive?

Does a given flowchart have any loops?

How should wires be attached to various electrical outlets so that all are connected using the least amount of wire?

In this chapter we will study algorithms to answer most of these questions. The remainder of this section is devoted to definitions, general remarks, and a discussion of data structures useful for representing graphs and digraphs in a computer program. Many statements and definitions apply to both graphs and digraphs, and we will use the same notation for both to minimize repetition. An edge $\{v, w\}$ in a graph or (v, w) in a digraph will be written vw. (For a graph, of course, $vw = wv$.)

A *subgraph* of a graph or digraph $G = (V, E)$ is a graph (or digraph) $G' = (V', E')$ such that $V' \subseteq V$ and $E' \subseteq E$. A *complete graph* is a graph with an edge between each pair of vertices. Vertices v and w are said to be *incident* with the edge vw and vice versa.

The edges of a graph or digraph $G = (V, E)$ induce a relation called the adjacency relation, A, on the set of vertices. Let v and w be elements of V. Then vAw (read "w is *adjacent* to v") if and only if vw is in E. In other words, vAw means that w can be reached from v by moving along an edge of G. If G is a graph, the relation A is symmetric. (That is, wAv if and only if vAw.)

Consider Fig. 4.1 again and suppose we wish to travel by airplane from Los Angeles (LA) to Fresno (FRES). There is an edge {LA, FRES} that is one possible route, but there are others. We could go from LA to SAC to SF to FRES, or we could go from LA to SD to FRES. These are all "paths" from LA to FRES in the graph. The concept of a path is very useful in many applications, including some (like this one) that involve routing people, telephone (or electronic) messages, automobile traffic, liquids or gases in pipes, etc., and others in which paths represent abstract properties. (See Exercise 4.2.) Formally, a *path from v to w* in a graph or digraph $G = (V, E)$ is a sequence of edges $v_0 v_1, v_1 v_2, \ldots, v_{k-1} v_k$, such that $v_0 = v$, $v_k = w$, and $v_0, v_1, \ldots, v_k$ are all distinct. The length of the path is k. A vertex v alone is considered to be a path of length zero from v to itself. The path {SD, FRES}, {FRES, SF}, {SF, SAC} is shown in Fig. 4.5. We will denote a path by listing the sequence of vertices through which it passes (but remember that the length of a path is the number of edges traversed). Thus the path in Fig. 4.5 is SD, FRES, SF, SAC, and has length three. A graph is *connected* if for each pair of vertices, v and w, there is a path from v to w.

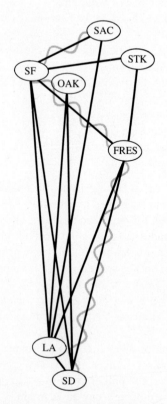

Figure 4.5   A path from SD to SAC.

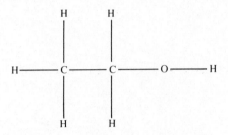

Figure 4.6 A tree: an alcohol molecule.

A *cycle* in a graph or digraph $G = (V, E)$ is like a path $v_0, v_1, \ldots, v_k$, with $k \geq 2$, except that $v_k = v_0$. A graph or digraph is *acyclic* if it has no cycles. A *tree* may be defined as a connected, acyclic graph; see Fig. 4.6. Note that with this definition of a tree, no node is singled out as the root. A *rooted tree* is a tree with one vertex designated as the root. The parent and child relations often used with trees can be derived once a root is specified.

If a graph is not connected, it may be partitioned into separate connected pieces: a *connected component* of a graph G is a maximal connected subgraph of G. The graph in Fig. 4.7 has three connected components.

In many applications of graphs and digraphs it is natural to associate a number with each edge. The numbers represent costs or benefits derived from using the particular edge in some way. Consider Fig. 4.1 once again and suppose that we want to fly from SD to SAC. There is no nonstop flight, but there are several routes or paths that could be used. Which is best? To answer this question we need a standard by which to judge the various paths. Some possible standards are

1. the number of stops,
2. the total ticket cost, and
3. the total flying time.

After choosing a standard, we could assign to each edge in the graph the cost (in stops, money, or time) of traveling along that edge. The total cost of a particular path is the sum of the costs of the edges traversed by that route. Figure 4.8 shows the airline graph with the (hypothetical) cost of a plane ticket written beside each

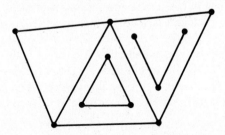

Figure 4.7 A graph with three connected components.

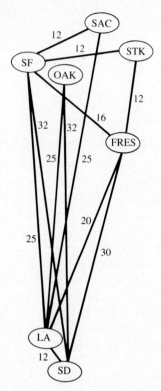

Figure 4.8 A weighted graph showing airline fares.

edge.[1] The reader may verify that the cheapest way to get from SD to SAC is to make one stop in LA. The general problem of finding "best" paths is studied in Section 4.3.

Figure 4.9, which shows some of the streets in a city, might be used to study the flow of automobile traffic. The number assigned to an edge indicates the amount of traffic that can flow along that section of the street in a certain time interval. The number is determined by the type and size of road, the speed limit, the number of traffic lights between the intersections shown in the graph as vertices (assuming that not every street is shown in the graph), and various other factors.

The assignment of numbers to edges occurs often enough in applications to merit a definition. A *weighted graph* (or *weighted digraph*) is a triple (V, E, W) where (V, E) is a graph (or digraph) and W is a function from E into $\mathbf{Z}^+$, the positive integers. For an edge e, $W(e)$ is called the *weight* of e. The weights in some applications will correspond to costs or undesirable aspects of an edge, whereas in other problems the weights are capacities or other beneficial properties of the edges. (The

---

[1] When writing the first edition of this book, I corrected these numbers every few months to reflect the effects of inflation and changing oil prices. I soon gave up.

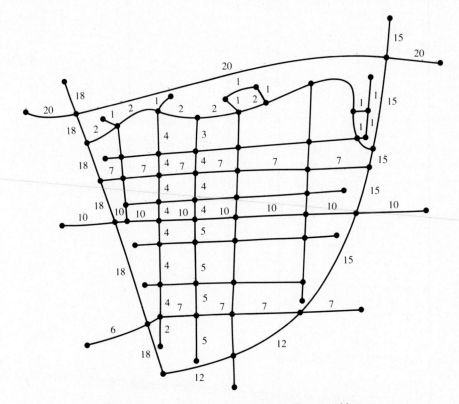

Figure 4.9   A street map showing traffic capacities.

terminology varies with the application; thus the terms cost, length, or capacity may be used instead of weight.) Sometimes it is desirable to allow negative or noninteger weights, but one must be careful when choosing algorithms since the correctness of some of them depends on restricting the weights to nonnegative integers.

4.1.2 Computer Representation of Graphs and Digraphs

We have seen two ways of representing a graph or digraph on paper: by drawing a picture in which vertices are represented by points and edges as lines, and by listing the vertices and edges. For solving problems in programs, we need other representations. Let $G = (V, E)$ be a graph or digraph with $|V| = n$, $|E| = m$, and $V = \{v_1, v_2, \ldots, v_n\}$.

G can be represented by an $n \times n$ matrix $A = (a_{ij})$, called the *adjacency matrix* for G. A is defined by

$$a_{ij} = \begin{cases} 1 & \text{if } v_i v_j \in E \\ 0 & \text{otherwise} \end{cases} \quad \text{for } 1 \leq i, j \leq n.$$

The adjacency matrix for a graph is symmetric (and only half of it need be stored). If $G = (V, E, W)$ is a weighted graph or digraph, the weights can be stored in the adjacency matrix by modifying its definition as follows:

$$a_{ij} = \begin{cases} W(v_i v_j) & \text{if } v_i v_j \in E \\ c & \text{otherwise} \end{cases} \quad \text{for } 1 \le i, j \le n,$$

where c is a constant whose value depends on the interpretation of the weights and the problem to be solved. If the weights are thought of as costs, ∞ (or some very high number) may be chosen for c because the cost of traversing a nonexistent edge is prohibitively high. If the weights are capacities, a choice of $c = 0$ is usually appropriate since nothing can move along an edge that is not there. See Figs. 4.10(a), 4.10(b), 4.11(a), and 4.11(b) for examples.

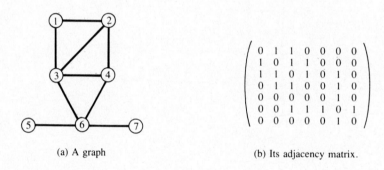

(a) A graph

(b) Its adjacency matrix.

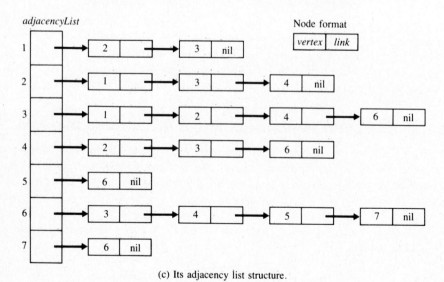

(c) Its adjacency list structure.

Figure 4.10 Representations for a graph.

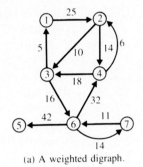

(a) A weighted digraph.

$$\begin{pmatrix} 0 & 25 & \infty & \infty & \infty & \infty & \infty \\ \infty & 0 & 10 & 14 & \infty & \infty & \infty \\ 5 & \infty & 0 & \infty & \infty & 16 & \infty \\ \infty & 6 & 18 & 0 & \infty & \infty & \infty \\ \infty & \infty & \infty & \infty & 0 & \infty & \infty \\ \infty & \infty & \infty & 32 & 42 & 0 & 14 \\ \infty & \infty & \infty & \infty & \infty & 11 & 0 \end{pmatrix}$$

(b) Its adjacency matrix.

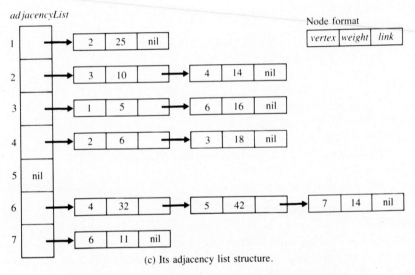

(c) Its adjacency list structure.

Figure 4.11 Representations for a weighted digraph.

Algorithms for solving some problems on graphs and digraphs require that every edge be examined and processed in some way at least once. If an adjacency matrix representation is used, we may as well think of a graph or digraph as having edges between all pairs of distinct vertices, because many algorithms would examine each entry in the matrix to determine which edges really exist. Since the number of edges is $n(n-1)/2$, in a graph, or $n(n-1)$ in a digraph, the complexity of such algorithms will be in $\Omega(n^2)$.

An alternative to the adjacency matrix representation is a data structure containing, for each vertex v, a linked list indicating which vertices are adjacent to v. The data in the adjacency lists will vary with the problem, but there is a fairly standard basic structure that is useful for many algorithms. The nodes in the linked lists have fields *vertex* and *link*, where *vertex* contains a vertex number and *link* contains a pointer. Each such node represents an edge in the graph or digraph. In particular, suppose we number the vertices so that $V = \{1, 2, \ldots, n\}$. Then, if a node containing, say, 2, in its *vertex* field is on the adjacency list for 7, it represents the edge 72. We use a pointer array, *adjacencyLists*, for the n listheads, one for the adjacency list of each vertex. This data structure for a graph is illustrated by the example in Fig. 4.10. Each edge is represented twice; that is, if *vw* is an edge, there is a node for *w* on the adjacency list for *v* and a node for *v* on the adjacency list for *w*. Thus there are $2m$ linked list nodes and n listheads. For a digraph each edge is represented exactly once. If the graph or digraph is weighted, a weight field is included in each node. Figure 4.11 illustrates the structure for a weighted digraph. The declarations for this structure are

```
const
    MAXVERTICES = { some appropriate value };
type
    VertexType = 1..MAXVERTICES;
    NodePointer = pointer to EdgeNode;
    EdgeNode = record
        vertex: VertexType;
        weight: integer;
        link: NodePointer
    end { EdgeNode };
    HeaderArray = array[1..MAXVERTICES] of NodePointer;
var
    adjacencyList: HeaderArray;
```

Additional fields may be added to the list heads or the linked list nodes as required by the algorithms to be used. The merit of this structure is that edges that do not exist in G do not exist in the representation either. If G is sparse (has few edges), it can be processed quickly. Note that if the nodes within an adjacency list appear in a different order, the structure still represents the same graph or digraph, but an algorithm using the list will encounter the nodes in a different order and may behave very differently. An algorithm should not assume any particular ordering (unless, of course, it constructs the list itself in a special way).

4.2
A Minimum Spanning Tree Algorithm

In this section and the next we present two similar algorithms of E. W. Dijkstra — one for finding minimum-spanning trees and one for finding shortest paths. (The minimum spanning tree algorithm was discovered independently by R. C. Prim.)

4.2.1 Definitions and an Overview of the Algorithm

The first problem we will study is the problem of finding a minimum spanning tree for a weighted graph (not directed).

A *spanning tree* for a graph, $G = (V, E)$, is a subgraph of G that is a tree and contains all the vertices of G. In a weighted graph the weight of a subgraph is the sum of the weights of the edges in the subgraph. A *minimum spanning tree* for a weighted graph is a spanning tree with minimum weight. There are many situations in which minimum spanning trees must be found. Whenever one wants to find the cheapest way to connect a set of terminals, be they cities, electrical terminals, computers, or factories, by using, say, roads, wires, or telephone lines, a solution is a minimum spanning tree for the graph with an edge for each possible connection weighted by the cost of that connection. Finding minimum spanning trees is also an important subproblem in various routing algorithms, that is, algorithms for finding efficient paths through a graph that visit every vertex (or every edge).

As the simple example in Figure 4.12 shows, a weighted graph may have more than one minimum spanning tree.

If G is not connected, then it cannot have any spanning tree, much less a minimum one: It would have a spanning forest. For simplicity in describing the minimum spanning tree algorithm, we will assume for a while that G is connected. It will be easy to extend the algorithm and the justification for it to finding minimum spanning forests.

The Dijkstra/Prim algorithm begins by selecting an arbitrary starting vertex, and then "branches out" from the part of the tree constructed so far by choosing a new vertex and edge at each iteration. During the course of the algorithm, the vertices may be thought of as divided into three (disjoint) categories as follows:

Tree vertices: in the tree constructed so far.
Fringe vertices: not in the tree, but adjacent to some vertex in the tree.
Unseen vertices: all others.

The key step in the algorithm is the selection of a vertex from the fringe and an incident edge. Actually, since the weights are on the edges, the focus of the choice is on the edge, not the vertex. The Dijkstra/Prim algorithm always chooses an edge from a tree vertex to a fringe vertex of minimum weight. The general structure of the algorithm can be described as follows.

Select an arbitrary vertex to start the tree;
while there are fringe vertices **do**
    select an edge of minimum weight between a tree vertex and
        a fringe vertex;
    add the selected edge and the fringe vertex to the tree
end

Figure 4.13 illustrates one iteration of the loop.

This algorithm is an example of a "greedy" algorithm. *Greedy algorithms* are algorithms for optimization problems (that is, where some quantity is to be minimized or maximized); they make locally optimal choices at each step in the hope that these choices will produce a globally optimal solution. This approach works for many problems but not for others. We will see other greedy algorithms later in this

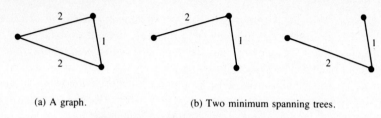

(a) A graph. (b) Two minimum spanning trees.

Figure 4.12 Minimum spanning trees.

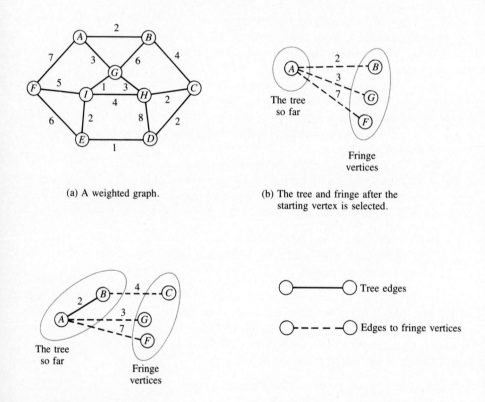

(a) A weighted graph.

(b) The tree and fringe after the starting vertex is selected.

(c) After selecting an edge and vertex. *BG* is not shown because *AG* is a better choice (has lower weight) to reach *G*.

Figure 4.13 One iteration of the loop.

book, including, in Chapter 8, another greedy approach to finding minimum spanning trees.

Can we be sure that this strategy will yield a minimum spanning tree? Is being greedy in the short term a good long-term strategy? In this case, yes; an easy induction proof works. Since the algorithm will start with no edges selected, the edges selected initially form a subset of the edges in some minimum spanning tree. We prove that, after adding a minimum edge from the tree to a fringe vertex, the set of selected edges is still contained in a minimum spanning tree.

Theorem 4.1   Let $G = (V, E, W)$ be a connected, weighted graph and let $E' \subseteq E$ be a subset of the edges in some minimum spanning tree $T = (V, E_T)$ for G. Let V' be the vertices incident with edges in E'. If xy is an edge of minimum weight such that $x \in V'$ and $y \notin V'$, then $E' \cup \{xy\}$ is a subset of a minimum spanning tree.

Proof. If xy is in E_T, the conclusion follows. Suppose xy is not in E_T. There is a path from x to y in T since trees are connected. Let vw be the first edge in that path with exactly one vertex in V', say, v. (See Fig. 4.14.) Let $E_{T'} = E_T - \{vw\} \cup \{xy\}$. The reader may verify that $E' \cup \{xy\} \subseteq E_{T'}$ and that T' (that is, $(V, E_{T'})$) is a spanning tree. Since v is in V' and w is not in V', by the choice of xy, $W(xy) \leq W(vw)$, so $W(T') \leq W(T)$ and T' is a minimum spanning tree.   □

After each iteration of the algorithm's loop, there may be new fringe vertices, and the set of edges from which the next selection is made will change. Figure 4.13(c) suggests that we need not consider all edges between tree vertices and fringe vertices. After AB was chosen, BG became a potential edge choice, but it is discarded because AG has lower weight and would be a better choice for reaching G. If BG had lower weight than AG, then AG could be discarded. For each fringe vertex, we need keep track of only one edge to it from the tree — the one with lowest weight. We will call such edges *candidate edges*.

We can now expand the outline of the algorithm. (Later we will consider data structures and implementation details and rewrite the algorithm in more detail.)

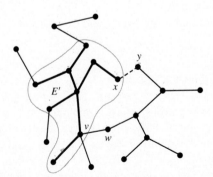

Figure 4.14 Minimum spanning tree T for Theorem 4.1. Edges shown as solid lines are in T. The heavy edges are also in E'. There may be many other edges in the graph that are not shown.

Algorithm 4.1   Minimum spanning tree (Dijkstra/Prim)

Input: $G = (V, E, W)$, a weighted graph.

Output: The edges in a minimum spanning tree.

1.   { Initialization }
    Let x be an arbitrary vertex.
    $V_T := \{x\}$; $E_T := \varnothing$;
    stuck := false;

2.   { Main loop; x has just been brought into the tree.
    Update fringe and candidates. Then add one vertex and edge. }
    while $V_T \neq V$ **and not** *stuck* **do**

3.   { Replace some candidate edges. }
      for each fringe vertex y adjacent to x **do**
        if $W(xy) < W($the candidate edge e incident with $y)$ **then**
          xy replaces e as the candidate edge for y;
        end { if }
      end { for };

4.   { Find new fringe vertices and candidate edges. }
      for each unseen y adjacent to x **do**
        y is now a fringe vertex;
        xy is now a candidate
      end { for };

5.   { Ready to choose next edge. }
      if there are no candidates **then** *stuck* := true {no spanning tree};
      else

6.   { Choose next edge. }
      Find a candidate edge, e, with minimum weight;
      $x :=$ the fringe vertex incident with e;
      Add x and e to the tree;
      { x and e are no longer fringe and candidate. }
      end { if }
    end { while }

    The algorithm terminates with $V_T = V$, and the edges in E_T form a minimum spanning tree if and only if G is connected. If it terminates for lack of candidate edges (*stuck* is true), the edges in E_T form a minimum spanning tree for a connected component of G. With slight modification, the tree construction can be "restarted" to find a forest of minimum spanning trees, one for each connected component of G. Figure 4.15 contains an example of the action of the algorithm.

    Can we determine how much time this algorithm requires before we consider the details of an implementation? The question should prompt another one from the reader: What basic operation should we choose to count? In steps 3 and 6 weights are compared, but in step 4 there are no comparisons, although the **for** loop there

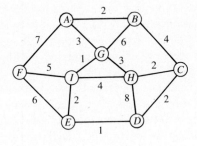

(a) A weighted graph.

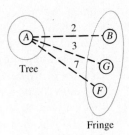

(b) After selection of
the starting vertex.

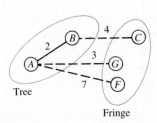

(c) *BG* was considered but did not
replace *AG* as a candidate.

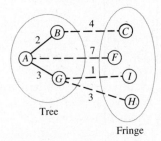

(d) After *AG* is selected and fringe and
candidates updated.

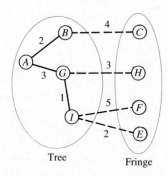

(e) *IF* has replaced *AF* as a candidate.

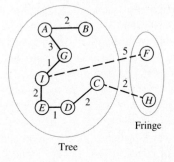

(f) After several more passes. The two candidate
edges will be put in the tree.

Figure 4.15 An example for the minimum spanning tree algorithm.

may do a lot of work. In several steps the "status" of vertices and edges is modified. The number of comparisons of weights is a good measure of work done only if the time spent on other operations is proportional to the number of comparisons. Is it? To answer this question, we have to examine the bookkeeping carefully, but the specific operations and the number of them that would be carried out depend on how the algorithm keeps track of fringe vertices and candidate edges, and how such lines as "**for** each fringe vertex y adjacent to x" are implemented. For example, maintaining a list of fringe vertices and testing each entry for adjacency to x may be much more, or much less, efficient than traversing an adjacency list for x and checking each entry to see if it is on the fringe. Hence we will defer the timing analysis until after we study data structures and implementation details. (It turns out that, even with a good implementation, there are examples of graphs where the running time is of higher order than the number of comparisons done.)

4.2.2 Implementation

We want to choose a data structure that stores only information that is really needed, and stores it in such a way that the operations required by the algorithm can be done quickly. Let $n = |V|$ and $m = |E|$.

The **for** loops in steps 3 and 4 may suggest that each fringe vertex and each unseen vertex be considered in turn and tested for adjacency to x. Suppose we use a linked list representation of these sets. Adding and deleting vertices would be easy with links. Even before considering other details, we can now derive a lower bound on the total number of operations done in steps 3 and 4. Since there are $n-1$ fringe and unseen vertices when the algorithm starts and only one is put in the tree at each iteration of the main loop, traversing lists of fringe and unseen vertices $n-1$ times will require $\Omega(n^2)$ operations. Suppose, on the other hand, that the graph is represented by linked adjacency lists roughly as described in Section 4.1. We could implement steps 3 and 4 by traversing the adjacency list for x and checking whether each y on the list is a fringe or unseen vertex. (A *status* array indexed by the vertices could be used to remember where each vertex is at any time.) Since each vertex in V plays the role of x in the algorithm at most once, each adjacency list would be traversed at most once. The number of operations required to traverse these lists is proportional to the number of edges, m, in the graph. Although m can be as large as $(n^2-n)/2$ (in a complete graph), for many graphs on which the algorithm may be used, it is much smaller. So, although traversing adjacency lists in steps 3 and 4 will not give a significant reduction in work for graphs with a large number of edges, it is better for some inputs and is the implementation we will use. The data structure is illustrated in Fig. 4.16; the rest of the features are explained in the next few paragraphs. We will need to store several data for each vertex. A common approach is to use an array of records, one for each vertex. Our main concern in presenting the algorithms here is clarity and readability. Because the notation for a field in a record in an array is bulky (e.g., *vertexData*[*v*].*status*) compared with the notation for individual arrays (e.g., *status*[*v*]), we will use separate arrays. The declarations appear with the algorithm.

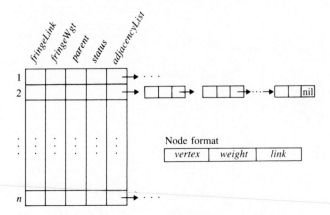

Figure 4.16   Data structure for Algorithm 4.2.

How should we keep track of the candidate edges? We must be able to find the candidate edge for a given fringe vertex (step 3), replace edges (step 3), and find a candidate with minimum weight (step 6). Our representation should permit implementation of these operations without searching lists or arrays unnecessarily. There is one candidate for each fringe vertex; we will link the fringe vertices and store data describing each candidate edge with its associated fringe vertex. We use arrays *fringeLink*, *parent*, and *fringeWgt*. For a fringe vertex v, *fringeLink*[v] contains the link to (actually, the number of) the next fringe vertex, and *parent*[v] contains the vertex that is the parent of v in the tree formed by the tree edges and candidate edges, taking the first vertex chosen (at step 1) as the root. In other words, *parent*[v] is the tree vertex incident with v's candidate edge; *fringeWgt*[v] is the weight of that edge. The value for *fringeWgt*[v] is copied from a *weight* field in an adjacency list so it can be found easily when needed in step 3. Now, replacing a candidate edge (step 3) requires only changing an entry in the *parent* and *fringeWgt* arrays. Finding (and deleting) a candidate edge with minimum weight can be done by traversing the list of fringe vertices via the *fringeLink*s and comparing the corresponding *fringeWgt*s.

When the algorithm terminates, the tree edges are {{z,*parent*[z]} | all z except the root}, and the *fringeWgt* array contains the weights of the selected edges. Figure 4.17 shows the data structure at an intermediate point in the execution of the algorithm on the example in Fig. 4.15 (specifically at the point illustrated in Fig. 4.15(d)). To make the figure a little easier to read, we show the vertex names as letters, as in Fig. 4.15.

Instead of testing "$V_T \neq V$" in step 2, which would require a scan of the *status* array, we use a counter *edgeCount* for edges in E_T; a tree with n vertices must have exactly $n-1$ edges.

Algorithm 4.1 is rewritten here as Algorithm 4.2 to show more of the detail.

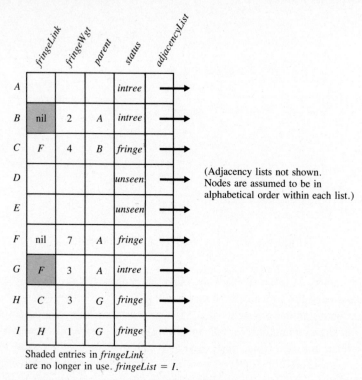

The table in the figure:

	fringeLink	fringeWgt	parent	status	adjacencyList
A				intree	→
B	nil	2	A	intree	→
C	F	4	B	fringe	→
D				unseen	→
E				unseen	→
F	nil	7	A	fringe	→
G	F	3	A	intree	→
H	C	3	G	fringe	→
I	H	1	G	fringe	→

(Adjacency lists not shown. Nodes are assumed to be in alphabetical order within each list.)

Shaded entries in *fringeLink* are no longer in use. *fringeList* = *I*.

Figure 4.17   Minimum spanning tree data structure for the situation in Fig. 4.15(d).

Algorithm 4.2   Minimum Spanning Tree

Input: $G = (V, E, W)$, a weighted graph with $V = \{1, \ldots, n\}$, represented by an adjacency list structure.

Output: A list of edges in a minimum spanning tree for G.

Comment: This algorithm assumes the following adjacency list structure.

```
const
    MAXVERTICES = { some appropriate value };

type
    VertexType = 1..MAXVERTICES;
    { Adjacency list structure for the graph }
    NodePointer = pointer to EdgeNode;

    EdgeNode = record
        vertex: VertexType;
        weight: integer;
        link: NodePointer
    end { EdgeNode };
```

HeaderArray = **array**[1..MAXVERTICES] **of** *NodePointer*;

procedure *MinSpanningTree* (*adjacencyList*: *HeaderArray*; *n*: *integer*);

type
    VertexData = **array**[*VertexType*] **of** *integer*;
    StatusType = (*intree, fringe, unseen*);

var
    status: **array**[*VertexType*] **of** *StatusType*;
    parent: *VertexData*;
    fringeLink: *VertexData*;
    fringeWgt: *VertexData*;
    ptr: *NodePointer*;   { for traversing adjacency lists }
    x, y : *VertexType*;
    fringeList: *integer*;   { first fringe vertex }
    edgeCount: *integer*;
    stuck: *boolean*;

begin
    { Initialization }
    x := 1; *status*[1] := *intree*;
    edgeCount := 0;
    fringeList := 0;
    for *y* := 2 **to** *n* **do** *status*[*y*] := *unseen* **end**;
    stuck := *false*;

    while *edgeCount* < *n*−1 **and not** *stuck* **do**

        { Traverse the adjacency list for *x*. }
        ptr := *adjacencyList*[*x*];
        while *ptr* ≠ **nil do**
            y := *ptr*↑.*vertex*;
            if *status*[*y*] = *fringe* **and** *ptr*↑.*weight* < *fringeWgt*[*y*] **then**
                { Replace *y*'s candidate edge by *xy*. }
                parent[*y*] := *x*;
                fringeWgt[*y*] := *ptr*↑.*weight*
            end { if *y* is a fringe vertex };
            if *status*[*y*] = *unseen* **then**
                { *y* is now in the fringe and *xy* a candidate. }
                status[*y*] := *fringe*;
                link *y* onto the beginning of the fringe list;
                parent[*y*] := *x*;
                fringeWgt[*y*] := *ptr*↑.*weight*
            end { if *y* was *unseen* };
            ptr := *ptr*↑.link
        end { while ptr ≠ nil };

{ Choose the next vertex and edge for the tree. }
if *fringeList* = 0 **then** *stuck* := *true*
else
     traverse the fringe list to find a candidate with minimum weight;
     x := the fringe vertex incident with the edge;
     remove *x* from the fringe list;
     status[*x*] := *intree*; *edgeCount* := *edgeCount*+1
end { if }

end { while edgeCount < *n*−1 and not stuck };

{ Output the tree. }
for *x* := 2 **to** *n* **do**
     output(*x*,*parent*[*x*])
end { for }
end { *MinSpanningTree* }

4.2.3   Analysis (Time and Space)

Let $n = |V|$ and $m = |E|$. The number of initialization operations performed is linear in n. The number of operations (comparisons of weights, changes in pointer values, etc.) done for each edge when traversing the adjacency lists in the "**while** $ptr \neq$ **nil**" loop is independent of n and m, so the total time required for the work done in all iterations of the traversal loop is in $\Theta(m)$. The test to see if the fringe list is empty is executed at most $n-1$ times. So far it looks as though the running time of the algorithm may be linear in m (since, usually, $m > n$). However, as many as (roughly) $n^2/2$ comparisons may be done (in total, for all iterations of the outer loop) to find minimum candidate edges, even if the number of edges is smaller. There may be $n-1$ candidates after the first pass through the loop; finding one with minimum weight would take $n-2$ comparisons. Both these numbers decrease by one on each subsequent pass, and the total number of operations is in $\Theta(n^2)$. Thus the worst-case running time, as well as the worst-case number of comparisons done, is in $\Theta(m+n^2) = \Theta(n^2)$. (The reader is encouraged to investigate ways to reduce the work performed to find the minimum candidates, but see Exercises 4.6 – 4.8.)

The data structure in Fig. 4.16 uses $4n$ cells in addition to those in the adjacency list representation of the graph. This is more extra space than is used by any of the algorithms we have studied so far, and it may seem like quite a lot. However, it allows a time-efficient implementation of the algorithm. (It would be worse if the extra space requirement were in $\Theta(m)$ since for many graphs $\Theta(m) = \Theta(n^2)$.)

4.2.4   Lower Bound

How much work is essential to find a minimum spanning tree? We claim that any minimum spanning tree algorithm requires time that is in $\Omega(m)$ in the worst case because it must examine, or process in some way, every edge in the graph. To see this, let G be a connected, weighted graph where each edge has weight at least 2, and suppose there were an algorithm that did not do anything at all to an edge, $e = xy$, in G. Then e is not in the output, T, of the algorithm. T contains a path from x to y so there is a cycle in G consisting of e and the edges in that path. Change the weight of e to 1. This could not change the action of the algorithm because it never examined e. Now the tree obtained by removing one of the edges in the cycle and using e instead is a spanning tree with lower weight than T. Therefore the algorithm is not correct.

4.3
A Shortest-Path Algorithm

4.3.1 Background

In Section 4.1 we briefly considered the problem of finding the best route between two cities on a map of airline routes (Fig. 4.8). Using as our criterion the price of the plane tickets, we observed that the best — i.e., cheapest — way to get from San Diego to Sacramento was to make one stop in Los Angeles. This is one instance, or application, of a very common problem on a weighted graph or digraph: finding a "shortest" path between two specified vertices. The weight, or length, of a path $v_0, v_1, \ldots, v_k$ in a weighted graph or digraph $G = (V, E, W)$ is $\sum_{i=0}^{k-1} W(v_i v_{i+1})$ — in other words, the sum of the weights of the edges in the path. If the path is called P we denote its weight by $W(P)$. A path from v to w is a *shortest path* from v to w if there is no path from v to w with lower weight. (It is conventional, alas, to mix the terminology of weight and length.) Observe that shortest paths are not necessarily unique.

How did we determine the shortest path from SD to SAC in Fig. 4.8? In fact, the reader probably used a very unalgorithmic method full of assumptions, such as that the fares are proportional to the distance between the cities and that the map is drawn approximately to scale. Then the reader probably picked a route that "looked" short. This is hardly an algorithm we would expect to program for a computer. We mention it to answer the above question honestly; people generally use very unrigorous ways to solve problems, especially on very small sets of data. In practice, the problem of finding a shortest, or cheapest, path between two vertices in a graph or digraph arises in applications in which V may contain several hundred vertices. An algorithm could consider all possible paths and compare their weights, but that could take a very long time. In this section we study Dijkstra's shortest-path algorithm; it is very similar in approach and timing to the minimum spanning tree algorithm in the previous section.

4.3.2 The Algorithm

We are given a weighted graph or digraph $G = (V, E, W)$ and two specified vertices v and w; the problem is to find a shortest path from v to w. Before proceeding, we should consider whether we need a new algorithm at all. Suppose we use the minimum spanning tree algorithm, starting at v. Will the path to w in the tree constructed by the algorithm always be a shortest path from v to w? Consider the path from A to C in the minimum spanning tree in Fig. 4.15. It is *not* a shortest path; the path A,B,C is shorter.

The *distance* from a vertex x to a vertex y, denoted $d(x,y)$, is the weight of a shortest path from x to y. Dijkstra's shortest-path algorithm will find shortest paths from v to the other vertices in order of increasing distance from v. It stops when it reaches w (though it can be modified to find the set of shortest paths from v to each other vertex if desired). The algorithm, like the minimum spanning tree algorithm in Section 4.2, starts at one vertex (v) and "branches out" by selecting certain edges that lead to new vertices. Also like the minimum spanning tree algorithm, it is greedy; it always chooses an edge to a vertex that appears to be closest. The vertices are again divided into three categories as follows:

Tree vertices: in the tree constructed so far.
Fringe vertices: not in the tree, but adjacent to some vertex in the tree.
Unseen vertices: all others.

Also, as in the minimum spanning tree algorithm, we keep track of only one candidate edge (the best found so far) for each fringe vertex. For each fringe vertex z, there is at least one path $v(=v_0), v_1, \ldots, v_k, z$ such that all the vertices except z are already in the tree. The candidate edge for z is the edge $v_k z$ from a shortest path of this form.

Whether or not G is a digraph, it is helpful to think of the tree and candidate edges as having an orientation; the tail of an edge is the vertex closer to v. Candidate edges go from a tree vertex to a fringe vertex. These edges will always be written to reflect this orientation; in other words, if we write xy, we are assuming that x is closer to v than y is. We will refer to x as $tail(xy)$ and y as $head(xy)$ even if G is not a directed graph.

Given the situation in Fig. 4.18(c), the next step is to select a candidate edge and fringe vertex. We choose a candidate edge e for which $d(v,tail(e))+W(e)$ is minimum. This is the weight of the path obtained by adjoining e to the known shortest path to $tail(e)$.

Since the quantity $d(v,tail(e))+W(e)$ for a candidate edge e may be used repeatedly, it can be computed once and saved. To compute it efficiently when e first becomes a candidate, we also save $d(v,y)$ for each y in the tree. Thus we use an array $dist$ as follows:

$dist[y] = d(v,y)$ for y in the tree, and
$dist[z] = d(v,y)+W(yz)$ for z on the fringe, where yz is the candidate edge to z.

After a vertex and the corresponding candidate edge are selected, the information in the data structure must be updated. In Fig. 4.18(d) the vertex I and the edge GI have just been selected. The candidate edge for F was AF, but now AF must be replaced by IF because IF yields a shorter path to F. We must also recompute $dist[F]$. The vertex E, which was unseen, is now on the fringe because it is adjacent

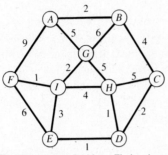

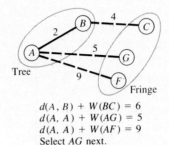

$d(A, B) + W(BC) = 6$
$d(A, A) + W(AG) = 5$
$d(A, A) + W(AF) = 9$
Select AG next.

(a) The Graph. Problem: Find a shortest path from A to H.

(b) An intermediate step.

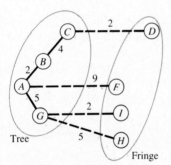

$d(A, C) + W(CD) = 8$
$d(A, A) + W(AF) = 9$
$d(A, G) + W(GI) = 7$
$d(A, G) + W(GH) = 10$
Select GI next.

(c) An intermediate step. (CH was considered but not chosen to replace GH as a candidate.)

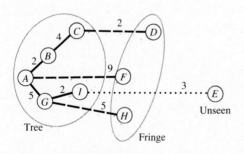

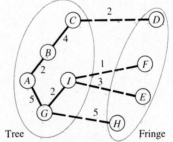

(d) GI is selected.

(e) AF was replaced bu IF as a candidate.

Figure 4.18 An example for the shortest-path algorithm.

to I, now in the tree. The edge IE becomes a candidate. (See Fig. 4.18(e).) Values of *dist* for new fringe vertices must be computed.

Does this method work? The questionable step is the selection of the next fringe vertex and candidate edge. For an arbitrary candidate e, $d(v, tail(e))+W(e)$ is not necessarily equal to $d(v, head(e))$ because shortest paths to $head(e)$ might not pass through $tail(e)$. (In Fig. 4.18, for example, the shortest path to H does not go through G, although GH is a candidate in Figs. 4.18(c), (d), and (e).) We claim that, if e is chosen by minimizing $d(v, tail(e))+W(e)$ over all candidates, then e does give a shortest path. This claim is proved in the following theorem.

Theorem 4.2   Let $G = (V, E, W)$ be a weighted graph or digraph with weights in Z^+. Let V' be a subset of V and let v be a member of V'. If e is chosen to minimize $d(v, tail(e))+W(e)$ over all edges with one vertex in V' and one in $V-V'$, then the path consisting of e adjoined to the end of a shortest path from v to $tail(e)$ is a shortest path from v to $head(e)$.

Proof.   Look at Fig. 4.19. Suppose e is chosen as indicated. Let $e = yz$, where y is in V', and let $v, x_1, \ldots, x_r, y$ be a shortest path from v to y. Let $P = v, x_1, \ldots, x_r, y, z$. $W(P) = d(v, y) + W(e)$. Let $v, z_1, \ldots, z_l, \ldots, z$ be any path from v to z; call it P'. We must show that $W(P) \le W(P')$. Let z_l be the first vertex in P' that is not in V'. (z_l may be z. If $l = 1$, interpret z_0 as v. In the algorithm, $z_{l-1}z_l$ would be a candidate edge.) $W(P) = d(v, y) + W(e) \le d(v, z_{l-1}) + W(z_{l-1}z_l)$ (by the choice of e) $\le W(P')$ since $v, z_1, \ldots, z_l$ is part of the path P'. $\square$

If there is a path from v to w at all, then w will be a leaf in the tree grown from v. There is no way to tell which of the tree edges are in the path to w until the algorithm terminates, so all of the paths that branch out from v are retained by using *parent* as in the minimum spanning tree algorithm.

The shortest-path algorithm uses virtually the same data structure as the minimum spanning tree algorithm; see Fig. 4.16. The only change is that *dist* replaces *fringeWgt*. For each fringe vertex z, *dist*[z] is the weight of the path from v

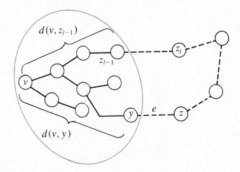

Figure 4.19   For the proof of Theorem 4.2.

to z using tree edges and the candidate edge for z. The other vertex of the candidate edge is, of course, $parent[z]$.

Algorithm 4.3   Shortest Path

Input: $G = (V, E, W)$, a weighted graph or digraph; $v, w \in V$. $V = \{1, \ldots, n\}$. G is represented by an adjacency list structure (as described in Algorithm 4.2).

Output: A shortest path from v to w.

```
procedure ShortestPath (adjacencyList: HeaderArray;
                            n: integer; v, w : VertexType);

type
      VertexData = array[VertexType] of integer;
      StatusType = (intree, fringe, unseen);

var
      status: array[VertexType] of StatusType;
      parent: VertexData;
      fringeLink: VertexData;
      dist: VertexData;
      ptr: NodePointer;                { for traversing adjacency lists }
      x, y : VertexType;
      fringeList: integer;             { first fringe vertex }
      stuck: boolean;

begin
      { Initialization }
      status[v] := intree;
      dist[v] := 0;
      parent[v] := 0;
      fringeList := 0;
      for y in V−{v} do status[y] := unseen end;
      x := v;
      stuck := false;

      while x ≠ w and not stuck do
            { Traverse the adjacency list for x. }
            ptr := adjacencyList[x];
            while ptr ≠ nil do
                y := ptr↑.vertex;
                if status[y] = fringe and dist[x]+ptr↑.weight < dist[y] then
                    { Replace y's candidate edge by xy. }
                    parent[y] := x;
                    dist[y] := dist[x]+ptr↑.weight
                end { if y is on the fringe };
```

        if *status*[*y*] = *unseen* **then**
            { *y* is now in the fringe, and *xy* a candidate. }
            status[*y*] := *fringe*;
            link *y* onto the beginning of the fringe list;
            parent[*y*] := *x*;
            dist[*y*] := *dist*[*x*]+*ptr*↑.*weight*
        end { if *y* was unseen };
        ptr := *ptr*↑.*link*
      end { while ptr ≠ nil };

      { Choose the next vertex and edge. }
      if *fringeList* = 0 **then** *stuck* := *true*
      else
        traverse the fringe list to find a vertex with minimum *dist*;
        x := this vertex;
        remove *x* from the fringe list;
        status[*x*] := *intree*
      end { if }

    end { while *x* ≠ *w* and not stuck };

    { Output the path. The vertices will be listed in reverse order,
    i.e., from *w* to *v*. }
    while *x* ≠ 0 **do**
      output (*x*);
      x := *parent*[*x*]
    end { while }
  end { *ShortestPath* }

Like Algorithm 4.2, Algorithm 4.3 runs in $\Theta(n^2)$ time in the worst case.

4.4
Traversing Graphs and Digraphs

4.4.1   Depth-first and Breadth-first Searches

Most algorithms for solving problems on a graph or digraph examine or process each vertex or edge. In the two algorithms considered so far, the order in which vertices and edges were considered was a fundamental part of the method used to solve the problem. Certain other strategies, or orders, for processing vertices and edges provide particularly useful and efficient methods for solving problems. Breadth-first and depth-first search are two such traversal strategies.

    Depth-first search is a generalization of preorder traversal of trees. The starting vertex may be determined by the problem or may be chosen arbitrarily. When each new vertex, say *v*, is visited, a path is followed as far as possible, "visiting" or processing all the vertices along the way, until a "dead end" is reached. A *dead end* is a vertex such that all its neighbors (vertices adjacent to it) have already been visited. At a dead end we back up along the last edge traversed and branch out in another direction. This has the effect of visiting all vertices in one subgraph adjacent to v before going on to a new subgraph adjacent to v, much as a preorder traversal visits all vertices in one subtree before going to the next subtree. (See Fig. 4.20.) The value of depth-first search was illustrated by J. E. Hopcroft and R. E. Tarjan, who developed many important algorithms that use it; several are presented in the rest of this chapter.

In a breadth-first search, vertices are visited in order of increasing distance from the starting point, say v, where distance is simply the number of edges in a shortest path. The central step of the breadth-first search, beginning with $d = 0$ and repeated until no new vertices are found, is to consider in turn each vertex, x, of distance d from v and, by examining all edges incident with x, find and process all vertices of distance $d+1$ from v. See Fig. 4.21 for examples of depth-first and breadth-first traversals.

These descriptions of the two traversal methods are somewhat ambiguous. For example, if two vertices are adjacent to v, which will be visited first? The answer

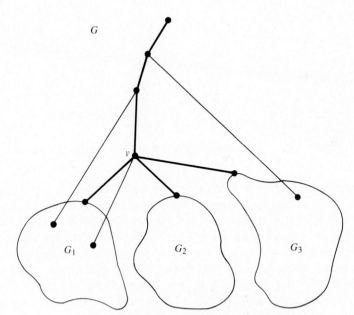

G_1 is completely traversed before exploring G_2, then G_3. Since G may not be a tree, there may be edges from the subgraphs to previously visited vertices.

Figure 4.20 Depth-first search.

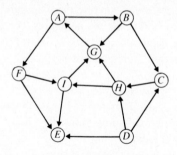

(a) A digraph.

Edges are numbered in the order traversed.

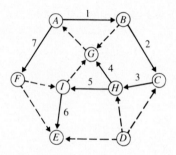

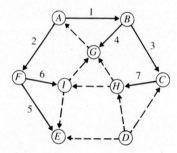

(b) Depth-first search beginning at *A*; order in
which vertices are visited: *A B C H G I E F*

(c) Breadth-first search beginning at *A*; order
in which vertices are visited: *A B F C G E I H*

Figure 4.21   Examples of depth-first and breadth-first searches.

will depend on implementation details, for example, the way in which the vertices
are numbered or arranged in the representation of *G*. An efficient implementation for
either method must keep a list of vertices that have been visited but whose adjacent
vertices have not yet all been visited. Note that when a depth-first search backs up
from a dead end, it is supposed to branch out from the most recently visited vertex
before pursuing new paths from vertices that were visited earlier. Thus the list of
vertices from which some paths remain to be traversed must be a stack. On the other
hand, in a breadth-first search, in order to ensure that vertices close to *v* are visited
before those farther away, the list must be a queue. Algorithms for both methods are
presented next. In both algorithms, vertices are marked (using a one-bit mark field,
or whatever is convenient) when first examined to prevent repeated work. The
instruction "visit *x*" is used to indicate when the desired processing of the vertex *x* is
done.

Algorithm 4.4   Depth-first Search

Input: $G = (V, E)$, a graph or digraph represented by an adjacency list structure as described in Section 4.1.2 with $V = \{1, 2, \ldots, n\}$; $v \in V$, the vertex from which the search begins.

Comment: For a stack S, we assume that the function call $Top(S)$ returns the value of the top item on S (without popping it).

> **procedure** *DepthFirstSearch (adjacencyList: HeaderList; v: VertexType)*;
> **var**
>   *S: Stack*;
>   *w: VertexType*;
>
> **begin**
>   initialize S to be empty;
>   visit, mark, and stack v;
>   **while** S is nonempty **do**
>   **while** there is an unmarked vertex w adjacent to $Top(S)$ **do**
>   visit, mark, and stack w
>   **end** { while there's an unmarked vertex ... };
>   Pop S
>   **end** { while S is nonempty }
> **end** { DepthFirstSearch }

Since a depth-first search does not back up from a vertex x until every edge from x has been examined, it has a very simple recursive description. We will use the recursive form in all the depth-first search applications.

Algorithm 4.5   Depth-first Search (Recursive)

> **procedure** *DFS(v: VertexType)*;
> **var**
>   *w: VertexType*;
>
> **begin**
>   visit and mark v;
>   **while** there is an unmarked vertex w adjacent to v **do**
> $DFS(w)$
>   **end** { while }
> **end** { DFS }

Algorithm 4.6   Breadth-first Search

Input: $G = (V, E)$, a graph or digraph represented by an adjacency list structure as described in Section 4.1.2 with $V = \{1, 2, \ldots, n\}$; $v \in V$, the vertex from which the search begins.

Comment: For a queue Q, we assume that the function call *RemoveFromQ(Q)* returns the value of the front item on Q and removes that item from Q.

> **procedure** *BreadthFirstSearch* (*adjacencyList*: HeaderList; *v*: VertexType);
> **var**
>   *Q*: *Queue*;
>   *w*: *VertexType*;
> **begin**
>   initialize Q to be empty;
>   visit and mark v; insert v in Q.
>   **while** Q is nonempty **do**
> x := *RemoveFromQ(Q)*;
>   **for** each unmarked vertex w adjacent to x **do**
>   visit and mark w;
>   insert w in Q
>   **end** { for }
>   **end** { while Q is nonempty }
> **end** { BreadthFirstSearch }

Many variations and extensions may be made to these algorithms, depending on what they are used for. It is often necessary, for example, to do some sort of processing on each edge. The descriptions of the algorithms do not mention edges explicitly, but of course the implementation of the lines that require finding an unmarked vertex adjacent to a given vertex, say x, would involve examining edges incident with x, and the necessary processing of edges would be done there. In Section 4.4.5 we will consider how to fit other kinds of processing into the depth-first search skeleton.

The preceding algorithms will visit only those vertices that can be reached by a path from v. If it is necessary to visit every vertex in G, instructions must be added to find an unmarked vertex after the traversal from v is complete and to begin again from the new starting point. All vertices must be unmarked before the first traversal, but the marks would not be erased before each succeeding one. In Fig. 4.21, for example, if a traversal were started at A all vertices except D would be visited. A second traversal beginning at D would visit D, then discover that all adjacent vertices are marked, and quit. Details for these extensions and for the implementation of depth-first search will be shown in Section 4.4.3, where we consider the problem of finding connected components of graphs. Implementation and analysis of breadth-first search are similar and are left as exercises.

4.4.2 Depth-first Search and Recursion

We have seen that depth-first search can be simply described by a recursive algorithm. In fact, there is a fundamental connection between recursion and depth-first search. In a recursive program, the call structure can be diagrammed as a rooted tree in which each vertex other than the root represents a recursive call to the program. The order in which the calls are executed corresponds to a depth-first traversal of the

tree. Consider, for example, the recursive definition of the Fibonacci numbers:
$F_0=0$, $F_1=1$, and for $n\geq2$, $F_n=F_{n-1}+F_{n-2}$. The call structure for a recursive computation of F_6 is shown in Fig. 4.22. Each vertex is labeled with the current value of
n, i.e., the n for which the subtree rooted at that vertex computes F_n. The order of
the execution of the recursive calls is indicated in the figure; it is the familiar
preorder. (The reader should be aware that it is extremely inefficient to compute the
Fibonacci numbers recursively; this example is used only to illustrate the connection
between depth-first traversal and recursion.)

Thus the logical structure of the solutions to a number of interesting problems
solved by recursive algorithms is a depth-first traversal of a tree. The tree is not
always explicitly part of the problem, nor is it represented explicitly as a data

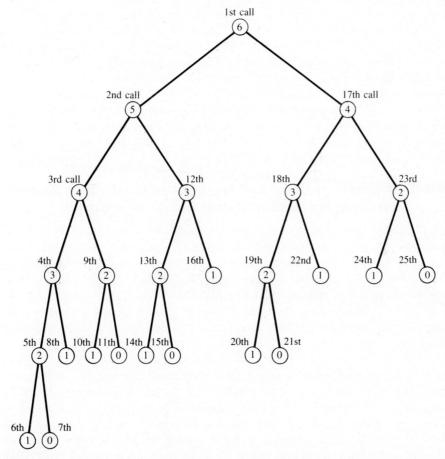

Figure 4.22 Call structure for recursive computation of Fibonacci numbers. Nodes are
labeled with the index of the Fibonacci number to be computed.

structure. Its relation to the solution of a problem is similar to the relation of a flowchart (a digraph) to a program; it gives an insight into what is going on.

Consider the problem of placing eight queens on a chessboard so that none is under attack by any other; in other words, so that none can reach another by moving along one row, column, or diagonal. It is not obvious that this can be done. We try as follows: Place a queen in the first (leftmost) square of the first (topmost) row. Then continue to place queens in each successive vacant row in the first column that is not under attack by any queen already on the board. Do this until all eight queens are on the board or all of the squares in the next vacant row are under attack. If the latter case occurs (which it does in the sixth row; see Fig. 4.23), go back to the previous row, move the queen there as few places as possible farther to the right so that it is still not under attack, and then proceed as before.

What tree is involved in this problem, and in what sense are we doing a depth-first search of it? The tree is shown in Fig. 4.23. Each vertex (other than the root) is labeled by a position on the chessboard. For $1 \le i \le 8$, the vertices at level i are labeled with board positions in row i. The children of a vertex v at level i are all board positions in row $i+1$ that would not be under attack if there were queens in all board positions along the path from the root to v; in other words, the children are all the safe squares in the next row. In terms of the tree, the problem is to find a path from the root to a leaf of length eight. As an exercise, the reader may write a recursive program for the queens problem such that the order in which the recursive calls are executed corresponds to a depth-first search. If there actually is a solution, only part of the tree in Fig. 4.23 is traversed. (Depth-first search, when used in a problem like this one, is also called back-track search.)

4.4.3   Implementation and Analysis of Depth-first Search: Finding Connected Components of a Graph

Let $G = (V, E)$ be a graph (not directed) with $n = |V|$ and $m = |E|$. A *connected component* of G is a maximal connected subgraph, i.e., a connected subgraph that is not contained in any larger connected subgraph. (The graph in Fig. 4.7, for example, has three connected components.) The problem of finding the connected components of a graph can be solved by using depth-first search with very little embellishment. We may start with an arbitrary vertex, do a depth-first search to find all other vertices (and edges) in the same component, and then, if some vertices remain, choose one and repeat.

We use the recursive version of depth-first search, Algorithm 4.5. Various parts of the algorithm could require a lot of work if we choose a poor implementation. The loop is controlled by the condition "while there is an unmarked vertex w adjacent to v." How can we determine efficiently if there is such a w? Certainly, we should use linked adjacency lists to represent the graph so that we can traverse the list for v looking for an unmarked w, but the **while** condition must be tested each time the search backs up to v to branch out again after a recursive call to *DFS*. Do we have to retraverse the adjacency lists (even part way) each time? Fortunately, no.

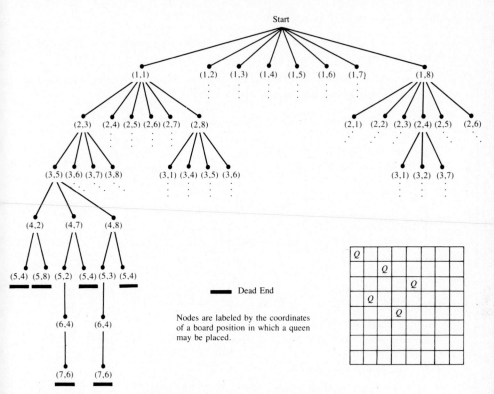

Figure 4.23 The eight-queens problem.

Marked vertices are never unmarked, so the search for an unmarked w adjacent to v can begin where the previous search through v's adjacency list ended. Thus, throughout the algorithm, the adjacency lists are traversed only once. A local pointer is used to keep track of our place in an adjacency list. (This means there is a pointer saved on the stack to keep track of our place in each adjacency list that has been partly, but not completely, traversed at any time.)

The problem of finding an unmarked vertex from which to start a new depth-first search can be handled similarly; in other words, instead of checking through the list of vertices from the beginning each time a depth-first search is completed, we start wherever we left off the previous time. If the partition of the graph into connected components is to be recorded in the data structure for later use, it can be done by marking each vertex and/or edge with the number of the component to which it belongs or by linking or making a separate list of the vertices and/or edges in each component. The particular method chosen would depend on how the information is to be used later.

The connected component algorithm is presented next, using a depth-first search procedure that makes the implementation explicit.

Algorithm 4.7   Connected Components

Input: $G = (V, E)$, a graph (not directed) represented by the adjacency list structure described in Section 4.1.2 with $V = \{1, 2, \ldots, n\}$.

Output: Lists of edges in each connected component. Also each vertex is numbered to indicate which component it is in.

```
procedure ConnectedComponents (adjacencyList: HeaderList; n: integer);
var
    mark: array[VertexType] of integer;
    { Each vertex will be marked with the number of the component it is in. }
    v: VertexType;
    componentNumber: integer;

procedure DFS(v: VertexType);
{ Does a depth-first search beginning at the vertex v }
var
    w: VertexType;
    ptr: NodePointer;
begin
    mark[v] := componentNumber;
    ptr := adjacencyList[v];
    while ptr ≠ nil do
        w := ptr↑.vertex;
        output (v,w);
        if mark[w] = 0 then DFS(w) end
        ptr := ptr↑.link
    end { while }
end { DFS }

begin { ConnectedComponents }
    { Initialize mark array. }
    for v := 1 to n do mark[v] := 0 end;
    { Find and number the connected components. }
    componentNumber := 0;
    for v := 1 to n do
        if mark[v] = 0 then
            componentNumber := componentNumber+1;
            output heading for a new component;
            DFS(v)
        end { if v was unmarked }
    end { for }
end { ConnectedComponents }
```

The number of operations done by *ConnectedComponents*, excluding the calls to *DFS*, is clearly linear in *n*. In *DFS(v)*, the number of instructions executed is proportional to the number of links traversed, since the instruction "*ptr := ptr↑.link*" is executed once each time through the **while** loop. Since the the adjacency lists are traversed only once, the complexity of the depth-first search, and hence the connected component algorithm, is in $\Theta(\max(n, m))$, which is usually $\Theta(m)$.

The space used by the adjacency list structure is in $\Theta(n+m)$. Additional space is used for the *mark* array and the stack, both linear in n, so the total amount of space used is in $\Theta(n+m)$.

Observe that the output of the connected components algorithm described will contain two copies of each edge because every edge is encountered twice. (See Fig. 4.10.) Here this may be a minor annoyance, but in some problems it is critical that an edge not be traversed twice. The reader should be able to modify the algorithm (and, if necessary, the data structure) to avoid the duplication in the output list without changing the order of the complexity of the algorithm.

4.4.4 Depth-first Search Trees

The edges that lead to new, i.e., unmarked, vertices during a depth-first search of a graph or digraph G form a rooted tree called a depth-first search tree. If not all of the vertices can be reached from the starting vertex (the root), then a complete traversal of G partitions the vertices into several trees. For an undirected graph, the search provides an orientation for each of its edges; they are oriented in the direction in which they are traversed. (If G is directed, its edges may be traversed only in the direction of their preassigned orientation.)

We say that a vertex v is an *ancestor* of a vertex w in a tree if v is on the path from the root to w; v is a proper ancestor of w if $v \neq w$. If v is a (proper) ancestor of w, then w is a (proper) descendant of v.

An edge of G that is traversed from a vertex to one of its ancestors in a depth-first search tree is called a *back edge*. If G is undirected, each of its edges will be a tree edge or a back edge. If G is a digraph, depth-first search partitions its edges into several classes: tree edges, back edges, edges that go from a vertex to one of its descendants other than a child, and edges called *cross edges* between two vertices such that neither is a descendant of the other. See Fig. 4.24 for illustrations. Note that the head and tail of a cross edge may be in two different trees. The reader should prove that there can be no cross edges or descendant edges if G is undirected. The distinctions between the various types of edges are important in some applications of depth-first search — in particular, in the algorithms studied in Sections 4.5 and 4.6.

4.4.5 A Generalized Depth-first Search Skeleton

Depth-first search provides the structure for many elegant and efficient algorithms. A depth-first search encounters each vertex several times: when the vertex is first visited and becomes part of the depth-first search tree, then several more times when the search backs up *to* it and attempts to branch out in a different direction, and finally, after the last of these encounters, when the search backs up *from* the vertex and does not pass through it or any of its descendants again. Depending on the

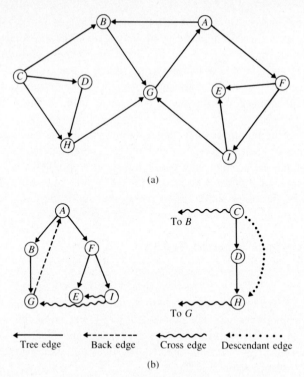

(a)

(b)

| Tree edge | Back edge | Cross edge | Descendant edge |

Figure 4.24 (a) A digraph. (b) Depth-first search trees for the digraph.

problem to be solved, an algorithm will process the vertices differently when they are encountered at various stages of the traversal. Many algorithms will also do some computation for the edges: perhaps for each edge, or perhaps only for edges in the depth-first search tree, or perhaps different kinds of computation for the different kinds of edges. The following skeleton algorithm shows exactly where the processing would be done for each kind of edge and for each kind of encounter with the vertices.

Algorithm 4.8 General Depth-first Search Skeleton

Input: $G = (V, E)$, a graph or digraph represented by the adjacency list structure described in Section 4.1.2 with $V = \{1, 2, \ldots, n\}$.

> **var**
> > *mark*: **array**[*VertexType*] **of** *integer*;
> > *markValue*: *integer*;
>
> **procedure** *DFS* (*v*: *VertexType*);
> { Does a depth-first search beginning at the vertex *v*, marking
> the vertices with *markValue*. }

var
    w: *VertexType;*
    ptr: *NodePointer;*

begin

    { Process vertex when first encountered (like preorder). }

    mark[*v*] := *markValue;*

    ptr := *adjacencyList*[*v*];
    while *ptr* ≠ **nil do**
      w := *ptr*↑.*vertex;*

        { Processing for every edge.
         (If *G* is undirected, each edge is encountered
         twice; an algorithm may have to distinguish the
         two encounters.)   }

      if *mark*[*w*] = 0 { unmarked } **then**

         { Processing for tree edges, *vw*. }

         DFS(*w*);

         { Processing when backing up to *v* (like inorder) }

      else

         { Processing for nontree edges.
         (If *G* is undirected, an algorithm may have
         to distinguish the case where *w* is the
         parent of *v*.)   }

      end { if };
      ptr := *ptr*↑.*link*
    end { while };

    { Processing when backing up from *v* (like postorder) }

end { DFS }

For an exercise, the reader should write an algorithm to determine if a graph (undirected) has a cycle. (This will require distinguishing between a nontree edge and an encounter with a tree edge "backwards," i.e., from a vertex to its parent.)

In some applications of depth-first search, we may need to know which vertices are on the path from the root to the current vertex (*v*). They are exactly the vertices that are on the stack. For some algorithms we need to know the order in which vertices are encountered for the first time. We can simply number the vertices as they are encountered by incrementing *markValue*. The number assigned to a vertex in this way is called its *depth-first search number*.

4.5
Biconnected Components of
a Graph

4.5.1   Articulation Points and Biconnected Components

In Section 4.1 we raised these questions:

If one city's airport is closed by bad weather, can you still fly between any other pair of cities?

If one computer in a network goes down, can messages be sent between any other pair of computers in the network?

In this section we consider undirected graphs only. As a graph problem, the question is:

If any one vertex (and the edges incident with it) is removed from a connected graph, is the remaining subgraph still connected?

This question is important in graphs representing all kinds of communication or transportation networks. It is also important to find those vertices, if any, whose removal can disconnect the graph.

Formally, a vertex v is an *articulation point* (also called a *cutpoint*) for a graph if there are distinct vertices w and x (distinct from v also) such that v is in every path from w to x. Clearly, the removal of an articulation point would leave an unconnected graph, so a connected graph is *biconnected* if and only if it has no articulation points. A *biconnected component* of a graph is a maximal biconnected subgraph, that is, a biconnected subgraph not contained in any larger biconnected subgraph. Figure 4.25 illustrates biconnected components. Observe that, although the biconnected components partition the edges into disjoint sets, they do not partition the vertices; some vertices are in more than one component. (Which vertices are these?)

There is an alternative characterization of biconnected components, in terms of an equivalence relation on the edges, that is sometimes useful. Two edges e and e' are equivalent if $e = e'$ or if there is a cycle containing both e and e'. Then each subgraph consisting of the edges in one equivalence class and the incident vertices is a biconnected component. (Verifying that the relation described is indeed an equivalence relation and verifying that it characterizes the biconnected components are left as exercises.)

The applications that motivate the study of biconnectivity should suggest a dual problem to the reader: how to determine if there is an edge whose removal would disconnect a graph, and how to find such an edge if there is one. For example, if a railroad track is damaged, can trains still travel between any pair of stations? Relationships between the two problems are examined in Exercise 4.40.

The algorithm we will study for finding biconnected components uses the depth-first search skeleton of Section 4.4.5 and the idea of a depth-first search tree from Section 4.4.4. During the search, information will be computed and saved so

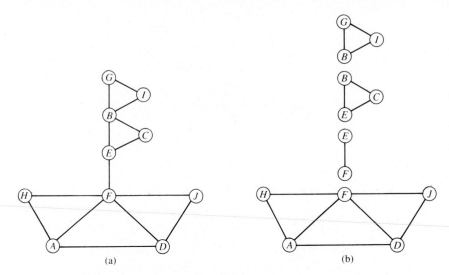

Figure 4.25   (a) A graph. (b) Its biconnected components.

that the edges (and, implicitly, the incident vertices) can be divided into biconnected
components as the search progresses. What information must be saved? How is it
used to determine the biconnected components? Several wrong answers to these
questions seem reasonable until they are examined carefully. Two edges are in the
same component if they are in a cycle, and every cycle must include at least one
back edge. The reader should work on Exercise 4.32 before proceeding; it requires
looking at a number of examples to determine relationships between back edges and
biconnected components.

From now on we will use the shorter term "bicomponent" in place of "biconnected component."

4.5.2   The Bicomponent Algorithm

Processing of vertices may be done when a vertex is *first visited*, when the search
backs up *to* it, and/or when the search backs up *from* it. The bicomponent algorithm
tests to see if a vertex in the tree is an articulation point each time the search backs
up to it. Suppose the search is backing up to v from w. If there is no back edge
from any vertex in the subtree rooted at w to a proper ancestor of v, then v must be
on every path in G from the root to w and is therefore an articulation point. See Fig.
4.26 for illustration. (The careful reader will note that this argument is not valid if v
is the root.) The subtree rooted at w, along with all back edges leading from it and
along with the edge vw, can be separated from the rest of the graph at v, but it is not
necessarily one bicomponent; it may be a union of several. We ensure that bicomponents are properly separated by removing each one as soon as it is detected. Vertices at the outer extremities of the tree are tested for articulation points before vertices closer to the root, ensuring that when an articulation point is found, the subtree

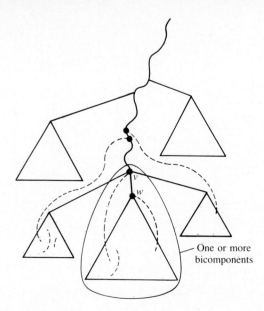

Figure 4.26 An articulation point *v* in a depth-first search tree. Every path from the root to *w* passes through *v*.

in question (along with the additional edges mentioned above) forms one bicomponent.

The discussion suggests that the algorithm must keep track of how far back in the tree one can get from each vertex by following tree edges (implicitly directed away from the root) and certain back edges. This information will be stored in an array *back*. The vertices will be numbered in the order in which they are first visited. These numbers, stored in an array *dfsNumber*, replace the marks used earlier. Values of *back* will be these vertex numbers. For a vertex *v*, *back*[*v*] may be assigned (or modified) when the search is going forward and a back edge from *v* is encountered (as in Fig. 4.27(b) with *v* = *F* and in Fig. 4.27(c) with *v* = *C*) and when the search backs up to *v* (as in Fig. 4.27(d) with *v* = *B*), since any vertex that can be reached from a child of *v* can also be reached from *v*. Determining which of two vertices is farther back in the tree is easy: If *v* is a proper ancestor of *w*, then *dfsNumber*[*v*] < *dfsNumber*[*w*]. Thus we can formulate the following rules for setting *back*[*v*]:

1. When proceeding forward from *v* and a back edge *vw* is detected, *back*[*v*] := min(*back*[*v*], *dfsNumber*[*w*]).

2. When backing up from *w* to *v*, *back*[*v*] := min(*back*[*v*], *back*[*w*]).

(These rules imply that *back*[*v*] must be properly initialized; *back*[*v*] will initially be assigned *dfsNumber*[*v*], but see Exercise 4.35.)

The condition tested to detect a bicomponent when backing up from *w* to *v* is *back*[*w*] ≥ *dfsNumber*[*v*]. (This condition is tested but not satisfied in Figs. 4.27(d)

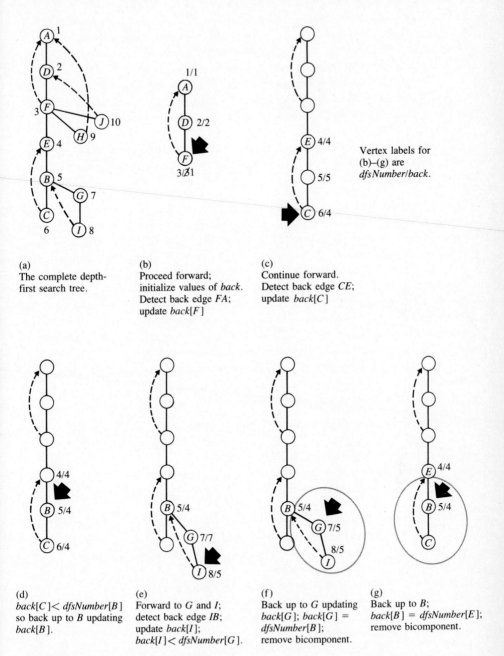

(a)
The complete depth-first search tree.

(b)
Proceed forward; initialize values of *back*. Detect back edge *FA*; update *back[F]*

(c)
Continue forward. Detect back edge *CE*; update *back[C]*

Vertex labels for (b)–(g) are *dfsNumber/back*.

(d)
back[C]< dfsNumber[B] so back up to *B* updating *back[B]*.

(e)
Forward to *G* and *I*; detect back edge *IB*; update *back[I]*; *back[I]< dfsNumber[G]*.

(f)
Back up to *G* updating *back[G]*; *back[G]* = *dfsNumber[B]*; remove bicomponent.

(g)
Back up to *B*; *back[B]* = *dfsNumber[E]*; remove bicomponent.

Figure 4.27 The action of the bicomponent algorithm on the graph in Fig. 4.25 (detecting the first two bicomponents).

and 4.27(e); it is satisfied in Figs. 4.27(f) and 4.27(g).) When the test is satisfied, v is an articulation point (except perhaps if v is the root of the tree); a complete bicomponent has been found and may be removed from further consideration. When this occurs, rule 2 above for resetting $back[v]$ may be skipped.

The problem of exactly when and how to test for bicomponents is subtle but critical to the correctness of an algorithm. (See Exercises 4.37–4.39.) The essence of the correctness argument is contained in the following theorem.

Theorem 4.3 In a depth-first search tree, a vertex v, other than the root, is an articulation point if and only if v is not a leaf and some subtree of v has no back edge incident with a proper ancestor of v.

Proof. Suppose that v, a vertex other than the root, is an articulation point. Then there are vertices x and y such that v, x, and y are distinct and v is on every path from x to y. At least one of x and y must be a proper descendant of v, since otherwise there would be a path between them using (undirected) edges in the tree without going through v. Thus v is not a leaf. Now suppose every subtree of v has a back edge to a proper ancestor of v; we claim that this contradicts the assumption that v is an articulation point. There are two cases: when only one of x and y is a descendant of v, and when both are descendants of v. For the first case, paths between x and y that do not use v are illustrated in Fig. 4.28. We leave the latter case as an exercise for the reader.

The remaining half of the proof is also left as an exercise. □

Theorem 4.3 does not tell us under what conditions the root is an articulation point. See Exercise 4.34.

We can now outline the work to be done in the depth-first search:

```
procedure BicompDFS(v: VertexType); { outline }
begin
    number v and initialize back[v];
    while there is an untraversed edge vw incident with v do
        if w is unmarked then
            BicompDFS(w);
            { Now backing up to v }
            if back[w] ≥ dfsNumber[v] then
                output a new bicomponent { the subtree rooted at w
                and incident edges };
            else { haven't found a new bicomponent }
                back[v] := min(back[v], back[w])
            end { of backing up from w to v }
        else { w is already in the tree }
            back[v] := min(dfsNumber[w], back[v])
        end { of processing w };
    end { while };
end { BicompDFS }
```

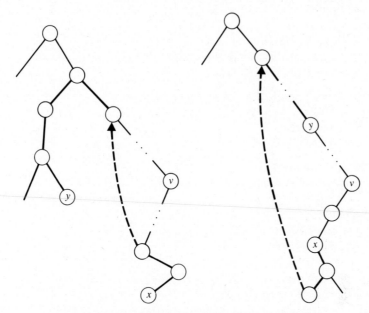

Figure 4.28   Examples for the proof of Theorem 4.3.

The algorithm must keep track of the edges traversed during the search so that those in one bicomponent can easily be identified and removed from further consideration at the appropriate time. As the example in Fig. 4.27 illustrates, when a bicomponent is detected, its edges are the edges most recently processed. Thus edges are stacked on *EdgeStack* as they are encountered. When a bicomponent is detected when backing up from, say, *w* to *v*, the edges in that bicomponent are the edges from the top of *EdgeStack* down to (and including) *vw*. These edges may then be popped.

Each adjacency list will be scanned exactly once, but every edge of *G* is in two adjacency lists and is encountered twice. Stacking an edge the second time it is encountered can result in incorrect output — some edges may be put in two different bicomponents. The algorithm avoids stacking edges that will cause this problem.

Algorithm 4.9   Biconnected Components

Input: $G = (V, E)$, a connected graph (not directed) represented by linked adjacency lists with $V = \{1, 2, \ldots, n\}$.

Output: Lists of the edges in each biconnected component of *G*.

    procedure *Bicomponents* (*adjacencyList*: *HeaderList*; *n*: *integer*);
    var
        dfsNumber: **array**[*VertexType*] **of** *integer*;
        back: **array**[*VertexType*] **of** *integer*;

```
    dfn: integer;
    v: VertexType;
    EdgeStack: Stack;
    { We assume that Top is a function that returns the top item on a
    stack (without popping it). }

procedure BicompDFS(v: VertexType);
var
    w: VertexType;
    ptr: NodePointer;

begin { BicompDFS }
    { Process vertex when first encountered. }
    dfn := dfn+1;
    dfsNumber[v] := dfn;  back[v] := dfn;
    ptr := adjacencyList[v];
    while ptr ≠ nil do
        w := ptr↑.vertex;
        if dfsNumber[w] < dfsNumber[v] then
            push vw on EdgeStack
            { else wv was a backedge already examined }
        end { if };
        if dfsNumber[w] = 0 { unmarked } then
            BicompDFS(w);
            { Now backing up to v }
            if back[w] ≥ dfsNumber[v] then
                output a heading for a new bicomponent;
                repeat
                    output Top(EdgeStack);
                    pop EdgeStack
                until vw is popped;
            else { haven't found a new bicomponent }
                back[v] := min(back[v], back[w])
            end { of backing up from w to v }
        else { w is already in the tree }
            back[v] := min(dfsNumber[w], back[v])
        end { of processing w };
        ptr := ptr↑.link;
    end { while };
end { BicompDFS }

begin { Bicomponents }
    for v := 1 to n do dfsNumber[v] := 0;
    dfn := 0;
    BicompDFS(1)
end { Bicomponents }
```

4.5.3   Analysis

As usual, $n = |V|$ and $m = |E|$. The initialization in *Bicomponents* includes $\Theta(n)$ operations. *BicompDFS* is the depth-first search skeleton with appropriate processing of vertices and edges added. The depth-first search skeleton takes time in $\Theta(\max(n, m)) = \Theta(m)$. (Since G is connected, $m \geq n-1$.) Thus if the amount of processing for each vertex and edge is bounded by a constant, the complexity of *Bicomponents* is in $\Theta(m)$. It is easy to see that this is the case. The needed observation is nontrivial only when the search backs up from w to v. Sometimes the **repeat** loop popping edges from *EdgeStack* is executed, sometimes not, and the number of edges popped each time varies. But each edge is stacked and popped at most twice (yes, some edges may be stacked when encountered from both directions), so overall the amount of work done is in $\Theta(m)$.

The amount of space used is $\Theta(n+m)$.

4.5.4   Generalizations

The prefix "bi" means "two." Informally speaking, a biconnected graph has two vertex-disjoint paths between any pair of vertices (see Exercise 4.29). We can define triconnectivity (and, in general, k-connectivity) to denote the property of having three (in general, k) vertex-disjoint paths between any pair of vertices. An efficient algorithm that uses depth-first search to find the triconnected components of a graph has been developed (see the notes and references at the end of the chapter), but it is much more complicated than the algorithm for bicomponents.

4.6
Strongly Connected Components of a Digraph

4.6.1   Definitions

A graph (undirected) is connected if and only if there is a path between each pair of vertices. Connectedness for digraphs may be defined in either of two ways, depending on whether or not we require that edges be traversed only from tail to head. A digraph $G = (V, E)$ is *strongly connected* if, for each pair of vertices v and w, there is a path from v to w (and hence by interchanging the roles of v and w in the definition, there is a path from w to v as well). That is, edges must be followed in the direction of their "arrows." G is *weakly connected* if, for each pair of vertices v and w, there is a sequence of vertices $v_0, v_1, \ldots, v_k$ such that $v_0 = v$, $v_k = w$, and for $i = 0, 1, \ldots, k-1$ either $v_i v_{i+1} \in E$ or $v_{i+1} v_i \in E$. That is, for weak connectivity, the direction of the arrows is ignored. We will focus on strong connectivity.

A *strongly connected component* (hereinafter called a *strong component*) of a digraph is a maximal strongly connected subgraph. We may give an alternative definition in terms of an equivalence relation, S, on the vertices. For v and w in V, let vSw if and only if there is a path from v to w and a path from w to v. Then a

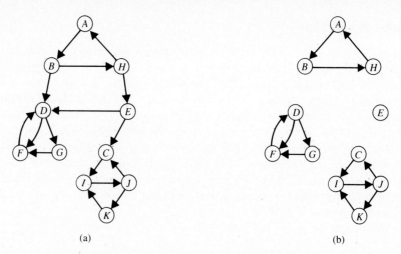

Figure 4.29 (a) A digraph. (b) Its strong components.

strong component consists of one equivalence class, C, along with all edges vw such that v and w are in C. See the example in Fig. 4.29. We will sometimes use the term strong component to refer only to the vertex set C; the meaning should be clear from the context.

The strong components of a digraph can each be collapsed to a point yielding a new digraph that has no cycles. Let $S_1, S_2, \ldots, S_p$ be the strong components of G. The *condensation* of G is the digraph $G' = (V', E')$, where V' has p elements, $s_1, s_2, \ldots, s_p$, and $s_i s_j$ is in E' if and only if there is an edge in E from some vertex in S_i to some vertex in S_j. See Fig. 4.30 for an example. Solutions to some problems on digraphs can be simplified by treating the strong components and the condensation separately, taking advantage of the special properties of each (in other words, that the former are strongly connected and the latter acyclic). (Consider the relationship of the strong components and the condensation of a program flowchart to the loop structure of the program. Also see Section 8.1.)

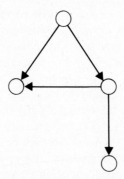

Figure 4.30 The condensation of the digraph in Fig. 4.29.

4.6.2 The Strong Component Algorithm

We will study an algorithm for finding strong components that is very similar in structure to the biconnected component algorithm in Section 4.5. See Exercise 4.50 for another algorithm.

Here again we use the depth-first search skeleton. In an undirected graph the set of vertices in a depth-first search tree is exactly the set of vertices in one connected component of the graph and is independent of the particular starting point chosen for the search. Also, as we observed earlier, every edge in that component will be a tree edge or a back edge. With a digraph, where each edge may be traversed in only one direction, the situation is more complex. The partition induced on the vertex set by forming depth-first search trees for a digraph is not invariant; it depends on the starting points chosen for each search. (Consider the tree that would result from starting a depth-first search of the digraph in Fig. 4.29 at E.) However, it is true that the set of vertices in any depth-first search tree is a union of strong components, so we may choose an arbitrary starting point, do a depth-first search, and by doing some computation similar to that done in the bicomponent algorithm, divide the tree into strong components.

Let $T = (V_T, E_T)$ be a depth-first search tree for the digraph $G = (V, E)$. Recall that, aside from tree edges and back edges, G may have cross edges and edges that lead from a vertex in V_T to one of its descendants. (See Section 4.4.4 and Fig. 4.24.)

Assuming that branches are drawn (traversed) in left-to-right order, all cross edges point in the same direction — to the left. This fact can be stated formally if we define the relation L (read "is to the left of") on V_T as follows: For v and w in V_T, vLw if and only if w is encountered after v in the search and w is not a descendant of v. Lemma 4.4 is used in the arguments that follow, which justify the algorithm. Its proof is easy and is left as an exercise.

Lemma 4.4   If v and w are in V_T and vw is in E, then wLv or either v or w is a descendant of the other. (The only possibility excluded is that vLw.)

For v in V_T let $oldest[v]$ be the oldest (closest to the root) ancestor of v in T that can be reached by following tree edges, back edges, and cross edges. The strong components in T can be characterized as follows.

Theorem 4.5   For v and w in V_T, v and w are in the same strong component if and only if $oldest[v] = oldest[w]$.

The proof is left as an exercise.

Thus an algorithm could be devised that does depth-first searches of a digraph and computes the values of $oldest$. However, it is possible to determine the strong components with less information; $oldest$ need not be computed explicitly for each vertex. The algorithm numbers vertices when they are first visited (storing the numbers in an array $dfsNumber$) and uses an array low such that, for a vertex v in T, $low[v]$ is the lowest-numbered vertex (not necessarily an ancestor of v) that is known to be in the same strong component as v and that can be reached by tree edges and one cross or back edge. When v is first encountered, $low[v]$ is initialized to $dfsNumber[v]$. When the algorithm backs up from v, it need only determine if it has just completed traversing a strong component. It does this by testing whether or not $low[v] = dfsNumber[v]$, for if after completely searching the subtree rooted at v no way has been found to reach a proper ancestor of v, there can be none. (Consider $v = D$ in Fig. 4.31. This argument uses the fact that cross edges go only to the left, i.e., to parts of T already examined. There can be no cross edge from the subtree leading to the right to nodes not yet explored.) Whenever the algorithm backs up from a vertex v such that $low[v] = dfsNumber[v]$, it removes all of the vertices in the subtree rooted at v. Each such set of vertices (along with the appropriate edges) is one strong component. (Can you prove this statement?)

When the algorithm backs up from a vertex v such that $low[v] < dfsNumber[v]$, the value of $low[v]$ is carried back to the parent of v, as when it backs up from J to I in Fig. 4.31. Values of low are, of course, also updated when the search is proceeding forward. Suppose that, when proceeding forward from a vertex v, an edge leading to a numbered vertex w is encountered. If w is a descendant of v, then $low[v] \leq dfsNumber[w]$ so $low[v]$ need not be changed. If vw is a back edge, $low[v]$ is assigned $\min(low[v], dfsNumber[w])$. If vw is a cross edge, things get a bit tricky. It is sometimes necessary and sometimes incorrect for $low[v]$ to be assigned $\min(low[v], dfsNumber[w])$. For example, in Figure 4.31, when the cross edge GF is

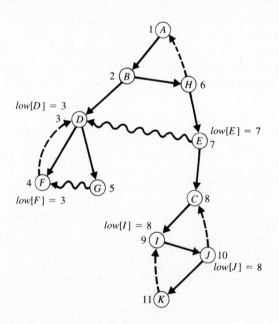

Vertices are numbered in the order visited.

Figure 4.31

encountered, $low[G]$ must be assigned $dfsNumber[F]$ so that G will not be incorrectly considered as a strong component by itself. On the other hand, if $low[E]$ is assigned $min(low[E], dfsNumber[D])$ when the cross edge ED is encountered, E would incorrectly be put in the same strong component as A, B, and H. The following theorem provides the criterion for doing the assignment.

Theorem 4.6   Suppose that, when a cross edge vw is encountered during a depth-first search, $low[v]$ is assigned $min(low[v], dfsNumber[w])$ if and only if w has not already been put in another strong component. This is exactly the case when v and w are in the same strong component, and the algorithm works correctly.

Proof.   Certainly if the strong component containing w has already been detected and removed, v is not in it. (This claim assumes that the algorithm has worked correctly so far; it can be formalized by using induction.) Suppose that w has not yet been removed from the tree. We will show that there is a path from w to v. Let $w_1 = w$, and let w_{i+1} be the vertex whose number is $low[w_i]$, for each i until for some k, $low[w_k] = dfsNumber[w_k]$. (See Fig. 4.32.) By the definition of low, there is a path from w to w_k. Either $w_k L v$ or w_k is an ancestor of v. We will show that the latter must be true by assuming that $w_k L v$ and deriving a contradiction. Consider

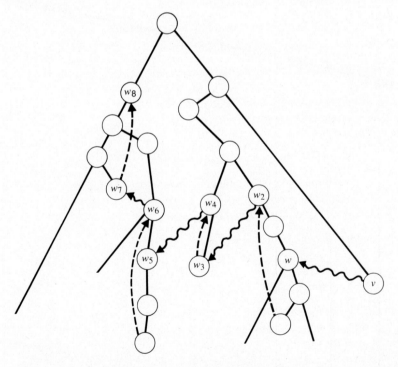

Figure 4.32   Illustration for the proof of Theorem 4.6. $w_k = w_8$; $w_i = w_4$.

the sequence $v = w_0$, $w = w_1$, $w_2, \ldots, w_k$. Let i be the largest index such that w_k is not an ancestor of w_i. Such an i exists since $w_k L v$; i could be 0. (In Fig. 4.32 $i = 4$.) Then $w_k L w_i$, and the search backed up from w_k before encountering w_i. Since $low[w_k] = dfsNumber[w_k]$, w_k and all of its descendants, including w_{i+1}, were removed from the tree. But then $low[w_i]$ would not have been assigned $dfsNumber[w_{i+1}]$. This contradicts the choice of w_{i+1}; thus w_k must be an ancestor of v. Then there is a path, using tree edges, from w_k to v. Adjoining this path to the one from w to w_k (which may use cross edges, back edges, and tree edges) gives a path from w to v. Hence w and v are in the same strong component and the assignment $low[v] := \min(low[v], dfsNumber[w])$ should be done.   □

The algorithm keeps track of the vertices to be put in the next strong component by pushing each vertex onto a stack SC as it backs up from that vertex. When the root of a strong component is detected at say, v, the vertices in the component, i.e., those in the subtree rooted at v, are the topmost vertices on SC, and they are popped.

The Boolean array *removed* indicates which vertices have been removed from the tree, so that cross edges may be handled properly.

Algorithm 4.10   Strongly Connected Components

Input: $G = (V, E)$, a digraph represented by linked adjacency lists.

Output: Lists of vertices in each strong component.

```
procedure StrongComponents (adjacencyList: HeaderList; n: integer);
var
    dfsNumber: array[VertexType] of integer;
    low: array[VertexType] of integer;
    dfn: integer;
    v: VertexType;
    SC: Stack;
    { We assume that Top is a function that returns the top item on a
    stack (without popping it). }
    removed: array[VertexType] of boolean;

procedure SCompDFS (v: VertexType);
var
    w: VertexType;
    ptr: NodePointer;

begin { SCompDFS }
    { Process vertex when first encountered. }
    dfn := dfn+1;
    dfsNumber[v] := dfn;  low[v] := dfn;
    removed[v] := false;
```

```
    ptr := adjacencyList[v];
    while ptr ≠ nil do
        w := ptr↑.vertex;
        if dfsNumber[w] = 0 { unmarked } then
            SCompDFS(w);
            { Now backing up from w to v }
            low[v] := min(low[v], low[w])
        else { w was already encountered }
            if not removed[w] then
                low[v] := min(dfsNumber[w], low[v])
            end { if w is still in the tree }
        end { of processing w };
        ptr := ptr↑.link;
    end { while };

    { Now backing up from v }
    if low[v] = dfsNumber[v] then
        output a heading for a new strong component;
        removed[v] := true; output v;
        while SC is nonempty and dfsNumber[Top(SC)] > dfsNumber[v] do
            output Top(SC);
            removed[Top(SC)] := true;
            pop SC
        end { while vertices from SC are in current strong component }
    else { haven't found a new strong component }
        push v onto SC
    end { backing up from v }

end { ScompDFS }

begin { StrongComponents }
    for v := 1 to n do dfsNumber[v] := 0 end;
    dfn := 0;
    for v := 1 to n do
        if dfsNumber[v] = 0 then SCompDFS(v) end
    end { for }
end { StrongComponents }
```

4.6.3 Analysis

Much of the analysis of the bicomponent algorithm carries over with small changes to the strong-component algorithm. The initialization and overhead in the body of the *StrongComponents* procedure is in $\Theta(n)$. The total number of steps executed in all calls to the depth-first search procedure is in $\Theta(\max(n, m))$, so the complexity of the strong-component algorithm is in $\Theta(\max(n, m))$. The extra space used for various arrays is in $\Theta(n)$. The recursion stack also uses $\Theta(n)$ space. So, including the adjacency lists, the space used is in $\Theta(n+m)$.

Exercises

Section 4.1: Definitions and Representations

4.1. Euler paths

a)   A popular game among grade-school children is to draw the following figure without picking up the pencil and without retracing a line. Try it.

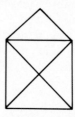

b)   Figure 4.33 provides a similar but slightly harder problem. It shows a river with two islands connected to each other and to the banks by seven bridges. The problem is to determine if there is a way to take a walk starting on either bank of the river or on either island and crossing each bridge exactly once. (No swimming allowed.) Try it.

c)   The problems in (a) and (b) may be studied abstractly by examining the following graphs. G_2 is obtained by representing each bank and island as a vertex and each bridge as an edge. (Some pairs of vertices are connected by two edges, but this departure from the definition of a graph will not cause trouble here.) The general problem is: Given a graph (with multiple edges between pairs of vertices permitted), find a path through the graph that traverses each edge exactly once. Such a path is called an Euler path. (The term "path" is used in a broader sense here than in the text; it may pass through a vertex more than once.) This problem is solvable for G_1 but not for G_2 that is, there is no way to walk across each bridge exactly once. Find a necessary and sufficient condition for a graph to have an Euler path.

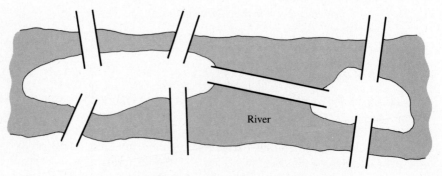

River

Figure 4.33   The Köenigsberg bridges.

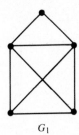
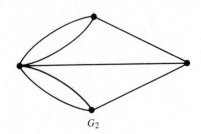

G_1 G_2

4.2. Suppose that a digraph G represents a binary relation R. Describe a condition on G that holds if and only if R is transitive.

Section 4.2: A Minimum Spanning Tree Algorithm

("Minimum spanning tree" is abbreviated "MST" in the exercises.)

*4.3. Prove that if the weights on the edges of a connected graph are distinct, then there is a unique minimum spanning tree.

4.4. Describe a sequence of connected weighted graphs G_n such that G_n has n vertices and the running time of the MST algorithm (Algorithm 4.2) for these graphs is linear in n.

4.5. Describe a sequence of connected weighted graphs G_n such that G_n has n vertices and the MST algorithm does no comparisons of weights when G_n is the input. (The algorithm will require time at least proportional to n because it must succeed in finding a minimum spanning tree.)

4.6. Let $G = (V, E, W)$ where $V = \{v_1, v_2, \ldots, v_n\}$, $E = \{v_1 v_i \mid i = 2, \ldots, n\}$, and for $i = 2, \ldots, n$, $W(v_1 v_i) = 1$. With this G as input, how many comparisons of edge weights would be done by the MST algorithm, in total, to find minimum candidate edges? (Working through this problem may suggest to the reader that saving information about the ordering of the weights of candidate edges could decrease the number of comparisons. The next two exercises suggest that it may not be easy.)

4.7. a)   How many comparisons of edge weights would be done by the MST algorithm, in total, if the input is a complete graph with n vertices?

b)   Suppose the vertices are $v_1, \ldots, v_n$, and $W(v_i v_j) = n+1-i$ for $1 \le i < j \le n$. How many of the edges are candidate edges at some time during the execution of the algorithm?

*4.8. Consider storing candidate edges in a min-heap (a heap where each node is smaller than its children). Estimate the number of comparisons of edge weights that would be done by the MST algorithm. Remember to consider the work that would have to be done on the heap when a candidate edge is replaced by another.

4.9. Complete the proof of Theorem 4.1 by showing that T' is a spanning tree.

4.10. Prove or disprove: The MST algorithm will work correctly even if weights may be negative.

Section 4.3: A Shortest-Path Algorithm

4.11. For the graph in Fig. 4.34, indicate which edges would be in the minimum spanning tree constructed by the MST algorithm (Algorithm 4.2) and which would be in the tree constructed by the shortest-path algorithm (Algorithm 4.3) after finding a shortest path from v_1 to v_6.

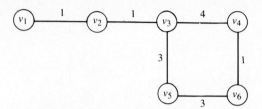

Figure 4.34 Graph for Exercise 4.11.

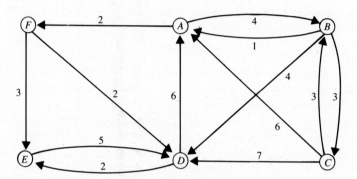

Figure 4.35 Digraph for Exercise 4.13.

4.12. Will Dijkstra's shortest-path algorithm (Algorithm 4.3) work correctly if weights may be negative? Justify your answer by an argument or a counterexample.

4.13. Here are the adjacency lists (with edge weights in parentheses) for a digraph. For convenience, the digraph is also shown in Fig. 4.35.

> A: B(4), F(2)
> B: A(1), C(3), D(4)
> C: A(6), B(3), D(7)
> D: A(6), E(2)
> E: D(5)
> F: D(2), E(3)

a) This digraph has *three* shortest paths from C to E (i.e., all with the *same* total weight). Find them. (List the sequence of vertices in each path.)

b) Which of these paths is the one that would be found by Dijkstra's shortest-path algorithm? (Give a convincing explanation or show the main steps of the algorithm.)

*4.14. Let $G = (V, E)$ be a graph or digraph, and let v and w be distinct vertices. As the previous exercise suggests, there can be more than one shortest path from v to w. Explain how to modify Dijkstra's shortest-path algorithm to determine the number of different shortest paths from v to w.

*4.15. Consider the problem of finding just the distance, but not a shortest path, from v to w in a weighted graph or digraph. Outline a modified version of the shortest-path algorithm to do this with the aim of eliminating as much work and extra space usage as possible. Indicate what changes, if any, you would make in the data structure used by the algorithm, and indicate what work or space you would eliminate.

4.16. Some graph algorithms are written with the assumption that the input is always a complete graph (where an edge has weight ∞ or 0 to indicate its absence from the graph for which the user really wants to solve the problem). Such algorithms are usually shorter and "cleaner" because there are fewer cases to consider. In the algorithms in Sections 4.2 and 4.3, for example, there would be no unseen vertices since all vertices would be adjacent to vertices in the tree constructed so far.

    a)   With the aim of simplifying as much as possible, rewrite the shortest-path algorithm with the assumptions that $G = (V, E, W)$ is a complete graph and that W maps E into $\mathbf{Z}^+ \cup \{\infty\}$. Describe any changes you would make in the data structures used.

    b)   Compare your algorithm and data structures with those in the text, using the criteria of simplicity, time (worst case and other cases), and space usage (for graphs with many edges that have weight ∞ and for graphs with few).

Section 4.4: Traversing Graphs and Digraphs

4.17. Is it always, sometimes, or never true that the order in which vertices are first visited (added to the tree) in the minimum spanning tree algorithm (Algorithm 4.2) is the same as the order in which the vertices would be visited by breadth-first or depth-first search? Justify your answer with examples or a proof.

4.18. Do the previous exercise for the shortest-path algorithm (Algorithm 4.3).

4.19. Give an example of a graph in which a depth-first search backs up from a vertex before all the vertices that can be reached from it via one or more edges are marked.

4.20. Write a depth-first search algorithm for an undirected graph such that the output is a list of the edges encountered, with each edge appearing once. (Describe your changes, if any, in the adjacency list structure.)

4.21. Write a depth-first search algorithm whose output is a list of the edges in the depth-first search tree.

4.22. Prove that if G is a connected graph (undirected), each of its edges either is in the depth-first search tree or is a back edge.

4.23. Prove that when a breadth-first search is done on a graph (undirected), every edge in the graph is either a tree edge or a cross edge. (A cross edge is an edge between two vertices such that neither is a descendant of the other.)

4.24. Let G be a connected graph, and let v be a vertex in G. Let T_D be a depth-first search tree formed by doing a depth-first search of G starting at v. Let T_B be the breadth-first search tree formed by doing a breadth-first search of G starting at v. Is it always true that $\text{depth}(T_D) \geq \text{depth}(T_B)$? Give a clear argument or a counterexample.

4.25. a)   Write an algorithm to determine if a graph has a cycle.

    b)   Write an algorithm to determine if a digraph has a cycle. How, if at all, does this algorithm differ from the previous one?

    c)   If you used depth-first search for the preceding algorithms, try to write algorithms for the same problems using breadth-first search, and vice versa. Do you

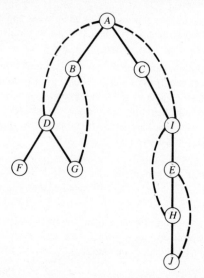

Figure 4.36   Depth-first search tree for Exercise 4.28.

see any strong reasons to prefer either of the search strategies for these problems?

4.26. Describe an algorithm to determine if a graph $G = (V,E)$, with $n = |V|$ and $m = |E|$, is a tree. Would you use the same algorithm if you could assume that the graph is connected? If not, describe one that uses that assumption as well.

*4.27. Consider the problem of finding the length of a shortest cycle in a graph (undirected). Here is a proposed solution that is not correct. Show why it does not always work. (Look for a fundamental flaw in the strategy, not a detail.)

When a back edge, say vw, is encountered during a depth-first search, it forms a cycle with the tree edges from w to v. The length of the cycle is $level[v]-level[w]+1$. Do a depth-first search, keeping track of the level of each vertex. Each time a back edge is encountered, compute the cycle length and save it if it is smaller than the shortest one previously seen.

Section 4.5: Biconnected Components of a Graph

4.28. List the articulation points in the graph with the depth-first search tree shown in Fig. 4.36.

4.29. Is the following property on a graph $G = (V, E)$ necessary and sufficient for G to be biconnected? Prove your answer.

For each pair of distinct vertices v and w in V, there are two distinct paths from v to w that have no vertices in common except v and w.

4.30. Show that for a graph $G = (V, E)$, the following relation, R, on E is an equivalence relation: eRe' if and only if $e = e'$ or there is a a cycle containing e and e'. How many equivalence classes are there in the following graph?

4.31. Show that a subgraph of a graph G consisting of the edges in one equivalence class of the relation R (described in the previous exercise) and the incident vertices is a maximal biconnected subgraph of G.

*4.32. The following two definitions of functions on the vertices in a depth-first search tree of a graph are attempts to provide necessary and/or sufficient conditions for two vertices to be in the same biconnected component of the graph. Show by exhibiting counterexamples that these attempts fail, i.e., that $old_i(v) = old_i(w)$ is neither necessary nor sufficient for v and w to be in the same biconnected component ($i = 1, 2$).

   a) $old_1(x) =$ the "oldest" — i.e., closest to the root — ancestor of x that can be reached by following tree edges (away from the root) and back edges, or x itself if no such path leads to an ancestor of x.

   b) $old_2(x) =$ the oldest ancestor of x that can be reached by following tree edges (away from the root) and *one* back edge, or x itself if no such path leads to an ancestor of x.

4.33. Complete the proof of Theorem 4.3.

4.34. Find a necessary and sufficient condition for the root of a depth-first search tree for a connected graph to be an articulation point. Prove it.

4.35. a) Would the bicomponent algorithm work properly if for each vertex v, $back[v]$ were initialized to the number of the parent of v in the depth-first search tree? If so, explain why; if not, give an example in which it does not work.

   b) What if $back[v]$ were initialized to ∞ for all v?

4.36. Give an example of a graph that shows that the bicomponent algorithm may produce incorrect answers if no attempt is made to avoid stacking an edge the second time it is encountered in the adjacency list structure (i.e., if the test "**if** $dfsNumber[w] < dfsNumber[v]$" near the beginning of the **while** loop in Algorithm 4.9 were omitted).

4.37. The test to detect a bicomponent when backing up from w to v is $back[w] \geq dfsNumber[v]$. Can the case $back[w] > dfsNumber[v]$ ever occur when the algorithm is executed? If so, explain in what circumstances it occurs; if not, explain why not.

4.38. Would the bicomponent algorithm work properly if the test for a bicomponent were changed to $back[v] \geq dfsNumber[v]$? If so, explain why; if not, give an example in which it does not work.

4.39. In the bicomponent algorithm, when a back edge vw is encountered, $back[v]$ is assigned $min(back[v], dfsNumber[w])$. Would the algorithm work correctly if $back[v]$ were assigned $min(back[v], back[w])$ instead? If so, explain why; if not, give an example in which it does not work.

4.40. A connected graph is *edge biconnected* if there is no edge whose removal disconnects the graph. Which, if either, of the following statements is true? Give a proof or counterexample for each.

a)   A biconnected graph is edge biconnected.

b)   An edge biconnected graph is biconnected.

4.41. Suppose that G is a connected graph. An edge e whose removal disconnects the graph is called a *bridge*. For example, the edge EF in Fig. 4.25 is a bridge. Give an algorithm for finding the bridges in a graph.

Section 4.6: Strongly Connected Components of a Digraph

4.42. Prove that the condensation of a digraph is acyclic.

4.43. Prove that the set of vertices in a depth-first search tree for a digraph is a union of strong components.

4.44. Prove Lemma 4.4.

4.45. Prove Theorem 4.5.

4.46. We observed that if vw is in G and w is a descendant of v in a depth-first search tree, it does not matter whether or not $low[v]$ is assigned $min(low[v], dfsNumber[w])$ when the edge vw is encountered. What does the algorithm actually do in such a case?

4.47. Draw a diagram showing the strong components of the digraph in Fig. 4.24(a).

4.48. Find the strong components of the digraph in Fig. 4.29 by carefully following the steps of the algorithm.

4.49. Extend or modify the strong-component algorithm so that it outputs a list of all the edges, as well as the vertices, in each strong component. Try to minimize the amount of extra time used to do so.

*4.50. We present here another algorithm for finding strong components. It has the advantage of being simpler to explain and the disadvantage of requiring two complete depth-first searches instead of one. It works as follows:

1.   Do a complete depth-first search of G numbering each vertex as the search backs up from it. (Hence the vertices are numbered in postorder.) (By "complete" we mean that every vertex must be visited; there may be more than one depth-first search tree. The short "driver" routine of Algorithm 4.10 takes care of this.)

2.   Construct a new graph G_R in which the direction of each edge in G is reversed.

3.   Do a complete depth-first search on G_R starting with the highest numbered vertex (using the postorder numbers assigned in Step 1). Whenever a new depth-first search tree is begun, always begin with the highest-numbered unvisited vertex. The strong components are exactly the sets of vertices in each depth-first search tree.

a)   Prove the correctness of the algorithm. In other words, prove that two vertices v and w are in the same strong component of G if and only if they are in the same depth-first search tree constructed for G_R in step 3.

b)   What is the complexity of this algorithm? (Sketch out some of the relevant implementation details to justify your answer.)

Additional Problems

4.51. Formulate the following problem as a graph or digraph problem. Indicate clearly what graph or digraph you would use and what general problem you would want to solve to obtain the desired answer.

A job consists of a series of tasks $S_1, S_2, \ldots, S_k$ where the amount of time required by task S_i is t_i for $1 \le i \le k$. A list of pairs (S_i, S_j) is given such that, if (S_i, S_j) is on the list, task S_i must be completed before task S_j is begun. Pairs that would be implied by transitivity are *not* on the list. For example, if (S_i, S_j) and (S_j, S_p) are given, then (S_i, S_p) is not. Assume that there is an unlimited number of processors available so that several tasks can be done at the same time. How much time is needed to complete the job?

It may be useful to include dummy tasks B, for "begin," and E, for "end," each of which takes no time. (Remember, you need only state the problem as a graph or digraph problem; you do not have to give an algorithm to solve it.)

4.52. We mentioned in Section 4.1 that, if a graph or digraph is represented by an adjacency matrix, then almost any algorithm that operates on the graph will have worst-case complexity in $\Omega(n^2)$, where n is the number of vertices. There are, however, some problems that can be solved quickly, even when the adjacency matrix is used. Here is one.

a)   Let $G = (V, E)$ be a digraph with n vertices. A vertex s is called a *sink* if, for every v in V such that $s \ne v$, there is an edge vs, and there are no edges of the form sv. Give an algorithm to determine whether or not G has a sink, assuming that G is given by its $n \times n$ adjacency matrix. (Note: The common definition of a sink does not require an edge to s from every other vertex, but here we do.)

b)   How many matrix entries are examined by your algorithm in the worst case? It is easy to give an algorithm that looks at $\Theta(n^2)$ entries, but there is a linear solution.

*4.53. Find the best lower bound you can for the number of adjacency matrix entries that must be examined to solve the problem described in the previous exercise. Prove that it is a lower bound. (Hint: You should easily be able to give a clear argument for $2n-2$. An adversary argument similar to the one in Section 3.3.3 can be used to get a stronger lower bound.)

4.54. Design an efficient algorithm to assign distinct integers 1 through n to the n vertices of an *acyclic* digraph so that $number[v] < number[w]$ if $v \ne w$ and there is a path from v to w. Your algorithm should run in $\Theta(n+m)$ time, where m is the number of edges. (This is called topological sorting.)

4.55. Design an efficient algorithm to find a path in a connected graph that goes through each edge exactly once in each direction. (Here, "path" means a sequence of vertices, $v_1, \ldots, v_k$, where v_{i+1} is adjacent to v_i $(1 \le i \le k-1)$; a vertex may appear in the path more than once.)

*4.56. An *Euler circuit* in a graph (undirected) is a circuit (i.e., a cycle that may go through some vertices more than once) that includes every edge exactly once. Give an algorithm that finds an Euler circuit in a graph, or tells that the graph does not have one.

4.57. Consider the following question:

Is there a vertex v in G such that every other vertex in G can be reached by a path from v?

If G is a graph, the question can be easily answered by a simple depth-first (or breadth-first) search and a check to see if every vertex was visited. Write an algorithm to solve the problem for a directed graph. What is the complexity of your algorithm?

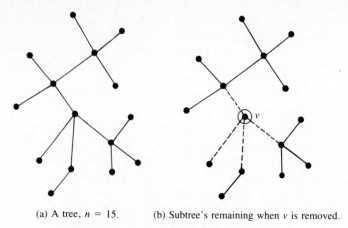

(a) A tree, $n = 15$.   (b) Subtree's remaining when v is removed.

Figure 4.37 Example for Exercise 4.60.

4.58. Suppose we want to find the shortest path from v to w in a graph G where the length of a path is simply the number of edges in the path (e.g., to plan an airline trip with the fewest stops). Which of the algorithms or traversal strategies from this chapter could you use? Which one would you use, and why?

4.59. A *bipartite* graph is a graph whose vertices can be partitioned into two subsets such that there is no edge between any two vertices in the same subset. Write an algorithm to determine if a graph is bipartite. What is the complexity of your algorithm?

*4.60. When a vertex and its incident edges are removed from a tree, a collection of subtrees remains. Write an algorithm that, given a graph that is a tree with n vertices, finds a vertex whose removal leaves no subtree with more than $n/2$ vertices. See Fig. 4.37 for an example. What is the complexity of your algorithm? (You should be able to get a linear solution.)

Programs

Each of the following program assignments requires a procedure that reads in a description of a graph or digraph and sets up the adjacency lists. Assume that the input contains the number of vertices and a list of pairs representing the edges, along with weights, if appropriate. Write this procedure so that, with small changes, it could be used for any of the problems.

Test data should be chosen so that all aspects of a program are tested. Include some of the examples in the text.

1.   The minimum spanning tree algorithm, Algorithm 4.2. Output should include the graph, the set of edges in the tree, along with their weights, and the total weight of the tree.

2.   The shortest-path algorithm, Algorithm 4.3. Output should include the graph (or digraph), the vertices between which a shortest path is sought, the edges in the path, along with their weights, and the total weight of the path.

3.   A breadth-first search algorithm to determine whether a graph has a cycle.

4.   A depth-first search algorithm to determine whether a graph has a cycle.

5.   The bicomponent algorithm, Algorithm 4.9.

6.   The strong-component algorithm described in Exercise 4.50.

Notes and References

The shortest-path and minimum spanning tree algorithms are from Dijkstra (1959), but that paper does not discuss implementation of the algorithms. (Prim's presentation of the minimum spanning tree algorithm is in Prim (1957).) The terminology for categorizing vertices in Sections 4.2 and 4.3 (e.g., *fringe vertex*) is from Sedgewick (1983). In some applications it is necessary to find a spanning tree with minimum weight among those that satisfy other criteria required by the problem, so it is useful to have an algorithm that generates spanning trees in order by weight so that each can be tested for the other criteria. Gabow (1977) presents algorithms that do this.

The adjacency list structure used in this chapter was suggested by Tarjan and is described, along with the algorithms, in Tarjan (1972) and Hopcroft and Tarjan (1973b). Hopcroft and Tarjan (1973a) presents an algorithm for finding the triconnected components of a graph. See Hopcroft and Tarjan (1974) for a very efficient algorithm to test graphs for planarity — another important problem for graphs. The strong component algorithm in Exercise 4.50 is based on Sharir (1981) and Aho, Hopcroft, and Ullman (1983).

See King and Smith-Thomas (1982) for optimal solutions to Exercises 4.52 and 4.53.

In Section 4.1 we listed several questions that might be asked about graphs and digraphs. One of the questions that we did not answer in this book is: How much of a commodity can flow from one vertex to another given capacities of the edges? This is the network flow problem; it has a rich variety of solutions and applications. The interested reader may consult Even (1979) and Ford and Fulkerson (1962).

Gibbons (1985) is a book on graph theory and algorithms; it covers topics in this chapter and many others. See also Even (1973 and 1979); Aho, Hopcroft and Ullman (1974); Deo (1974); Reingold, Nievergelt and Deo (1977); and Sedgewick (1983).

5

String Matching

5.1
The Problem and a
Straightforward Solution

In this chapter we study the problem of detecting the occurrence of a particular substring, called a pattern, in another string, called the text. The problem is usually presented in the context of character strings and arises often in text processing, and we will assume this context in our discussion and examples. However, the solutions presented also can be used in other contexts, e.g., matching a string of bytes containing graphical data, machine code, or other data, and matching a sublist of a linked list. Let P be the pattern and T the text, and let m be the length of (i.e., the number of characters in) P and n the length of T. We will assume that n is fairly large relative to m. In the algorithms we will use $P.length$ and $T.length$ to denote the lengths. Characters in P and T are denoted by subscripted p's and t's, respectively.

The reader should think about the problem and write out (or at least outline) an algorithm to solve it before proceeding. This algorithm will probably be very similar to the first one we present here, which is fairly straightforward.

Starting at the beginning of each string, we compare characters, one after the other, until either the pattern is exhausted or a mismatch is found. In the former case we are done; a copy of the pattern has been found in the text. In the latter case we start again, comparing the first pattern character with the second text character. In general, when a mismatch is found, we (figuratively) slide the pattern one more place forward over the text and start again, comparing the first pattern character with the next text character.

Example 5.1　Straightforward string matching

Comparisons are done (in left-to-right order) on the pairs of characters indicated by arrows.

P : ABABC	ABABC	ABABC
↓↓↓↓↓	↓	↓↓↓↓
T : ABABABCCA	ABABABCCA	ABABABCCA
		Successful match

Observe that moving the pattern all the way past the point where the first mismatch occurred could fail to detect an occurrence of the pattern.　　　　　　■

Algorithm 5.1　Straightforward String-matching Algorithm

Input:　P and T, the pattern and text strings. If $P.length$ and $T.length$ are not known in advance, their explicit use in the algorithm can be replaced by end-of-string tests.

Output:　The index in T where a copy of P begins. The index will be $T.length+1$ if no match for P is found.

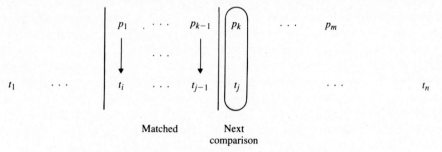

Figure 5.1 The general picture for Algorithm 5.1.

Comment: The general picture is shown in Fig. 5.1. The index variable i is not really needed in the algorithm since it can be computed from j and k (that is, $i = j-k+1$).

function *Match* $(P, T : String)$: *index*;
var
    i, j, k : *index*;
    { i is the current guess at where P begins in T;
     j is the index of the current character in T;
     k is the index of the current character in P.   }
begin
    $i := 1; j := 1;\ \ k := 1$;
    while $j \le T.length$ **and** $k \le P.length$ **do**
       if $t_j = p_k$ **then**
          $j := j+1;\ \ k := k+1$
       else { slide pattern forward and start over }
          $i := i+1;\ \ j := i;\ \ k := 1$
       end { if }
    end { while };
    if $k > P.length$ **then** *Match* $:= i$ { match found }
    else *Match* $:= j$ { $j = T.length+1$, no match }
    end { if }
end { Match }

Analysis

We will count the character comparisons done by our string-matching algorithms. This is certainly reasonable for Algorithm 5.1 given its simple loop structure. There are a few easy cases. If the pattern appears at the beginning of the text, m comparisons are done. If p_1 is not in T at all, n comparisons are done. What is the worst case? The number of comparisons would be maximized if for each value of i — that is, each possible starting place for P in T — all but the last character of P matched the corresponding text characters. Thus the number of character comparisons in the worst case is at most mn, and the complexity of the algorithm is in $O(mn)$. To show that the worst case *requires* (roughly) mn comparisons (i.e., to show that the worst-case complexity is in $\Theta(mn)$), we must show that the situation described can really occur, i.e., that P and T can be constructed so that all characters in P but the last one match corresponding characters beginning anywhere in T. (For some algorithms, inputs that require a lot of work at one step may require very little work at another step. Thus adding up the maximum possible work at each step gives an upper bound but not necessarily an exact value for the work done in the worst case.) Let $P = 'AA \ldots AB'$ ($m-1$ A's followed by a B) and $T = 'A \ldots A'$ (n A's).

This worst-case example is not one that occurs often in natural language text. In fact, Algorithm 5.1 works quite well on the average for natural language. In some empirical studies the algorithm did only about 1.1 character comparisons for each character in T (up to the point where a match was found or to the end of T if no match was found). Thus few characters in the text had to be examined more than once.

Algorithm 5.1 has a property that in some applications is undesirable: It may often be necessary to back up in the text string (the assignment "$j := i$" in the **while** loop). If the text is long, this may involve some tedious overhead. The algorithm we present in the next section was devised specifically to eliminate the need to back up in the text. It turned out to be faster (in the worst case) as well.

5.2
The Knuth-Morris-Pratt Algorithm

5.2.1 Pattern Matching with Finite Automata

We first describe briefly, without formal algorithms, an approach to the pattern-matching problem that has some important good points but also some drawbacks. The construction used by the main algorithm of this section was suggested by the method we describe now and salvages some of its advantages while eliminating the disadvantages.

Given a pattern P, it is possible to construct a *finite automaton* that can be used to scan the text for a copy of P very quickly. A finite automaton can easily be interpreted as a special kind of machine or flowchart, and a knowledge of automata theory is not necessary to understand this method. Let Σ be the alphabet, or set of characters, from which the characters in P and T may be chosen, and let $\alpha = |\Sigma|$. The flowchart, or finite automaton, has three types of nodes:

A *start* node.

A *stop* node, which means "Stop; a match was found." It is denoted by $*$.

Some *read* nodes, which mean "Read the next text character. If there are no further characters in the text string, halt; there is no match."

The flowchart has α arrows leading out from each *read* node. Each arrow is labeled with a character from Σ. The arrow that matches the text character just read

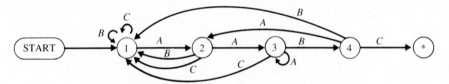

Figure 5.2   The finite automaton for $P =$ '*AABC*'.

is the arrow to be followed; i.e., it indicates which node to go to next. The reader should study the example in Fig. 5.2 to understand why the arrows point where they do. The *read* nodes serve as a sort of memory. For instance, if execution reaches the third *read* node, the last two characters read from the text were *A*'s. What preceded them is irrelevant. For a successful match, they must be followed immediately by a *B* and a *C*. If the next character is a *B* we can move on to node 4, which remembers that *AAB* has appeared. On the other hand, if the next character read at node 3 were a *C*, we would have to return to node 1 and wait for another *A* to begin the pattern.

Once the flowchart for the pattern is constructed, the text can be tested for an occurrence of the pattern by examining each text character only once, hence in $O(n)$ time. This is a big improvement over Algorithm 5.1, both in speed and in the fact that, once a text character has been examined, it never has to be reconsidered; there is no backing up in the text. The difficulty is constructing the finite automaton — that is, deciding where all the arrows go. There are well-known algorithms to construct the finite automaton to recognize a particular pattern, but in the worst case these algorithms require a lot of time. The difficulty arises from the fact that there is an arrow for each character in Σ leading out from each *read* node. It takes time to determine where each arrow should point, and space to represent $m\alpha$ arrows. Thus a better algorithm will have to eliminate some of the arrows.

5.2.2 The Knuth-Morris-Pratt Flowchart

When constructing the finite automaton for a pattern P, it is easy to put in the arrows that correspond to a successful match. For example, when drawing Fig. 5.2 for the pattern '*AABC*', the first step is to draw

The difficult part is inserting the rest of the arrows. The Knuth-Morris-Pratt algorithm (which, for brevity, will be called the KMP algorithm) also constructs a sort of flowchart to be used to scan the text. The KMP flowchart contains the easy arrows — i.e., the ones to follow if the desired character is read from the text — but it contains only one other arrow from each node, an arrow to be followed if the desired character was *not* read from the text. The arrows are called *success links* and *failure links*, respectively. The KMP flowchart differs from the finite automaton in several details: The character labels of the KMP flowchart are on the nodes rather than on

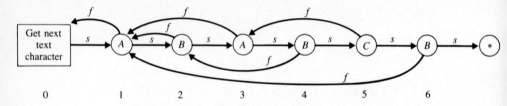

Figure 5.3   The KMP flowchart for $P = $ 'ABABCB'.

the arrows; the next character from the text is read only after a success link has been followed; the same text character is reconsidered if a failure link is followed; there is an extra node that causes a new text character to be read; and the starting point is the node that corresponds to the first pattern character. As in the finite automaton, if the * is reached, a copy of the pattern has been found; if the end of the text is reached elsewhere in the flowchart, the scan terminates unsuccessfully. This informal description of the scanning procedure should enable the reader to use the KMP flowchart in Fig. 5.3 to scan a text string. Try 'ACABAABABA' and refer to Table 5.1 if you have difficulty.

We now need a computer representation of the KMP flowchart, an algorithm to construct it (to determine how to set the failure links), a formal algorithm for the scan procedure, and an analysis of the two algorithms.

Table 5.1
Action of the KMP flowchart in Fig. 5.3 for the pattern 'ABABCB' on the text 'ACABAABABA'.

| KMP cell number | Text character being scanned | | Success (s) or failure (f) |
	Index	Character	
1	1	A	s
2	2	C	f
1	2	C	f
0	2	C	get next char.
1	3	A	s
2	4	B	s
3	5	A	s
4	6	A	f
2	6	A	f
1	6	A	s
2	7	B	s
3	8	A	s
4	9	B	s
5	10	A	f
3	10	A	s
4	11	none	failure

5.2.3   Construction of the Knuth-Morris-Pratt Flowchart

This flowchart representation is quite simple; it uses two arrays, one containing the characters of the pattern and one containing the failure links. The success links are implicit in the ordering of the array entries.

Let *fail* be the array of failure links; $fail[k]$ will be the index of the node pointed to by the failure link at the kth node, for $1 \le k \le m$. The special node that merely forces the next text character to be read is considered to be the zero-th node; $fail[1] = 0$. To see how to set the other failure links, we consider an example.

Example 5.2   Setting failure links for the KMP algorithm

Let $P =$ '*ABABABCB*' and suppose that the first six characters have matched six consecutive text characters as indicated:

```
P :           | A B A B A B | C B
              | ↓ ↓ ↓ ↓ ↓ ↓ |
T :     . . . | A B A B A B | x  . . .
```

Suppose that the next text character, x, is not a '*C*'. The next possible place where the pattern could begin in the text is at the third position shown, as follows:

```
P :                 | A B A B | A B C B
T :     . . .   A B | A B A B | x  . . .
```

The pattern is moved forward so that the longest initial segment that matches part of the text preceding x is lined up with that part of the text. Now x should be tested to see if it is an *A* to match the third *A* of the pattern. Thus the failure link for the node containing the *C* should point to the node containing the third *A*.   ■

The general picture is shown in Fig. 5.4. When a mismatch occurs, we want to slide *P* forward, but maintain the longest overlap of a prefix of *P* with a suffix of the part of the text that has matched the pattern so far. Thus, the current text character should be compared to p_r next; that is, $fail[k]$ should be r. But we want to construct the flowchart before we ever see *T*. How do we determine r without knowing *T*? The key observation is that when we do scan *T*, the part of *T* just scanned will have matched the part of *P* just scanned, so we need only find the longest overlap of a prefix of *P* with a suffix of the part of *P* just scanned. More precisely,

$fail[k]$ is the largest r (with $r < k$) such that $p_1 \ldots p_{r-1}$ matches $p_{k-r+1} \ldots p_{k-1}$.

Thus the failure links are determined by repetition within *P* itself.

An occurrence of the pattern could be missed if r were chosen too small. (Consider what would happen if in Example 5.2 the failure link for *C* were set to point to the second *A*, and if $x = A$ and is followed by *BCB* in the text.)

Although we have described the correct values for the failure links, we still do not have an algorithm to compute them efficiently. Suppose that the first $k-1$ failure

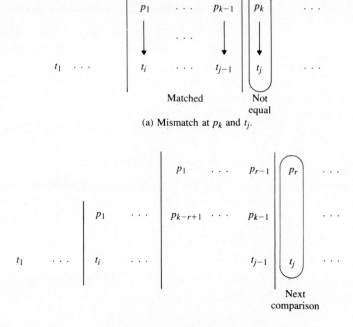

(a) Mismatch at p_k and t_j.

(b) Slide p to line up the longest prefix that
matches a suffix of the scanned characters.

Figure 5.4   Sliding the pattern for the KMP algorithm.

links have been computed. Then we have the picture in Fig. 5.5(a). To assign *fail*[*k*]
we need to match a suffix ending at p_{k-1}. The easy case is when $p_{k-1}=p_{fail\,[k-1]}$.
Then the two matching sequences in Fig. 5.5(a) can be extended by one more character, so in this case *fail*[*k*] is assigned *fail*[*k*−1]+1. In the example in Fig. 5.6,
fail[6] = 4 because $p_1p_2p_3$ matches $p_3p_4p_5$. Since $p_6=p_4$, *fail*[7] is assigned 5.

   What if $p_{k-1}\neq p_{fail\,[k-1]}$, as in Fig. 5.6 for $k = 8$? We must find an initial substring of *P* that matches a substring ending at p_{k-1}. In this case the match in Fig.
5.5(a) cannot be extended, so we look farther back. Let $r = fail[k-1]$ and let
$r' = fail[r]$. By the properties of the failure links we have the matches shown in Fig.
5.5(b). If $p_{r'}=p_{k-1}$, we have an initial substring to match a substring ending at p_{k-1}
and *fail*[*k*] should be r'+1. If $p_{r'}\neq p_{k-1}$, we must follow the failure link from node r'
and try again. This process is continued until we find a failure link r such that
$p_r =p_{k-1}$ or (as in Fig. 5.6 for $k = 8$) $r = 0$. In either case, *fail*[*k*] should be r+1.

Algorithm 5.2   KMP flowchart construction

Input: *P*, a string of characters. If the length of *P* is not known in advance, its
explicit use in the algorithm can be replaced by an end-of-string test.

Output: *fail*, the array of failure links.

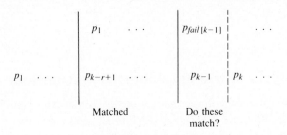

Matched

Do these match?

(a) By definition of *fail* [k − 1].

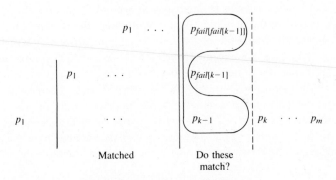

Matched

Do these match?

(b) Looking back for a match for p_{k-1}.

Figure 5.5 Computing *fail* links.

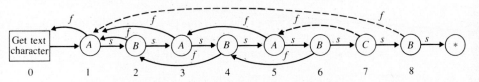

Figure 5.6 Computing *fail* links — an example.

```
procedure KMPsetup (P: String; var fail: IndexArray);
var
    k,r: integer;
begin
    fail[1] := 0;
    for k := 2 to P.length do
        r := fail[k−1];
        while r>0 and p_r ≠ p_{k−1} do¹
            r := fail[r]
        end { while };
```

---

[1] Reminder to Pascal programmers: By the semantics of our Boolean expressions, no reference to p_r is made if $r = 0$.

$fail[k] := r+1$
  end { for }
  end { KMPsetup }

5.2.4   Analysis of the Flowchart Construction

Let $m = P.length$. It is easy to see that the complexity of Algorithm 5.2 is in $O(m^2)$. The body of the **for** loop is executed $m-1$ times, and each time, the body of the **while** loop is executed at most m times because r starts somewhere in P and "jumps" backward, at worst to zero. But this analysis is not careful enough.

We will count character comparisons as we did for Algorithm 5.1. Since the comparison is part of the condition of the **while** loop, the running time of the algorithm is bounded by a multiple of the number of comparisons. (Actually, since the character comparison is not executed when $r = 0$, we should also note that $r = 0$ cannot occur more than $m-1$ times.)

We call a comparison "successful" if $p_r = p_{k-1}$ and "unsuccessful" otherwise. A successful comparison terminates the **while** loop so at most $m-1$ successful comparisons are done (for k from 2 to $P.length$). After every unsuccessful comparison r is decreased (since $fail[r] < r$), so we can bound the number of unsuccessful comparisons by determining how many times r can decrease. Observe the following:

1. r is initially assigned $fail[1] = 0$ (the first time the first statement in the **for** loop is executed).

2. r is increased by exactly 1 each subsequent time that the first statement in the **for** loop is executed. The relevant steps are:

    $fail[k] := r+1$ (at the end of the previous pass through the **for** loop);
    k is incremented by the **for** loop control;
    $r := fail[k-1]$ (the first instruction in the **for** loop).

    So the first statement in the **for** loop is actually assigning $r+1$ to r.

3. r is incremented $m-2$ times.

4. r is never negative.

Since r starts at 0 and is incremented by 1 $m-2$ times and is never negative, r cannot be decreased more than $m-2$ times. Thus the number of unsuccessful comparisons is at most $m-2$ and the total number of character comparisons is at most $2m-3$. Observe that, to count character comparisons, we actually counted the number of times the index r changed. The latter is another good measure of the work done by the algorithm. The important conclusion is that the complexity of the construction of the flowchart is linear in the length of the pattern.

5.2.5   The Knuth-Morris-Pratt Scan Algorithm

We have already informally described the procedure for using the KMP flowchart to scan the text. The algorithm follows.

Algorithm 5.3   KMP Scan Algorithm

Input: P and T, the pattern and text strings; *fail*, the array of failure links set up in Algorithm 5.2. If the length of T is not known in advance, its explicit use in the algorithm can be replaced by an end-of-string test. The length of P would have been found when setting up the *fail* array.

Output: The index in T where a copy of P begins. The index will be *T.length*+1 if no match for P is found.

```
function KMPmatch (P, T : String; fail: IndexArray) : index;

var
    j, k : index;
    { j indexes text characters;
      k indexes the pattern and fail array. }

begin
    j := 1; k := 1;

    while j ≤ T.length and k ≤ P.length do

        if k = 0 or tⱼ = pₖ then²
            j := j+1;  k := k+1
        else { follow fail arrow }
            k := fail[k]
        end { if }

    end { while };

    if k > P.length then KMPmatch := j−P.length  { match found }
    else KMPmatch := j  { j = T.length+1, no match }
    end { if }

end { KMPmatch }
```

The analysis of the scan algorithm uses an argument very similar to that used to analyze the algorithm to set up the failure links, and it is left to the reader. The number of character comparisons done by Algorithm 5.3 is at most $2n$, where $n = T.length$. Thus, the Knuth-Morris-Pratt pattern-matching algorithm, which is comprised of Algorithms 5.2 and 5.3, does $\Theta(n+m)$ operations in the worst case, a significant improvement over the $\Theta(mn)$ worst-case complexity of Algorithm 5.1. Some empirical studies have shown that the two algorithms do roughly the same number of character comparisons on the average (for natural language text), but the KMP algorithm never has to back up in the text.

---

[2] Reminder to Pascal programmers: By the semantics of our Boolean expressions, no reference to p_k is made if $k = 0$.

5.3
The Boyer-Moore Algorithm

5.3.1   The New Idea

> The chief defect of Henry King
> Was chewing little bits of string.
> Hillaire Belloc
> *Cautionary Tales* (1907)

As usual, let P be a pattern of length m and T a text string of length n. For both Algorithm 5.1 and KMP, if P is found beginning at, say, t_i, then each of the characters $t_1, t_2, \ldots, t_{i+m-1}$ has been examined (that is, has participated in at least one comparison). The key insight of the Boyer-Moore (from now on, BM) algorithm is that some characters may be skipped over entirely. In fact, our intuition suggests that the longer the pattern (the more information given the algorithm about what it has to find), the faster a good algorithm should be able to jump past places in the text where the pattern cannot appear.

The BM algorithm uses two heuristics for deciding how far the pattern may be slid over the text string, and it always scans the pattern from right to left. The first heuristic for sliding P is illustrated in the following example.

Example 5.3   Boyer-Moore's first heuristic

We are searching for the pattern *must* in a quotation from Oscar Wilde: "If you wish to understand others you must intensify your own individualism." The pattern is positioned over the string at each place where a potential match will be checked. The comparisons are indicated by arrows. The first four comparisons are as follows:

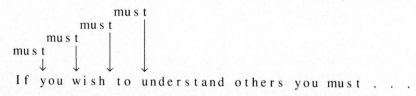

When the last character in *must*, the '*t*', is compared to the *y* in *you* we observe not only that there is no match at this position, but also, since there is no *y* at all in *must*, that there can be no match that overlaps the *y*. We may slide the pattern four places to the right. Similarly, after each of the next two comparisons we slide the pattern four places because there is no *w* or blank in *must*. At the fourth comparison we have a mismatch, but there is a *u* in the pattern, so we slide the pattern to line up the *u*'s and check for a match. (As always in BM, we start at the right end of the pattern.) The next several comparisons are:

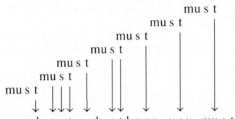

If you wish to understand others you must . . .

After the mismatch of the *u* in *must* and the *r* in *understand*, the pattern slides just far enough to pass the *r*. Similarly with the *s* in *must* and the *o* in *others*. The last comparison shown is a mismatch, but the text character *u* does appear in the pattern, so the pattern is slid over to line up the *u*'s. Four more comparisons (right to left) will confirm that a match has been found.   ∎

   In this example only 18 character comparisons are done, but since the match occurs at position 38 in *T*, the other algorithms would do at least 41 comparisons.

   The number of positions we can "jump" forward when there is a mismatch depends on the text character being read, say t_j. We will store these numbers in an array *charJump* indexed by the character set Σ. For controlling the scanning algorithm, it is more convenient to know the amount by which the text index *j* should be incremented to begin the next right-to-left scan of the pattern, rather than the amount by which *P* slides forward. As can be seen in Example 5.3 and Fig. 5.7, this jump in *j* may be larger than the distance *P* slides. If t_j does not appear in *P* at all, we can jump forward *m* places. Figure 5.7 illustrates how to calculate the jump for the case when t_j does occur in *P*. In fact, t_j may occur more than once in *P*. We need to make the smallest possible jump, lining up the rightmost instance of t_j in *P*; otherwise we might go past a copy of *P*. (We never want to slide the pattern backwards; when our current position in *P* is already left of the rightmost instance of t_j in *P*, $charJump[t_j]$ will not be useful.)

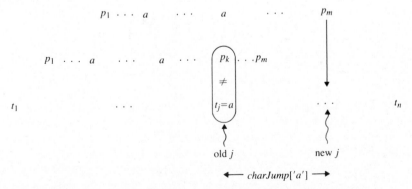

Figure 5.7   Sliding the pattern to line up a matching character.

Algorithm 5.4   Computing Jumps for the Boyer-Moore Algorithm

> **type**
>   *AlphabetArray* = **array**[char] **of** *integer*;
>
> **procedure** *ComputeJumps* (*P*: *String*; **var** *charJump*: *AlphabetArray*);
> **var**
>   *ch*: *char*;
>   *k*: *integer*;
>
> **begin**
>   **for** each *ch* in Σ **do**
>   *charJump*[*ch*] := *P.length*
>   **end** { for };
>   **for** *k* := 1 **to** *P.length* **do**
>   *charJump*[p_k] := *P.length*−*k*
>   **end** { for }
> **end** { ComputeJumps }

Clearly the amount of time used to compute the jumps is in $\Theta(|\Sigma|+m)$, where *m* is *P.length*.

5.3.2   And the "Old" Idea

Simply using *charJump* to skip through the text makes the Boyer-Moore algorithm run much faster than the Knuth-Morris-Pratt algorithm for many cases. Combining *charJump* with an idea similar to that of the *fail* arrows in the KMP algorithm can improve the algorithm further.

Example 5.4   Boyer-Moore's second heuristic

Suppose some (rightmost) segment of *P* has matched part of *T* before a mismatch occurs.

```
P :           b a t s a n d c | a t s |
                              |↓ ↓ ↓  |
T :  . . .              d | a t s |  . . .
                        ↑
                        j
```

The current text character is a '*d*'. Using *charJump*['*d*'], we would slide the pattern only one place right to line up its *d* over the *d* in *T*. However, we know that the letters in *T* to the right of the current position are '*ats*', the same letters that form the suffix of *P* that was just scanned. If we know that *P* does not have another instance of '*ats*', then we can slide *P* all the way past the '*ats*' in *T*. If *P* does have an earlier instance of '*ats*', we could slide *P* so that earlier '*ats*' lines up with the matched letters in *T*. For the previous example, the next position for a potential match is:

```
P :                      b │ a t s │ a n d c a t s

T :           . . .      d │ a t s │   . . .
                                        ↑
                                      new j
```

In order not to miss a potential match, if P has more than one substring that matches the matched suffix we line up the rightmost one (of course excluding the suffix itself). ∎

The general picture is shown in Fig. 5.8(a); the mismatch occurs at p_k and t_j. Figure 5.8(b) shows the pattern slid to the right to line up a substring with the matched suffix. We want $matchJump[k]$ to be the amount to increment j, the text position index, to begin the next right-to-left scan of the pattern after a mismatch has occurred at p_k. For $k < m$, we define $matchJump[k]$ as follows:

Let r be the largest index such that $p_r \ldots p_{r+m-k-1}$ matches $p_{k+1} \ldots p_m$ and $p_{r-1} \neq p_k$. (The condition $p_{r-1} \neq p_k$ is included because we already know that p_k does not match t_j; if $p_{r-1} = p_k$, then p_{r-1} will not match t_j either.)

$$matchJump[k] = m-r+1.$$

Sometimes there will not be a substring of P that matches the matched suffix $(p_{k+1} \ldots p_m)$. Then, as illustrated in Fig. 5.8(c), we line up the longest prefix of P that matches a suffix of P. If this prefix has, say, q characters, then $matchJump[k] = m-k+m-q$.

Example 5.5   Computing $matchJump$

Let $P =$ 'wowwow'. We compute the $matchJump$ values beginning at the right end of the pattern. See Fig. 5.9. Values already computed are shown below the pattern. The question mark indicates the position we are currently working on. At each step we slide the pattern to the right to line up a substring that matches a suffix. The character preceding the substring must be different from the character preceding the suffix. (For the rightmost pattern position, $matchJump$ is assigned 1. This value is not particularly important; $charJump$ will generally be larger.) Note that this example illustrates the values for $matchJump$, but not the actual steps carried out by the algorithm below to compute them. ∎

Algorithm 5.5   Computing Jumps Based on Partial Matches

```
procedure ComputeMatchJumps (P: String; var matchJump: IndexArray);
var   k, q, qq : integer;
      back: IndexArray;
begin
      for k := 1 to m do
            matchJump[k] := 2*m-k; { largest possible jump }
      end { for };
```

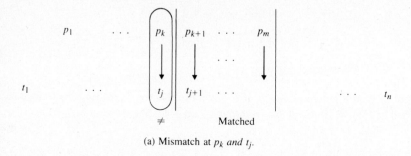

(a) Mismatch at p_k and t_j.

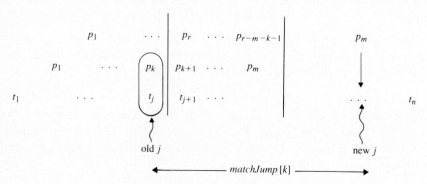

(b) Line up substring of p that matches the matched suffix (and satisfies $p_{r-1} \neq p_k$).

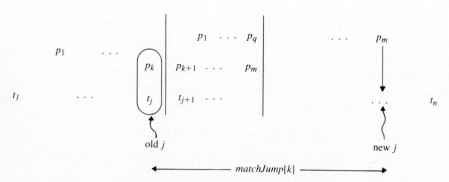

(c) If no substring of P matches $p_{k+1} \cdots p_m$, line up a prefix of P that gives a partial match.

Figure 5.8   Sliding the pattern to line up a matched substring.

$k := m; q := m+1;$
while $k>0$ **do**
    $back[k] := q;$
        while $q \leq m$ **and** $p_k \neq p_q$ **do**
            $matchJump[q] := \min(matchJump[q], m-k);$
            $q := back[q]$
        end;
        $k := k-1; \quad q := q-1;$
end { while $k>0$ };

for $k := 1$ **to** q **do**
    $matchJump[k] := \min(matchJump[k], m+q-k);$
end { for };

$qq := back[q];$
while $q \leq m$ **do**
    while $q \leq qq$ **do**
        $matchJump[q] := \min(matchJump[q], qq-q+m);$
        $q := q+1$
    end { while $q \leq qq$ };
    $qq := back[qq]$
end { while $q \leq m$ }
end { ComputeMatchJumps }

Note that the match of the first *ow* and the second *ow* was not used because both are preceded by a *w*; if a mismatch occurs at position 4 in the pattern, there is no point in lining up another *w* at that position when scanning the text.

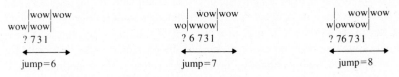

$P:$ w o w w o w
$matchJump:$ 8 7 6 7 3 1

Figure 5.9 Computing *matchJump* — an example.

5.3.3   The Boyer-Moore Scan Algorithm

Algorithm 5.6   The Boyer-Moore Scan Algorithm

Input: *P* and *T*, the pattern and text strings; *charJump* and *matchJump*, the arrays described in Secs. 5.3.1 and 5.3.2. If *T.length* is not known in advance, its explicit use in the algorithm can be replaced by an end-of-string test. The length of *P* would have been found to set up the two jump arrays.

Output: The index in *T* where a copy of *P* begins. The index will be *T.length*+1 if no match for *P* is found.

> **function** *BMmatch* (*P*, *T* : *String*; *charJump*: *AlphabetArray*;
> $\qquad\qquad\qquad$ *matchJump*: *IndexArray*) : *index*;
> **var**
> $\quad$ *j, k* : *index*;
> $\quad$ { *j* indexes text characters;
> $\qquad$ *k* indexes the pattern.   }
> **begin**
> $\quad$ *j* := *P.length*; *k* := *P.length*;
> $\quad$ **while** *j* ≤ *T.length* **and** *k* > 0 **do**
> $\qquad$ **if** $t_j = p_k$ **then**
> $\qquad\qquad$ *j* := *j*−1; *k* := *k*−1
> $\qquad$ **else** { slide *P* forward }
> $\qquad\qquad$ *j* := *j*+max(*charJump*[t_j], *matchJump*[*k*]);
> $\qquad\qquad$ *k* := *P.length*
> $\qquad$ **end** { if }
> $\quad$ **end** { while };
> $\quad$ **if** *k* = 0 **then** *BMmatch* := *j*+1   { match found }
> $\quad$ **else** *BMmatch* := *T.length*+1   { no match }
> $\quad$ **end** { if }
> **end** { BMmatch }

5.3.4   Remarks

The behavior of the Boyer-Moore algorithm depends on the size of the alphabet and the repetition within the strings. In empirical studies using natural language text and $m \geq 5$, the algorithm did only roughly 0.24 to 0.3 character comparisons per character in the text (up to the point of the match or the end of the text). In other words, it examined only roughly one-quarter to one-third of the characters. (See Fig. 5.10 for the results of one such study comparing the three algorithms. The experiments used 20 patterns of length *m* from 1 to 14. The text length was 5000.) For binary strings, BM does not do quite as well (*charMatch* does not help much); roughly 0.7 comparisons were done for each text character. In all these cases, the expected number of comparisons is *sublinear*, that is, bounded by *cn* for a constant $c < 1$. If the pattern is quite small ($m \leq 3$), then the overhead of preprocessing the pattern is not

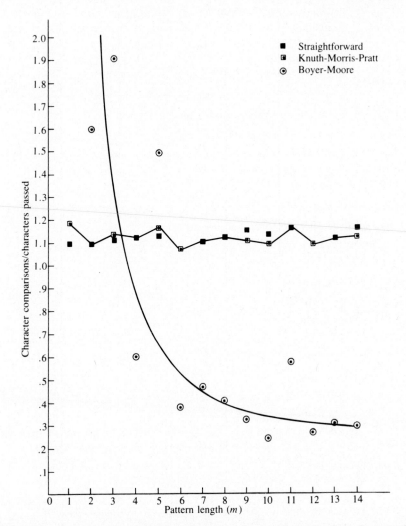

Figure 5.10 Comparison of string-matching algorithms. (From G. de V. Smit, "A Comparison of Three String Matching Algorithms," *Software — Practice and Experience*, vol. 12, Copyright 1982, John Wiley & Sons, Ltd. Reprinted by permission of John Wiley & Sons, Ltd.)

worthwhile; BM does more comparisons than the straightforward approach (Algorithm 5.1).

There are several improvements and modifications to the BM algorithm that make it run faster. (See the notes and references at the end of this chapter.) Like Algorithm 5.1, BM does some backing up in the text string (because it scans the pattern right-to-left).

Two extensions to the pattern-matching problem are often useful: Find *all* occurrences of the pattern in the text, and find any one of a finite set of patterns in the text.

Exercises

Section 5.1: The Problem and a Straightforward Solution

5.1. Rewrite Algorithm 5.1 eliminating the variable i.

5.2. Rewrite Algorithm 5.1 to work on inputs that are simply linked lists. Assume that each node has a *key* field and a *link* field, and assume that T and P are pointers to the first nodes of the two lists.

Section 5.2: The Knuth-Morris-Pratt Algorithm

5.3. Draw the finite automaton (flowchart) for the pattern 'ABAABA', where $\Sigma = \{A, B, C\}$.

5.4. Give the fail indexes used by the KMP algorithm for the following patterns:

  a)   *AAAB*

  b)   *AABAACAABABA*

  c)   *ABRACADABRA*

  d)   *ASTRACASTRA*

5.5. Give a pattern beginning with an *A* and using only letters from $\{A, B, C\}$ that would have the following fail indexes (for the KMP algorithm):

         0 1 1 2 3 4 2 2

5.6. Show that the KMP scan algorithm (Algorithm 5.3) does at most $2n$ character comparisons.

5.7. How will the KMP algorithms behave if the pattern and/or the text are null (have length zero)? Will they "crash"? If not, will their output be meaningful and correct?

5.8. Recall that the pattern $P = $ 'A...AB' ($m-1$ *A*'s followed by one *B*) and the text string $T = $ 'A...A' (n *A*'s) are a worst-case input for Algorithm 5.1.

  a)   Give the values of the *fail* indexes for P. Exactly how many character comparisons are done by *KMPsetup* (Algorithm 5.2) to compute them?

  b)   Exactly how many character comparisons are done by the KMP scan algorithm to scan T for an occurrence of P?

  c)   Given m, find a pattern Q with m letters such that *KMPsetup* does more character comparisons for Q than it does for the pattern P with m letters described above.

5.9. Prove that Algorithm 5.2 sets the KMP failure links so that

         fail[k] is the largest r (with $r < k$) such that $p_1 \ldots p_{r-1}$ matches $p_{k-r+1} \ldots p_{k-1}$.

*5.10. The strategy for setting the fail links for the KMP algorithm has a flaw that is illustrated by Fig. 5.3. If a mismatch occurs at the fourth character, a *B*, *fail*[4] points us back to another *B*, which of course will not match the current text character either. Modify Algorithm 5.2 so that *fail* values satisfy the condition stated in Section 5.2.3

(and repeated in the previous exercise) *and also* the condition that $p_r \neq p_k$. (Be careful; a common first guess does not work.)

5.11. Rewrite the KMP algorithms to work on inputs that are simply linked lists. Assume that each node has a *key* field and a *link* field, and assume that T and P are pointers to the first nodes of the lists.

Section 5.3: The Boyer-Moore Algorithm

5.12. List the values in the *charJump* array for the Boyer-Moore algorithm for the following patterns assuming that the alphabet is $\{A, B, \ldots, Z\}$.

    a)   ABRACADABRA

    b)   ASTRACASTRA

5.13. List the values in the *matchJump* array for the Boyer-Moore algorithm for the following patterns.

    a)   AAAB

    b)   AABAACAABABA

    c)   ABRACADABRA

    d)   ASTRACASTRA

5.14. As Example 5.3 showed, just using the *charJump* values, without using *matchJump*, can give a very fast scan. However, the statement

$$j := j + \max(charJump[t_j], matchJump[k])$$

in Algorithm 5.6 cannot simply be replaced by

$$j := j + charJump[t_j]$$

Why not? What other (small) change is needed to make the scan algorithm work?

5.15. Recall that the pattern $P = `A \ldots AB'$ ($m-1$ $A$'s followed by one $B$) and the text string $T = `A \ldots A'$ (n A's) are a worst-case input for Algorithm 5.1.

    a)   Give the values of the *charJump* and *matchJump* arrays for P.

    b)   Exactly how many character comparisons are done by the BM scan algorithm to scan T for an occurrence of P?

5.16. Give a formula relating *matchJump[k]* to the number of positions the pattern slides right when there is a mismatch at p_k.

5.17. Suppose that P and T are bitstrings.

    a)   Show the values in the *charJump* and *matchJump* arrays for the pattern 1101101011.

    b)   For bitstrings in general, which array, *charJump* or *matchJump*, will yield the longer "jumps"?

Additional Problems

5.18. Rewrite each of the three scan algorithms (Algorithms 5.1, 5.3, and 5.6) so that they find all occurrences of the pattern in the text.

5.19. P is a character string (of length m) consisting of letters and at most one asterisk ('*'). The asterisk is a "wild-card" character; it can match any sequence of zero or more characters. For example, if $P = `sun*day'$ and $T = `happysundaemonday'$, there is a match beginning at the '*s*' and ending at the second '*y*'; the asterisk "matches" *daemon*. Give an algorithm to find a match of *P* in a text string *T* (consisting of *n* characters), if there is one, and give an upper bound on the order of its worst-case time.

5.20. Let $X = x_1 x_2 \ldots x_n$ and $Y = y_1 y_2 \ldots y_n$ be two character strings. We say that *X* is a *cyclic shift* of *Y* if there is some *r* such that $X = y_{r+1} \ldots y_n y_1 \ldots y_r$. Give an $O(n)$ algorithm to determine whether *X* is a cyclic shift of *Y*.

5.21. a) Write an efficient algorithm to determine whether a (long) string of text contains 25 consecutive blanks. (Do not just give an exact copy of an algorithm in the text; customize it.)

   *b) Construct a worst-case (or near worst-case) example for your algorithm. How many character comparisons are done in this case?

   c) Suppose the text string contains ordinary English text in which blanks separate words and sentences but in which there is very rarely more than one blank together. If the text length is *n*, approximately how many character comparisons will your algorithm do?

*5.22. Investigate the problem of finding any one of a finite set of patterns in a text string. Can you extend any of the algorithms in this chapter to produce an algorithm that does better than scanning for each of the patterns separately?

Programs

Implement all three string-searching algorithms, including a counter for the number of character comparisons done; run a large set of test cases; and compare the results.

Notes and References

The main references for the algorithms presented here are Knuth, Morris, and Pratt (1977) and Boyer and Moore (1977). (The code in Algorithm 5.5 follows Smit (1982).) Guibas and Odlyzko (1977), Galil (1979), and Apostolico and Giancarlo (1986) present various worst-case linear versions of the Boyer-Moore algorithm. See also Aho and Corasick (1975). Boyer and Moore (1977) and Smit (1982) give empirical comparisons of the algorithms described in this chapter. The graph in Fig. 5.10 is from Smit.

6

Dynamic Programming

6.1
Introduction

Top-down algorithm design is natural and powerful. We think and plan in a general way first, then add more detail. We solve a high-level, complex problem by breaking it down to subproblems. Using recursion, we solve a large problem by breaking it down to smaller instances of the same problem. Divide and Conquer, a recursive algorithm design technique, proved especially useful for developing fast sorting algorithms. But as good as recursion is, it has its limitations.

The Fibonacci numbers provide a simple and dramatic example. The Fibonacci numbers are a sequence of integers defined by the recurrence $F_n = F_{n-1} + F_{n-2}$ for $n > 1$, with boundary values $F_0 = 0$ and $F_1 = 1$. They are defined recursively, and it may seem natural to compute them in a recursive function, but, as Fig. 4.22 illustrates, the recursive computation is extremely inefficient because it repeats a lot of work. The amount of work done by the recursive computation of F_n (measured by, say, the number of additions done, or the number of recursive calls to the function) is more than ϕ^n, where ϕ is the Golden Ratio, approximately 1.6. The amount of work is exponential in n, but F_n can be computed in $n-1$ iterations of a simple loop that starts at the "bottom," i.e., with 0 and 1, and repeatedly adds the two previous Fibonacci numbers to get the next one.

A dynamic programming algorithm stores the results, or solutions, for small subproblems and looks them up, rather than recomputing them, when it needs them later to solve larger subproblems. Thus dynamic programming is especially well suited to problems in which a recursive algorithm would solve many of the subproblems repeatedly. It is not easy to characterize dynamic programming algorithms formally, though they have some common features. These algorithms can best be recognized by seeing (and doing) several examples.

This chapter differs from most of the others in that we usually focus on one problem or application area and consider a variety of algorithms for it. When covering an application, we mention algorithm design techniques such as Divide and Conquer and greedy algorithms where they are useful. In this chapter, however, we focus on a technique, developing dynamic programming solutions for problems from different application areas.

6.2
Matrix Multiplication and Optimal Binary Search Trees

In this section we present two problems that, although quite different, have very similar solutions. They should serve as a good introduction to dynamic programming. (Note that both of the solutions presented here can be speeded up using techniques described in Yao (1982).)

6.2.1 Multiplying a Sequence of Matrices

Suppose that we want to determine the best order in which to carry out matrix multiplications when more than two matrices are to be multiplied together. We use the
ordinary matrix multiplication algorithm (Algorithm 1.2) each time we multiply two
matrices. Thus to multiply a $p \times q$ matrix and a $q \times r$ matrix, we do pqr element-
wise multiplications. There are two important observations to be made. One, we get
the same result no matter the order in which multiplications are done. That is,
matrix multiplication is *associative*: $A(BC) = (AB)C$. Second, the order can make a
big difference in the amount of work done. Consider the following example.

Example 6.1

We want to multiply

$$\begin{array}{ccccccc} A & \times & B & \times & C & \times & D \\ 30 \times 1 & & 1 \times 40 & & 40 \times 10 & & 10 \times 25 \end{array}$$

The following computations show how many multiplications are done for several
orderings.

$((AB)C)D$	$30 \times 1 \times 40 + 30 \times 40 \times 10 + 30 \times 10 \times 25 = 20{,}700$
$A(B(CD))$	$40 \times 10 \times 25 +\ \ 1 \times 40 \times 25 +\ \ 30 \times 1 \times 25 = 11{,}750$
$(AB)(CD)$	$30 \times 1 \times 40 + 40 \times 10 \times 25 + 30 \times 40 \times 25 = 41{,}200$
$A[(BC)D]$	$1 \times 40 \times 10 +\ \ 1 \times 10 \times 25 +\ \ 30 \times 1 \times 25 = 1400$ ∎

Suppose we are given matrices $A_1, A_2, \ldots, A_n$, where the dimensions of A_i
are $d_{i-1} \times d_i$. How should we compute

$$A_1 \times A_2 \times \cdots \times A_n$$

and what is the minimum number of element-wise multiplications needed? We shall
focus on the problem of finding the minimum number of multiplications; later we
will make the algorithm "remember" how the minimum is achieved.

We attempt to develop a Divide and Conquer algorithm. The subproblems we
have to solve are to determine the minimum number of multiplications needed to
compute the matrix product of some contiguous subsequence of the matrices. We
define the function M by

$M(i, j) =$ the minimum number of multiplications needed to compute
$A_i \times \cdots \times A_j$ for $1 \leq i \leq j \leq n$.

Suppose the matrices are factored as follows:

$$(A_i \times \cdots \times A_k) \times (A_{k+1} \times \cdots \times A_j)$$

The dimensions of the two factor matrices are $d_{i-1} \times d_k$ and $d_k \times d_j$. We can use
recursive calls to M to find out how many multiplications are needed for each factor.
But what is the best choice for k? Since we do not know the best place to split the

sequence of matrices into two factors, we minimize over all choices for k. Thus we have the following recurrence relation for M:

$$M(i,j) = \min_{i \leq k \leq j-1} \left[M(i,k) + M(k+1,j) + d_{i-1}d_k d_j \right] \quad \text{for } 1 \leq i < j \leq n \quad (6.1)$$

$$M(i,i) = 0.$$

In this formula, $M(i,k)$ is the minimum number of multiplications needed to compute $A_i \times \cdots \times A_k$ (considering all possible ways to factor it), and $M(k+1,j)$ is the minimum number of multiplications needed to compute $A_{k+1} \times \cdots \times A_j$ (considering all possible ways to factor it). For the last multiplication, i.e., $(A_i \cdots A_k) \times (A_{k+1} \cdots A_j)$, $d_{i-1}d_k d_j$ multiplications are done. Taking the minimum over all choices for k (between i and $j-1$) gives the minimum number of multiplications needed for the complete product $A_i \times \cdots \times A_j$. $M(i,i) = 0$ because $M(i,i)$ is, by definition, the number of multiplications needed to compute A_i.

It is a simple matter to write a recursive function subprogram for M and compute $M(1,n)$, but the function will be extremely inefficient. For an illustration of the repeated work, observe that the matrices will be factored in both of the following ways:

$$\left[A_i \times \cdots \times A_k \right] \times \left[(A_{k+1} \times \cdots \times A_{k'}) \times (A_{k'+1} \times \cdots \times A_j) \right]$$

$$\left[(A_i \times \cdots \times A_k) \times (A_{k+1} \times \cdots \times A_{k'}) \right] \times \left[A_{k'+1} \times \cdots \times A_j \right]$$

Computing $M(k+1, k')$ will be a subproblem when computing $M(k+1, j)$ and also when computing $M(i, k')$. It will also be a subproblem for many other computations. If M were computed this way, the total number of recursive calls would be exponential in n.

Divide and Conquer works well when we know (in some sense) which subproblems are really needed to solve the main problem, and when the subproblems do not "overlap" as the subsequences of matrices do. Here we should solve all the small subproblems and save the results to reuse later instead of recomputing.

How many subproblems are there in the matrix multiplication problem? Since we compute $M(i, j)$ only for $1 \leq i \leq j \leq n$, there are fewer than n^2. From now on we view M as an $n \times n$ (upper triangular) table (i.e., a two-dimensional array) instead of a recursive function. Now

$M[i, j] = $ the minimum number of multiplications needed to compute $A_i \times \cdots \times A_j$ for $1 \leq i \leq j \leq n$.

We need to compute and fill in entries in the table until we can determine the entry for $M[1, n]$. The diagonal entries are known; they are all zeros. The recurrence relation rewritten as a formula for the table entries is

$$M[i,j] = \min_{i \leq k \leq j-1} \left[M[i,k] + M[k+1,j] + d_{i-1}d_k d_j \right] \quad \text{for } 1 \leq i < j \leq n.$$

To compute $M[i,j]$ we need to know $M[i,k]$ and $M[k+1,j]$ for the indicated values of

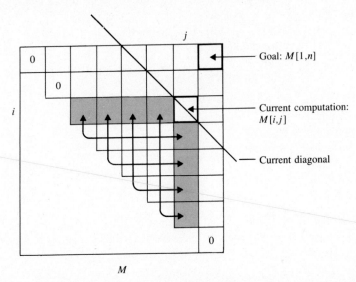

Figure 6.1 Computation of $M[i,j]$. The double-headed arrows show the pairs of entries used to compute $M[i,j]$. Each pair corresponds to one possible factorization.

k. These are the entries in row i to the left of column j and the entries in column j below row i. (See Fig. 6.1.) Thus we compute the table entries along diagonals starting just above the main diagonal and moving toward the upper right corner.

We originally wanted to know the best order in which to multiply the sequence of matrices, not just the minimum number of element-wise multiplications. So each time we determine the best way to factor a subsequence of matrices we will save the k that achieves the best factorization. That is, $factor[i,j] = k$ for which $M[i,k]+M[k+1,j]+d_{i-1}d_k d_j$ is minimum.

Algorithm 6.1   Factoring a Sequence of Matrices

Input: An array d containing the dimensions of the matrices; n, the number of matrices to be multiplied.

Output: A table *factor* that contains the best factorization for each subsequence of the matrices; *multiplications*, the number of multiplications done by the best factorization of the whole sequence.

       procedure *MatrixOrder* (*d*: *IntegerArray*; *n*: *integer*;
                     var *multiplications*: *integer*; **var** *factor*: *Matrix*);

    var
       M: *Matrix*;
       i, j, k, diagonal: *integer*;
    begin
       for $i := 1$ **to** n **do** $M[i,i] := 0$ **end**;

```
for diagonal := 1 to n−1 do
    { Diagonal #1 is just above the main diagonal. }
    for i := 1 to n−diagonal do
        j := i+diagonal;
        M[i,j] :=  min  (M[i,k]+M[k+1,j]+d_{i−1}d_k d_j);
                  i≤k≤j−1
        factor[i,j] := k that gave the minimum value for M[i,j]
    end { for i }
end { for diagonal };
multiplications := M[1,n]
end { MatrixOrder }
```

Example 6.2

For the sequence of matrices in Example 6.1, $d_0 = 30$, $d_1 = 1$, $d_2 = 40$, $d_3 = 10$, and $d_4 = 25$. *MatrixOrder* would produce the following tables. The best way of multiplying the matrices uses $M[1,4] = 1400$ multiplications.

$$
M = \begin{vmatrix} 0 & 1200 & 700 & 1400 \\ - & 0 & 400 & 650 \\ - & - & 0 & 10000 \\ - & - & - & 0 \end{vmatrix} \qquad factor = \begin{vmatrix} - & 1 & 1 & 1 \\ - & - & 2 & 3 \\ - & - & - & 3 \\ - & - & - & - \end{vmatrix}
$$

■

The following procedure, when called as *ShowOrder*(1, n), produces a fully parenthesized expression showing the order in which the matrices should be multiplied. The algorithm should be fairly clear if one realizes that the factor point k for the matrices $A_i, \ldots, A_j$ describes the *last*, not the first, multiplication to be done. Note that *ShowOrder* is recursive. Why is recursion, rather than dynamic programming, appropriate here?

Algorithm 6.2   The Order for Multiplying the Matrices

Input: i and j, the indexes of the first and last matrix in the subsequence $A_i, \ldots, A_j$. (The matrix *factor* is used by *ShowOrder*.)

Output: A parenthesized expression for $A_i \times \cdots \times A_j$.

```
procedure ShowOrder (i,j: integer);
var
    k: integer;
begin
    if i = j then  write('A', i)
    else
        k := factor[i,j];
        write( '(' );
        ShowOrder(i,k);
        write('*');
```

> *ShowOrder*($k+1, j$);
>   write(')')
>   **end** { if }
> **end** { ShowOrder }

For the *factor* table in Example 6.2, the output of *ShowOrder* is '(A1*((A2*A3)*A4))'.

Analysis

Algorithm 6.1 computes roughly $n^2/2$ entries of M. For each entry it finds the minimum of fewer than n expressions involving a few arithmetic operations. Thus the total number of steps is in $\Theta(n^3)$. This is far better than doing an exponential number of steps. Each call to *ShowOrder* (Algorithm 6.2) results in writing exactly one matrix name or asterisk, so there are $2n-1$ calls, and *ShowOrder* takes $\Theta(n)$ time.

Algorithm 6.1 uses the tables M and *factor*, hence uses $\Theta(n^2)$ extra space. A recursive solution would use only $\Theta(n)$ space (for the stack). The investment of the extra space to produce the much faster algorithm is worthwhile.

There is an $\Theta(n^2)$ algorithm for determining the best factorization of a sequence of matrices. See the notes and references at the end of this chapter.

6.2.2   Constructing Optimal Binary Search Trees

In this section we consider the problem of how to best arrange a set of keys (from some linearly ordered set) in a binary search tree to minimize the average search time if we know that some keys are looked up more often than others. In a binary search tree the keys at the nodes satisfy the following property: The key at each node is greater than all the keys in its left subtree and less than all keys in its right subtree. See Fig. 6.2. (Thus an inorder traversal of the tree would produce a sorted list of the keys.) To search for a particular key, we begin at the root and follow the left or right branch depending on whether the key sought is less than or greater than the key at the current node.

Algorithm 6.3   Searching for a Key in a Binary Search Tree

Input: *root*, a pointer to the root of the tree; and K, the key sought.

Output: *ptr*, a pointer to the node containing K; *ptr* will be **nil** if K is not in the tree.

> **procedure** *TreeSearch* (*root*: NodePtr; K: KeyType; **var** *ptr*: NodePtr);
> **var**
>   *found*: boolean;
> **begin**
>   *ptr* := *root*;
>   *found* := *false*;

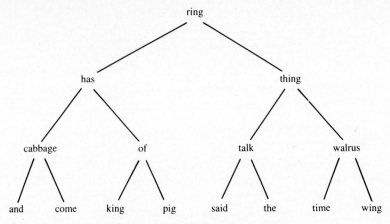

Figure 6.2 A binary search tree.

```
while not found and ptr ≠ nil do
    if ptr↑.key = K then found := true
    elsif K < ptr↑.key then ptr := ptr↑.leftChild
        else ptr := ptr↑.rightChild
    end { if }
end { while }
end { TreeSearch }
```

We use as our measure of work the number of key comparisons done, or the number of nodes of the tree that are examined, while searching for a key. (Although in the high-level language algorithm K is compared to a key in the tree twice, it is common to count it as one comparison, as it could be implemented in assembly language with one comparison that distinguishes the three results: $<$, $=$, and $>$.) Thus the number of comparisons done to find a key in the tree is one plus the level of the node containing the key. In the worst case (including cases in which K is not in the tree), the number of comparisons is the depth of the tree plus one. Suppose there are n nodes in the tree. Then, when implicitly assuming that that all keys are equally likely to be sought, it is best to keep the tree fairly full; the number of comparisons (in the worst case and on the average) is roughly $\lg n$. If the tree structure is arbitrary (and hence may consist of one long chain), the worst case is in $O(n)$.

Now let us assume that the keys are $K_1, K_2, \ldots, K_n$ and that the probability of each key being sought is $p_1, p_2, \ldots, p_n$, respectively. The probabilities would usually come from past experience or other knowledge about the application. Suppose we have arranged the keys in a binary search tree T; let c_i be the number of comparisons done by Algorithm 6.3 to locate K_i (i.e., the level of K_i plus one). The average number of nodes examined for T is

$$A(T) = \sum_{i=1}^{n} p_i c_i.$$

Example 6.3   Computing average search time

Table 6.1 shows a list of keys and data on the number of times each key was looked up in a (hypothetical) test. The probability for each key is computed from the data. (The data were chosen to make the computation easy; they are not particularly

Table 6.1
Data on the keys.

Key	Number of searches	Probability (p_i)
and	30	.15
cabbage	5	.025
come	10	.05
has	5	.025
king	10	.05
of	25	.125
pig	5	.025
ring	15	.075
said	15	.075
talk	10	.05
the	30	.15
thing	15	.075
time	10	.05
walrus	5	.025
wing	10	.05
	Total = 200	Total = 1.00

Table 6.2
Computation of average search time.

Key	Probability (p_i)	Comparisons (c_i)	$p_i \times c_i$
and	.15	4	.6
cabbage	.025	3	.075
come	.05	4	.2
has	.025	2	.05
king	.05	4	.2
of	.125	3	.375
pig	.025	4	.1
ring	.075	1	.075
said	.075	4	.3
talk	.05	3	.15
the	.15	4	.6
thing	.075	2	.15
time	.05	2	.1
walrus	.025	3	.075
wing	.05	4	.2
			Total = 3.25

realistic.) Now suppose that a binary search tree has been constructed as in Fig. 6.2. Table 6.2 shows the computation of the average search time.

The average search time is 3.25. It should seem pretty clear that this tree is not optimal. The two keys sought most often, *and* and *the*, are at the bottom level and thus require the maximum search time. Putting *and* at the root might not improve the average because, to maintain the order property of the tree, all the other keys would be in the right subtree. The prospects look better for putting *the* at the root. But let us tackle the problem systematically. ∎

We want to find a binary search tree for the keys $K_1, K_2, \ldots, K_n$ with search probabilities $p_1, p_2, \ldots, p_n$ that has minimum average search time. Assume that the keys have been sorted. If we choose K_k as the root for the tree, then $K_1, \ldots, K_{k-1}$

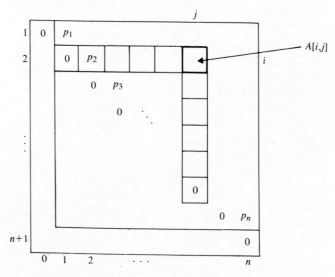

Figure 6.3 Choosing K_k as the root.

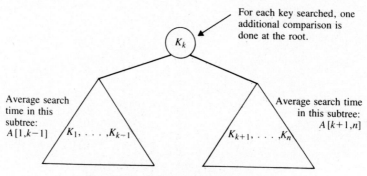

Figure 6.4 Computation of $A[i,j]$.

must go in the left subtree and $K_{k+1}, \ldots, K_n$ must go in the right subtree, and we now need optimal arrangements for the two subtrees. Since we do not know which is the best choice for the root, that is, how k should be chosen, we minimize over all choices. We could now write a recurrence relation for A, the minimum average search time, but, as with the matrix multiplication problem, we would observe that a lot of repeated work would be done by a recursive solution. The running time of the algorithm would be exponential. Instead we define A as an $n \times n$ table.

$A[i,j]$ = the minimum average search time for a binary search tree with keys $K_i < \ldots < K_j$, where $1 \le i \le j \le n$.

When a tree is constructed from two subtrees and a root K_k, one additional comparison (to the new root) will be done in a search for each key in the subtrees, so the average search time for the new tree is increased by $1 \times p_q$ for each key K_q in the subtrees. (See Fig. 6.3) We also need to count the contribution to the average made by searches for the root: $1 \times p_k$. The formulas for A are:

$$A[i,j] = \min_{i \le k \le j} \left[A[i,k-1] + \sum_{q=i}^{k-1} p_q + A[k+1,j] + \sum_{q=k+1}^{j} p_q + p_k \right]$$

$$= \min_{i \le k \le j} \left[A[i,k-1] + A[k+1,j] + \sum_{q=i}^{j} p_q \right]$$

$$= \min_{i \le k \le j} \left[A[i,k-1] + A[k+1,j] \right] + \sum_{q=i}^{j} p_q \qquad \text{for } i < j.$$

$$A[i,i] = p_i.$$

To compute $A[i,j]$, we need entries in row i to the left of $A[i,j]$ and entries in column j below $A[i,j]$. So, as for the matrix multiplication problem, we initialize the main diagonal entries, then compute entries along diagonals above the main diagonal until we reach $A[1,n]$. Note that it is also helpful to fill the diagonal just below the main diagonal with zeros because the formula for $A[i,j]$ refers to $A[i,i-1]$ and $A[j+1,j]$. (The zeros may be thought of as the average search time for empty subtrees.) (See Fig. 6.4.) To keep track of which key is chosen for the root at each step, we use the table *root*; *root*$[i,j]$ is the index of the key chosen as the root of the subtree containing $K_i, \ldots, K_j$.

Algorithm 6.4   Finding an Optimal Binary Search Tree

Input: An array p containing the probabilities for each key; n, the number of keys.

Output: A table *root* that describes how to arrange the keys; *searchTime*, the average search time for the optimal tree.

procedure *OptimalSearchTree* (*p*: RealArray; *n*: integer;
                var *searchTime*: real; **var** *root*: Matrix);
     var
        A: **array**[$1 .. n+1, 0 .. n$] **of** *real*;
        i, j, k, diagonal: integer;

```
begin
    { Initialization }
    for i := 1 to n do
        A[i, i] := p[i];
        A[i, i−1] := 0;
        root[i, i] := i
    end;  A[n+1, n] := 0;
    { Compute entries. }
    for diagonal := 1 to n−1 do
        { Diagonal #1 is just above the main diagonal. }
        for i := 1 to n−diagonal do
            j := i+diagonal;
            A[i, j] := min (A[i, k−1]+A[k+1, j]);
                     i≤k≤j
            root[i, j] := k that gave the minimum value for A[i, j];
            A[i, j] := A[i, j]+∑_{q=i}^{j} p [q ];
        end { for i }
    end { for diagonal };
    searchTime := A[1, n]
end { OptimalSearchTree }
```

$\sum_{q=i}^{j} p[q]$ need not be computed "from scratch" each time; we leave it to the reader to devise an efficient way to compute these sums. Also, if integer computation is faster or more convenient for any reason, data on past searches for the keys (as in the second column in Table 6.1) could be used directly instead of probabilities as weights for the keys. In any case, the amount of work done by Algorithm 6.4 is clearly in $\Theta(n^3)$.

6.3
Approximate String Matching

In Chapter 5 we studied several algorithms to find a copy of a character string called the *pattern* in another string called the *text*. Those algorithms searched for an exact copy of the pattern. However, in many applications we cannot expect an exact copy. A spelling corrector, for example, may search a dictionary for an entry that is similar to a given (misspelled) word. In speech recognition, samples may vary. Other applications in which close, but not exact, matches are sought include identifying sequences of amino acids and recognizing bird songs. In this section we will show a dynamic programming solution to the problem of finding an approximate match for a pattern in a string. As in Chapter 5, we will use character strings here, but the method is clearly applicable to strings of bytes of other kinds of data, for example, for speech recognition.

Let $P = p_1 p_2 \ldots p_m$ be the pattern and $T = t_1 t_2 \ldots t_n$ be the text. We assume that n is large relative to m.

Let k be a nonnegative integer. A *k-approximate match* is a match of P in T that has at most k differences. The differences may be any of the following three types:

1. The corresponding characters in P and T are different.
2. P is missing a character that appears in T.
3. T is missing a character that appears in P.

Example 6.4

The following match is a 3-approximate match. It has one of each of the permissible differences. (There are no blanks in P and T; the spaces are used to show the match more clearly.)

$$P = \qquad\qquad\text{un escessaraly}$$
$$\qquad\qquad\qquad\downarrow\ \downarrow \qquad\qquad\downarrow$$
$$T = \qquad .\ .\ .\quad \text{unne cessarily}\ .\ .\ . \qquad\blacksquare$$

The inputs for the problem are P, T, and k. For the dynamic programming solution, we define:

$D[i,j]$ = the minimum number of differences between $p_1 \ldots p_i$ and a segment of T ending at t_j.

There will be a k-approximate match ending at t_j for any j such that $D[m,j] \le k$. Thus if we want to find the first k-approximate match, we can stop as soon as we find an entry less than or equal to k in the last row of D. The rules for computing entries of D consider each of the possible differences that may occur at p_i and t_j, and, of course, the possibility that those two characters may match. $D[i,j]$ is the minimum of the following three values:

1. If $p_i = t_j$ then $D[i-1,j-1]$ else $D[i-1,j-1]+1$.
2. $D[i-1,j]+1$ (the case where p_i is missing from T).
3. $D[i,j-1]+1$ (the case where t_j is missing from P).

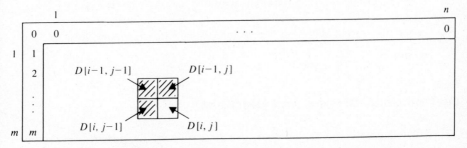

Figure 6.5 Computation of $D[i,j]$. The three shaded entries are used.

Table 6.3
The table D for Example 6.5.

	0	H	a	v	e		a		h	s	p	p	y	d	a	y	.
	0	0	0	0	0	0	0	0	0	0	0	0	0				
h 1	1	1	1	1	1	1	1	1	0	1	1	1	1				
a 2	2	2	1	2	2	2	1	2	1	1	2	2	2				
p 3	3	3	2	2	3	3	2	2	2	2	1	2	3				
p 4	4	4	3	3	3	4	3	3	3	3	2	1	2				
y 5	5	5	4	4	4	4	4	4	4	4	3	2	1				

Each entry depends on only the entries above it and to its left in the table (see Fig. 6.5), so the computation may be done in a natural row-by-row or column-by-column order. Since n may be much larger than m, it is more efficient to compute the entries of D column by column. To start the computation, we use a row 0, with $D[0,j] = 0$ for all j (intuitively, because a null section of the pattern differs in zero places from the null suffix of $t_1 \ldots t_j$), and a column 0, with $D[i,0] = i$ (because $p_1 \ldots p_i$ differs in i places from a null prefix of T).

Example 6.5   Computing the table D

Suppose $P =$ 'happy', $k = 1$, and T is the mistyped sentence 'Have a hsppy day.' Table 6.3 shows the values of D. The entries are computed column by column, and as soon as an entry in the fifth row is found to have the value 1, the computation terminates.   ∎

The work done to compute each entry of D is a small constant, so the total work done is in $O(nm)$. This is about as fast as the first, straightforward, algorithm we considered for exact pattern matching (Algorithm 5.1).

What about space? The space used by a dynamic programming algorithm for its table is often a reasonable price to pay for saving time. The table D in this algorithm, however, is m by n, and n is very large. Clearly the whole table does not have to be stored. Only entries from the current column and the previous one are needed, so the algorithm can be written using roughly $2m$ cells.

By now, writing out the algorithm should be an easy exercise, and we leave it to the reader.

There is a recent algorithm for k-approximate matching that runs in $O(kn)$ time.

6.4
Distances in Graphs and Digraphs

We studied Algorithm 4.3 which finds a shortest path and the distance between two specified vertices in a weighted graph or digraph. The algorithm used the linked adjacency list structure and ran in $\Theta(n^2)$ time in the worst case. Now we consider the following problem:

Given a weighted graph or digraph $G = (V, E, W)$ with $V = \{v_1, \ldots, v_n\}$, represented by the weight matrix with entries

$$w_{ij} = \begin{cases} W(v_i v_j) & \text{if } v_i v_j \in E \\ \infty & \text{if } i \neq j \text{ and } v_i v_j \notin E \\ 0 & \text{if } i = j, \end{cases}$$

compute the $n \times n$ matrix D defined by d_{ij} = the distance from v_i to v_j. (The distance is the weight of a minimum weight path.)

See Fig. 6.6 for an example.

One approach to computing D would be to use Algorithm 4.3 repeatedly, starting over at each vertex, but we can use dynamic programming to develop a more streamlined algorithm (eliminating the data structures used in Algorithm 4.3).

How do we compute $D[i, j]$? A shortest path may go through any of the other vertices in any order; we restrict the problem to make it simpler, then build a solution from the simpler subproblems. Let $D^{(k)}$ be the matrix with entries defined by

$D^{(k)}[i, j]$ = the weight of a shortest path from v_i to v_j using only vertices from $\{v_1, v_2, \ldots, v_k\}$ as intermediate vertices in the path.

Then $D^{(0)} = W$, and $D^{(n)} = D$, the goal of the problem. How do we compute $D^{(k)}$ from $D^{(k-1)}$? Two cases must be considered:

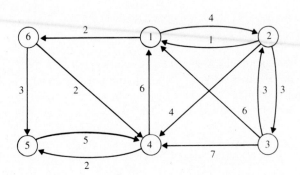

$$W = \begin{pmatrix} 0 & 4 & \infty & \infty & \infty & 2 \\ 1 & 0 & 3 & 4 & \infty & \infty \\ 6 & 3 & 0 & 7 & \infty & \infty \\ 6 & \infty & \infty & 0 & 2 & \infty \\ \infty & \infty & \infty & 5 & 0 & \infty \\ \infty & \infty & \infty & 2 & 3 & 0 \end{pmatrix} \qquad D = \begin{pmatrix} 0 & 4 & 7 & 4 & 5 & 2 \\ 1 & 0 & 3 & 4 & 6 & 3 \\ 4 & 3 & 0 & 7 & 9 & 6 \\ 6 & 10 & 13 & 0 & 2 & 8 \\ 11 & 15 & 18 & 5 & 0 & 13 \\ 8 & 12 & 15 & 2 & 3 & 0 \end{pmatrix}$$

Figure 6.6 The weight matrix and the distance matrix for a digraph.

Case 1: Among paths from v_i to v_j using only $\{v_1, v_2, \ldots, v_k\}$ as intermediate vertices, a shortest path does *not* use v_k. Then $D^{(k)}[i,j] = D^{(k-1)}[i,j]$.

Case 2: Among paths from v_i to v_j using only $\{v_1, v_2, \ldots, v_k\}$ as intermediate vertices, a shortest path *does* use v_k. Then the path can be broken into two segments, the segment from v_i to v_k and the segment from v_k to v_j (as in Fig. 6.7). The vertex v_k does not appear within either of these segments (i.e., it appears only as an endpoint), so all the intermediate vertices are in $\{v_1, v_2, \ldots, v_{k-1}\}$. The two segments must each be shortest paths between their respective endpoints (among paths using only vertices in $\{v_1, v_2, \ldots, v_{k-1}\}$ as intermediate vertices) because, if they were not, they could be replaced by shorter paths, contradicting the fact that the entire path from v_i to v_j is shortest. Thus in Case 2, $D^{(k)}[i,j] = D^{(k-1)}[i,k]+D^{(k-1)}[k,j]$.

Example 6.6

The computation of $D^{(6)}[4,3]$ for the digraph in Fig. 6.6 illustrates Case 1. $D^{(5)}[4,3]$ = 13 (because the best path from 4 to 3 using only $\{1,2,3,4,5\}$ is the path 4,1,2,3, which has weight 13). Now allowing the use of vertex 6 does not give a better path. $D^{(5)}[4,6] = 8$ (by the path 4,1,6), and $D^{(5)}[6,3] = 15$ (by the path 6,4,1,2,3). So $D^{(6)}[4,3] = D^{(5)}[4,3] = 13$.

Computing $D^{(6)}[1,5]$ illustrates Case 2. $D^{(5)}[1,5] = 10$ (because the best path from 1 to 5 using only $\{1,2,3,4,5\}$ is the path 1,2,4,5, which has weight 10). Allowing the use of vertex 6 gives a shorter path: 1,6,5, with weight 5. We get this by adding $D^{(5)}[1,6] = 2$ and $D^{(5)}[6,5] = 3$.   ∎

From the two possible cases, we conclude that

$$D^{(k)}[i,j] = \min(D^{(k-1)}[i,j], \; D^{(k-1)}[i,k]+D^{(k-1)}[k,j]).$$

The algorithm computes a sequence of matrices: $D^{(0)}, D^{(1)}, \ldots, D^{(n)}$. Since the computation of $D^{(k)}$ uses only $D^{(k-1)}$, we do not have to save the earlier matrices. It appears that we need only two $n \times n$ matrices. In fact, we need only one; the computation can all be done in the matrix D. The values in D at the end of the kth pass

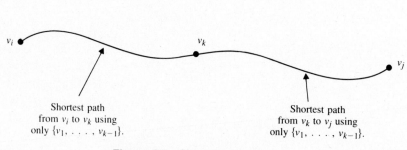

Shortest path from v_i to v_j
using only $\{v_1, \ldots, v_k\}$.

v_i v_k v_j

Shortest path
from v_i to v_k using
only $\{v_1, \ldots, v_{k-1}\}$.

Shortest path
from v_k to v_j using
only $\{v_1, \ldots, v_{k-1}\}$.

Figure 6.7   Shortest path uses v_k.

of the loop are the values of $D^{(k)}$. During the kth pass, the matrix contains entries from $D^{(k-1)}$ and $D^{(k)}$. It is easy to see, however, that the entries used when computing $D[i,j]$ have not yet been updated on the current pass, so they are entries from $D^{(k-1)}$ as they should be.

Algorithm 6.5   Distances in a Graph or Digraph

Input: W, the weight matrix for a graph or digraph with vertices $v_1, \ldots, v_n$; and n.

Output: D, an $n \times n$ matrix such that $D[i,j]$ = the distance from v_i to v_j.

```
procedure Distances ( W: Matrix; n: integer; var D: Matrix);
var
    i, j, k : integer;
begin
    D := W;  { i.e., copy W into D }
    for k := 1 to n do
    for i := 1 to n do
    for j := 1 to n do
        D[i,j] := min(D[i,j], D[i,k]+D[k,j])
    end  end end { for loops }
end { Distances }
```

Clearly Algorithm 6.5 does $\Theta(n^3)$ operations.

This algorithm is certainly short and simple, so our main idea of considering paths that use a restricted set of intermediate vertices was a good one. But this certainly was not obvious when we started. Instead of restricting the paths to use only vertices in $\{v_1, v_2, \ldots, v_k\}$, we might have restricted the paths to use *any* k intermediate vertices. We suggest that the reader consider why that choice would not have worked as well.

6.5
Some Comments on Developing a Dynamic Programming Algorithm

The essence of dynamic programming algorithms is that they trade space for speed by storing solutions to subproblems rather then recomputing them. Based on the examples in this chapter, we can make some general comments about how to develop a dynamic programming solution to a problem.

1.   It is usually useful to tackle the problem "top-down" as if we were going to develop a recursive algorithm; we figure out how to solve a large problem by finding and using solutions to smaller problems.

2.   If it appears that saving results from smaller problems can avoid repeated computation, define an appropriate table for saving results, and make a clear statement to characterize the entries in the table. Using this characterization, write a formula (or an algorithm) for computing the entries.

3. Figure out the order in which the table entries must be computed. The appropriate initialization can be determined at this point.

4. Determine how to find the solution to the problem from the data in the table. For problems such as those in Section 6.2, the answer is in a particular place in the table. For others, such as the k-approximate string-matching problem, we may need to scan part of the table to find an entry that satisfies some criterion. For some problems, again like those in Section 6.2, the table may serve as input for another algorithm to construct a solution if we want to show the optimal matrix factorization or construct the optimal binary search tree.

The first two steps are the most creative parts of developing the solution. Sometimes it is far from obvious how to break the problem down and what to put in the table. For the k-approximate matching problem, why did we define table entries with respect to a match of a segment of T ending at t_j instead of a segment begin-ning at t_j? Why did we define table entries as numbers of differences rather than indexes in T of where "good" matches occur? The answer to these questions is that the alternatives would not have worked as smoothly (if at all), but of course we cannot know that until we try them.

The solution to the distance problem (Section 6.4) used a technique that also helps in other problems. In problems that involve making an optimal selection from a set of objects (here, the vertices to traverse in a shortest path), it is often a good idea to order the objects and, at the kth iteration of the algorithm, allow a selection from any of the first k objects. This is just a guideline, not an absolute rule; the exercises include one problem for which this approach works and another one for which a different way of restricting the selections works better.

Experience with dynamic programming (and recursion) helps provide good intuition about what will work best for various problems.

The correctness of the formula for the table entries may also not be obvious. It may require some careful reasoning to show that the formula truly corresponds to the verbal description of the table entries.

Exercises

Section 6.2: Matrix Multiplication and Optimal Binary Search Trees

6.1. Show that if the recurrence relation in Equation 6.1 is used to compute the optimal number of element-wise multiplications for multiplying a sequence of matrices, then the number of recursive calls is bounded from below by an exponential function of n. (A similar argument would show that the corresponding recursive solution for the optimal binary search tree problem is also exponential.)

6.2. Suppose that the dimensions of the matrices A, B, C, and D are 20×2, 2×15, 15×40, and 40×4, respectively, and that we want to know how best to compute $A \times B \times C \times D$. Show the tables M and *factor* computed by Algorithm 6.1.

6.3. Let $A_1, \ldots, A_n$ be matrices where the dimensions of A_i are $d_{i-1} \times d_i$, for $i = 1, \ldots, n$. Here is a proposal for a greedy algorithm to determine the best order in which to perform the matrix multiplications to compute $A_1 \times A_2 \times \cdots \times A_n$:

At each step, choose the largest remaining dimension (from among $d_1, \ldots, d_{n-1}$), and multiply two adjacent matrices that share that dimension.

Observe that this strategy produces the optimal factorization for the matrices in Example 6.1.

a)   What is the order of the running time of this algorithm (only to determine the order in which to multiply the matrices, not including the actual multiplications)?

b)   Either give a convincing argument that this strategy will always minimize the number of multiplications, or give an example in which it does not do so.

6.4. Construct an example with only three or four matrices in which the worst factorization does at least 100 times as many element-wise multiplications as the best factorization.

6.5. a)   Compute the values in the matrix A and the matrix *root* in the dynamic programming algorithm for finding the optimal binary search tree for the following keys. (The probabilities appear in parentheses after each key.)

$$A \ (.20), \ B \ (.24), \ C \ (.16), \ D \ (.28), \ E \ (.04), \ F \ (.08)$$

b)   Draw the optimal tree.

6.6. Suppose that Algorithm 6.4 has been run for keys $K_1, \ldots, K_n$, with probabilities $p_1, \ldots, p_n$. Write an algorithm that uses the *root* table computed by Algorithm 6.4 to actually construct the optimal binary search tree. What is the order of the running time of your algorithm? (It should be in $O(n)$.)

6.7. Describe a straightforward greedy algorithm for the problem of constructing optimal binary search trees. Does it always produce the optimal tree? Justify your answer with an argument or a counterexample.

Section 6.3: Approximate String Matching

6.8. An algorithm for finding an exact string match need tell us only where the pattern begins or ends in the text. We can determine the unspecified end of the match in the text because we know the length of the pattern. This is not the case with approximate matching because we do not know how many characters are missing from the pattern or the text. Show how to modify the algorithm for detecting k-approximate matches so that it tells where the approximate match of the pattern begins in T.

6.9. Write out the algorithm for k-approximate matching.

Section 6.4: Distances in Graphs and Digraphs

6.10. Use Algorithm 6.5 to compute the distance matrix for the digraph whose adjacency matrix is

$$\begin{bmatrix} 0 & 2 & 4 & 3 \\ 3 & 0 & \infty & 3 \\ 5 & \infty & 0 & 3 \\ \infty & 1 & 4 & 0 \end{bmatrix}$$

6.11. a)   Use Algorithm 6.5 to compute the distance matrix for the digraph whose adjacency matrix is

$$\begin{pmatrix} 0 & 2 & 4 & 3 \\ 3 & 0 & \infty & 3 \\ 5 & \infty & 0 & -3 \\ \infty & -1 & 4 & 0 \end{pmatrix}$$

b) Give an explanation why this algorithm will work correctly even if some of the weights are negative, so long as there are no negative cycles. (A negative cycle is a cycle for which the sum of the weights of its edges is negative.)

6.12. The order of the **for** loops in Algorithm 6.5 is important. Show that if k varied in the innermost loop as shown below, the algorithm would not give the correct results.

$D := W;$ { i.e., copy W into D }
for $i := 1$ **to** n **do**
for $j := 1$ **to** n **do**
for $k := 1$ **to** n **do**
    $D[i,j] := \min(D[i,j], D[i,k]+D[k,j])$
end end end { for loops }

6.13. Give an algorithm to find the length of a shortest cycle in a graph.

Additional Problems

*6.14. The binomial coefficients are defined by

$$C(n,k) = C(n-1,k-1) + C(n-1,k) \text{ for } n > 0 \text{ and } k > 0$$

$$C(n,0) = 1 \text{ for } n \geq 0, \ C(0,k) = 0 \text{ for } k > 0.$$

$C(n,k)$ is also called "n choose k" and denoted $\begin{bmatrix} n \\ k \end{bmatrix}$. It is the number of ways to choose k objects from among n objects. Consider the following three ways to compute $C(n,k)$ for $n \geq k$.

1. A recursive function as suggested by the recurrence relation for $C(n,k)$.
2. A dynamic programming algorithm.
3. The formula $C(n,k) = n!/k!(n-k)!$.

a) Write out an outline of each method to make it clear you understand what work is to be done for each.

b) Compare the amount of work done by each method. Indicate what operations you are counting. Compare the amount of space used by each method. Does any of the three methods have any other strong advantages or disadvantages? (For example, is one of them more likely to cause an arithmetic overflow error?)

6.15. Let L be an array of n distinct integers. Give an algorithm to find the length of a longest increasing subsequence of entries in L. For example, if the entries are 11, 17, 5, 8, 6, 4, 7, 12, 3, a longest increasing subsequence is 5, 6, 7, 12. How much time does your algorithm take?

6.16. Two character strings may have many common substrings. For example, *photograph* and *tomography* have several common substrings of length one (i.e., single letters), and common substrings *ph*, *to*, and *ograph* (as well as all the substrings of *ograph*). The maximum common substring length is 6.

Let $X = x_1 x_2 \ldots x_n$ and $Y = y_1 y_2 \ldots y_m$ be two character strings. Give an algorithm to find the maximum common substring length for X and Y. What is the order of the running time of your algorithm? (Note: There is an $\Theta(nm)$ dynamic programming solution, and there are other $\Theta(nm)$, non-dynamic-programming solutions. Try to find two solutions.)

6.17. Suppose you are given three strings of characters: $X = x_1 x_2 \ldots x_m$, $Y = y_1 y_2 \ldots y_n$, and $Z = z_1 z_2 \ldots z_{m+n}$. Z is said to be a *shuffle* of X and Y if Z can be formed by interspersing the characters from X and Y in a way that maintains the left-to-right ordering of the characters from each string. For example, *cchocohilaptes* is a shuffle of *chocolate* and *chips*, but *chocochilatspe* is not. Devise a dynamic programming algorithm that takes as input X, Y, Z, m, and n, and determines whether Z is a shuffle of X and Y. (Hint: The values in the table you construct should be Boolean, not numeric.) What is the order of the running time of your algorithm?

6.18. The *Partition Problem* is:

> Given n integers, is there a way to partition the integers into two disjoint subsets so that the sums of the integers in each of the two subsets are equal? More formally, given $s_1, \ldots, s_n$, is there a subset I of $\{1, 2, \ldots, n\}$ such that
> $$\sum_{i \in I} s_i = \sum_{i \notin I} s_i ?$$

Give a dynamic programming algorithm for the Partition Problem. Estimate the running time of your algorithm.

6.19. Suppose you have n dollars to invest in any of m enterprises. Assume that n is an integer and that all investments must be in integer amounts. The table *return* describes the expected returns for individual investments. Specifically, *return*$[d, j]$ is the expected return for an investment of d dollars in enterprise j.

   a)   Write an algorithm to determine the maximum possible expected return for investing n dollars. (You may assume that the columns of *return* are nondecreasing; i.e., investing more money in one enterprise will not decrease your return from that enterprise.)

   b)   What is the order of the running time of your algorithm?

   c)   Expand your algorithm to determine the optimal investment plan (i.e., do whatever is needed to tell how much to invest in each enterprise). What is the order of the running time?

   d)   Suppose you cannot make the assumption made in part a? In other words, suppose that investing additional dollars in one enterprise can reduce your overall return from that enterprise. Either prove that your algorithm already works correctly in such cases, or give an example in which it does not compute the maximum possible return, and then show how to modify it so that it works correctly in general.

*6.20. Suppose that the denominations of the coins in a country are $c_1 > c_2 > \ldots > c_n$ (e.g., 50, 25, 10, 5, 1 for the United States). The problem to consider is:

> Given an integer a, what is the minimum number of coins needed to make a cents in change? (You may assume that $c_n = 1$, so that it is always possible to make change for any amount a.)

a) A greedy algorithm for this problem would use as many of the largest coin as possible, then as many of the next largest as possible, etc. For example, to ˷ ᴉke change for $1.43 (U.S.), use two half-dollars, one quarter, one dime, one ᴨᴉᴄᴋel, and three pennies. Prove that the greedy algorithm works for U.S. coins; i.e., it will make the change using the minimum possible number of coins.

b) Make up an example of denominations for a fictitious country's coin system in which the greedy algorithm will not give the minimum number of coins.

c) Give a dynamic programming algorithm to solve the problem. How fast is your algorithm?

6.21. Give an algorithm to determine how many different ways there are to give a cents in change using any coins from among pennies, nickels, dimes, quarters, and half-dollars. For example, there are six ways to give 17 cents change: a dime, a nickel, and two pennies; a dime and seven pennies; three nickels and two pennies; two nickels and seven pennies; one nickel and 12 pennies; and 17 pennies.

*6.22. Suppose that you have inherited the rights to 500 previously unreleased songs recorded by the popular group Raucous Rockers. You plan to release a set of five compact disks (numbered 1 through 5) with a selection of these songs. Each disk can hold a maximum of 60 minutes of music, and a song cannot overlap from one disk to the next. Since you are a classical music fan and have no way to judge the artistic merits of these songs, you decide on the following criteria for making the selection:

1. The songs will be recorded on the set of disks in the order of the dates they were written.

2. The number of songs included will be maximized.

Suppose you have a list of the lengths of the songs, $l_1, l_2, \ldots, l_{500}$, in order by the date they were written. (Each song is less than 60 minutes long.)

Give an algorithm to determine the maximum number of songs that can be included in the set satisfying the given criteria. (Hint: Let $T[i, j]$ be the minimum amount of time needed for any i songs selected from among the first j songs. T should be interpreted to include the blank time, if any, at the end of a completed disk. In other words, if a selection of songs uses one disk plus the first fifteen minutes of a second disk, count the time for that selection as 75 minutes even if there are a few blank minutes at the end of the first disk.)

Programs

1. Write a program to construct an optimal binary search tree using Algorithm 6.4 and your solution to Exercise 6.6.

2. Write a program for the k-approximate matching algorithm, storing at most two columns at a time. Include the enhancements for Exercise 6.8.

Notes and References

Bellman (1957) and Bellman and Dreyfus (1962) are standard references for dynamic programming.

For a much more extensive discussion of optimal binary search trees, see Knuth (1973). Yao (1982) describes techniques for speeding up dynamic programming solutions, and contains $O(n^2)$ algorithms for the matrix multiplication and binary search tree problems covered in Section 6.2.

Section 6.3 is based on Wagner and Fischer (1974). Hall and Dowling (1980) is a survey of approximate string matching techniques. The $O(kn)$ algorithm for k-approximate string matching is in Landau and Vishkin (1986).

Algorithm 6.5, for finding distances in graphs and digraphs, is from Floyd (1962).

Exercise 6.22 was contributed by J. Frankle.

Bentley (1986) describes the use of dynamic programming to solve chess endgames with a specific set of pieces on the board by working backwards from all possible checkmate positions.

7

Polynomials and Matrices

7.1
Introduction

The problems examined in this chapter are polynomial evaluation (with and without preprocessing the coefficients), polynomial multiplication (as an illustration of the discrete Fourier transform), and multiplication of matrices and vectors. The operations usually used for such tasks are multiplication and addition. On older computers, multiplication took a lot more time than addition, and some of the algorithms presented "improve" upon the straightforward or most widely known methods by reducing the number of multiplications at the expense of some extra additions. Hence their value depends on the relative costs of the two operations. Other algorithms presented reduce the number of both operations (for large input sizes).

Many lower-bound results are stated without proof in this chapter. See the notes and references at the end of the chapter for further comment and references on these results.

7.2
Evaluating Polynomial Functions

7.2.1   Algorithms

Let $p(x) = a_n x^n + a_{n-1} x^{n-1} + \cdots + a_1 x + a_0$ be a polynomial with real coefficients and $n \geq 1$. Suppose that the coefficients $a_0, a_1, \ldots, a_n$ and x are given and that the problem is to evaluate $p(x)$. The number of multiplications and additions done may seem a reasonable measure of work, but some algorithms may use division and subtraction and do fewer multiplications and additions. Thus, particularly when discussing lower bounds, we will consider the total number of multiplications and divisions and the total number of additions and subtractions. The two types of operations will be denoted $*/$ and $\pm$, respectively.

The obvious way to solve the problem is to compute each term and add it to the sum of the others already computed. The following algorithm does this.

Algorithm 7.1   Polynomial Evaluation — Term by Term

Input: The coefficients of $p(x)$ in the array a; $n \geq 1$, the degree of p; and x.
Output: The value of $p(x)$.

> **function** *Poly* (*a: RealArray; n: integer; x: real*): *real*;
> **var**
>> *p, xpower : real*;
>> *i: integer*;
>
> **begin**
>> $p := a[0] + a[1] * x$;
>> *xpower* := *x*;

```
    for i := 2 to n do
        xpower := xpower*x;
        p := p+a[i]*xpower
    end { for };
    Poly := p
end { Poly }
```

Algorithm 7.1 does $2n-1$ multiplications and n additions.

Horner's Method

Is there a better way? Is there a way to compute $ab+ac$, given a, b, and c, with fewer than two multiplications? Yes, of course, by factoring it as $a(b+c)$. Similarly, the key to Horner's method for evaluating $p(x)$ is simply a particular factorization of p:

$$p(x) = (\cdots ((a_n x + a_{n-1})x + a_{n-2})x + \cdots + a_1)x + a_0.$$

The computation is done in a short loop with only n multiplications and n additions.

Algorithm 7.2   Polynomial Evaluation — Horner's Method

Input: a, n, and x as in Algorithm 7.1.

Output: The value of $p(x)$.

```
function HornerPoly (a: RealArray; n: integer; x: real): real;
var
    p: real;
    i: integer;
begin
    p := a[n];
    for i := n−1 to 0 by −1 do
        p := p*x+a[i]
    end { for };
    HornerPoly := p
end { HornerPoly }
```

Thus simply by factoring p we have cut the number of multiplications in half without increasing the number of additions. Can the number of multiplications be reduced further? Can the number of additions be reduced?

7.2.2   Lower Bounds for Polynomial Evaluation

Just as we used decision trees as an abstract model for establishing lower bounds for sorting (and other problems), we need a model for polynomial evaluation algorithms (and other related computation problems). Recall that the algorithms represented by decision trees worked for a fixed input size and had no loops. Here we use a model called *straight-line programs*. The programs perform a sequence of arithmetic operations; they do no looping and no branching. The operands may be inputs to the problem, I, some constants, C, and intermediate results computed from these. The constants may seem unnecessary — we did not use any in the two polynomial evaluation algorithms we examined — but allowing constants simplifies the lower-bound arguments, and any lower bound derived in a model that allows constants will be valid for a more restricted model that does not.

Formally, a straight-line program is a finite sequence of steps of the form

$$s_i = q \ o \ r$$

where q and r are inputs, constants, or the results of earlier steps; i.e., q and r are in $I \cup C \cup \{s_j \mid j < i\}$, and o is an arithmetic operator. The last step should compute $p(x)$. For the problem of evaluating a polynomial $p(x) = a_n x^n + a_{n-1} x^{n-1} + \cdots + a_1 x + a_0$, the input set is $I = \{x, a_0, a_1, \ldots, a_n\}$. The inputs should be thought of as indeterminates, that is, abstract symbols with no assumptions about their values.

Example 7.1 A straight-line program for Horner's method with $n = 2$

$$s_1 := a_2 * x$$
$$s_2 := s_1 + a_1$$
$$s_3 := s_2 * x$$
$$s_4 := s_3 + a_0$$

∎

The number of steps in a straight-line program is clearly a reasonable measure of the work it does. Most of the theorems count $*/$ and $\pm$ steps separately. We will illustrate the proof techniques by showing that, if divisions are not permitted, a straight-line program to evaluate a polynomial of degree n must do at least $n \pm$'s. It can be shown by a similar but more complicated argument that, if divisions are not permitted, at least n multiplications are needed. It is also known that, if divisions are permitted, at least $n */$'s are required. Thus Horner's method uses the optimal number of $*/$'s, and, since division takes at least as much time as multiplication, it uses the best mix of these two operators.

We say that a step $s_i := q \ o \ r$ *uses* an input α if and only if $q = \alpha$ or $r = \alpha$, or $q = s_j$ for some $j < i$ and s_j uses α, or $r = s_j$ for some $j < i$ and s_j uses α. In other words, if we "expand" s_i by replacing the results from earlier steps until only inputs and constants remain, then α appears in the expression. In Example 7.1, for example, s_3 uses a_2, a_1, and x.

Theorem 7.1 A straight-line program using only $*$, $+$, and $-$ to evaluate

$$p(x) = a_n x^n + a_{n-1} x^{n-1} + \cdots + a_1 x + a_0$$

where $a_0, \ldots, a_n$ and x are arbitrary inputs, must have at least $n \pm$ steps.

Lemma 7.2 A straight-line program (using only $*$, $+$, and $-$) to compute $a_0 + \ldots + a_n$ must have at least $n \pm$ steps.

Proof by induction on n. For $n = 0$, we observe that any program has at least zero $\pm$ steps. For $n > 0$, suppose that P is a program for $a_0 + \cdots + a_n$. The idea of the proof is to substitute 0 for a_n to produce a program that computes $a_0 + \cdots + a_{n-1}$, then to use the induction assumption. Let

$$s_i := q \; o \; r$$

be the first $\pm$ step that uses a_n. (There must be such a step; otherwise the result of the computation would be a multiple of a_n.) Since no previous $\pm$ step used a_n, q or r must be a_n itself or a multiple of a_n. If we substitute 0 for a_n we would have one of the following cases:

Case 1: $s_i := q \pm 0$
Case 2: $s_i := 0 + r$
Case 3: $s_i := 0 - r$

For Cases 1 and 2, eliminate this step from the program and substitute q or r, respectively, for s_i in all other steps where s_i appears. For Case 3, replace this step with

$$s_i := -1 * r.$$

In all cases, we have eliminated one $\pm$ step. Replace a_n by 0 in all steps where it appears. We now have a program that computes $a_0 + \cdots + a_{n-1}$. By the induction assumption, it has at least $n - 1$ $\pm$ steps. Therefore the original program P had at least $n \pm$ steps. $\square$

Proof of the theorem. Let P be a program to compute $a_n x^n + a_{n-1} x^{n-1} + \cdots + a_1 x + a_0$. Replace every reference to x with "1." This does not change the number of $\pm$ steps. The resulting program now computes $a_n + \cdots + a_0$, so it must have at least $n \pm$ steps. $\square$

7.2.3 Preprocessing of Coefficients

Preprocessing (also called preconditioning) some of the data in a problem means, informally, that some of the input is known in advance and that a specialized program may be written. Suppose that a problem has inputs I and I' and that we denote an algorithm for the problem by A. When we speak of preprocessing I, we mean finding an algorithm A_I with input I' that produces the same output as A with inputs I and I'. Thus the preprocessing problem has two parts: the algorithm A_I, which depends on I, and an algorithm that, with I as input, produces the algorithm A_I. Rigorously speaking, A and A_I solve different problems and, as we shall see, their complexities may differ.

In some situations one polynomial must be evaluated for a large number of different arguments. One example is a power series approximation of a function. In such cases, preprocessing the coefficients may reduce the number of $*/$'s required for each evaluation.

Let $p(x) = a_n x^n + a_{n-1} x^{n-1} + \cdots + a_1 x + a_0$, where $n = 2^k - 1$ for some $k \geq 1$. Thus, p has 2^k terms, some of which may be zero. The procedure for evaluating $p(x)$

described here assumes that p is monic, i.e., that $a_n = 1$. Extending the algorithm to the general case is left as an exercise.

If $n = 1$, then $p(x) = x + a_0$ and is evaluated by doing one addition. Suppose that $n > 1$ and that p is written as follows for some j and b:

$$p(x) = (x^j + b)q(x) + r(x),$$

where q and r are monic polynomials of degree $2^{k-1}-1$ (i.e., with half as many terms as p, counting zero terms, if any). Then $p(x)$ can be evaluated by carrying out the following steps:

1.　Evaluate $q(x)$ and $r(x)$.

2.　Compute x^j.

3.　Multiply $(x^j + b)$ by $q(x)$ and add $r(x)$.

Since q and r satisfy the same conditions as p — i.e., they are monic and their degree is $2^{k'}-1$ for some k' — the same scheme could be used recursively to evaluate them. How must j and b be chosen to ensure that q and r have the desired properties? Clearly $j = \text{degree}(p) - \text{degree}(q) = 2^k - 1 - (2^{k-1}-1) = 2^{k-1}$. Note that, since j is a power of 2, x^j can be computed fairly quickly. The correct value for b becomes clear when we divide $p(x)$ by $x^j + b$ to obtain $q(x)$, the quotient, and $r(x)$, the remainder. See Fig. 7.1. For r to be monic, $a_{2^{k-1}-1} - b$ must be 1, so $b = a_{2^{k-1}-1} - 1$. Thus the preprocessing algorithm factors p as follows:

$$p(x) = \left[x^{2^{k-1}} + (a_{2^{k-1}-1} - 1) \right] q(x) + r(x).$$

It factors q and r recursively by the same procedure. The factorization is complete when q and r have degree 1. The following example illustrates the entire procedure.

Example 7.2

Let $p(x) = x^7 + 6x^6 + 5x^5 + 4x^4 + 3x^3 + 2x^2 + x + 1$.

Then $k = 3$, $j = 2^{k-1} = 4$, $b = a_{2^{k-1}-1} - 1 = a_3 - 1 = 2$, and $x^j + b = x^4 + 2$. Figure 7.2(a) shows the computation of $q(x)$ and $r(x)$.

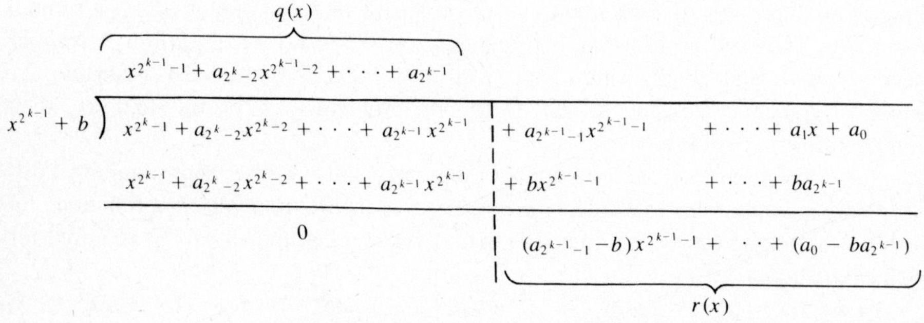

Figure 7.1　$p(x)$ divided by $x^j + b$.

$$
\begin{array}{r|l}
\multicolumn{2}{l}{\overbrace{x^3 + 6x^2 + 5x + 4}^{q(x)}} \\
x^4 + 2 \,\big)\; x^7 + 6x^6 + 5x^5 + 4x^4 & + 3x^3 + 2x^2 + x + 1 \\
x^7 + 6x^6 + 5x^5 + 4x^4 & + 2x^3 + 12x^2 + 10x + 8 \\
\hline
 & \underbrace{x^3 - 10x^2 - 9x - 7}_{r(x)}
\end{array}
$$

(a) Computing $q(x)$ and $r(x)$.

$q(x) = x^3 + 6x^2 + 5x + 4$

$k = 2;\ j = 2^{k-1} = 2$

$b = a_{2^{k-1}-1} - 1 = a_1 - 1 = 4$

$$
\begin{array}{r|l}
\multicolumn{2}{l}{\qquad x + 6} \\
x^2 + 4 \,\big)\; x^3 + 6x^2 & + 5x + 4 \\
x^3 + 6x^2 & + 4x + 24 \\
\hline
 & x - 20
\end{array}
$$

Thus

$q(x) = (x^2 + 4)(x + 6) + (x - 20)$

$r(x) = x^3 - 10x^2 - 9x - 7$

$k = 2;\ j = 2^{k-1} = 2$

$b = a_{2^{k-1}-1} - 1 = a_1 - 1 = -10$

$$
\begin{array}{r|l}
\multicolumn{2}{l}{\qquad x - 10} \\
x^2 - 10 \,\big)\; x^3 - 10x^2 & - 9x - 7 \\
x^3 - 10x^2 & - 10x + 100 \\
\hline
 & x - 107
\end{array}
$$

and

$r(x) = (x^2 - 10)(x - 10) + (x - 107).$

(b) Recursively factoring $q(x)$ and $r(x)$.

Figure 7.2 Example 7.2.

Thus $p(x) = (x^4+2)(x^3+6x^2+5x+4) + (x^3-10x^2-9x-7)$. Factor $q(x)$ and $r(x)$ in the same way, as shown in Fig. 7.2(b). Thus

$$p(x) = (x^4+2)[(x^2+4)(x+6)+(x-20)] + [(x^2-10)(x-10)+(x-107)].$$

Using this formula, evaluating $p(x)$ requires five multiplications: three that appear explicitly in the factorization and two to compute x^2 and x^4. Horner's method would have required seven. Observe, however, that ten additions (and subtractions) are done instead of seven. ■

Analysis of Polymonial Evaluation with Preprocessing of Coefficients

The number of operations done to evaluate $p(x)$ (after the preprocessing work has been done) may be counted easily by considering the three steps used to describe the procedure:

1. Evaluate $q(x)$ and $r(x)$ recursively.

   This suggests the use of a recurrence relation.

2. Compute x^j.

   The largest j used is 2^{k-1}; $x^2, x^4, x^8, \ldots, x^{2^{k-1}}$ may be computed by doing $k-1$ multiplications.

3. Multiply $(x^j + b)$ by $q(x)$ and add $r(x)$

   One multiplication and two additions.

Let $M(k)$ be the number of multiplications done to evaluate a monic polynomial of degree $2^k - 1$, *not* counting the powers of x (since they may be computed once and used as needed). Let $A(k)$ be the number of additions (and subtractions). Then:

$$M(1) = 0$$
$$M(k) = 2M(k-1) + 1 \qquad \text{for } k > 1$$

and

$$A(1) = 1$$
$$A(k) = 2A(k-1) + 2 \qquad \text{for } k > 1.$$

By expanding $M(k)$ a few times we see that $M(k) = 4M(k-2) + 2 + 1 = 8M(k-3) + 4 + 2 + 1 = \sum_{i=0}^{k-2} 2^i = 2^{k-1} - 1$. The total number of multiplications, then, is $2^{k-1} - 1 + k - 1$, the $k-1$ term for computing powers of x. Since $n = 2^k - 1$, the number of multiplications is $(n+1)/2 + \lg(n+1) - 2$, or roughly $n/2 + \lg n$. It is easy to show that $A(k) = (3n-1)/2$. (The reader should check that these formulas describe the number of operations done in the example.)

Whether or not eliminating $n/2 - \lg n$ multiplications by doing $n/2$ extra additions is a time-saver, we have illustrated an important point: Lower bounds that have been obtained for a problem without preprocessing, in this case n */'s for evaluating a polynomial of degree n, may no longer be valid. The particular operations permitted in the preprocessing (for example, division of polynomials, as in this case, or finding roots of polynomials) can also affect the number of operations required. A lower bound of $\lceil n/2 \rceil$ */'s has been established for polynomial evaluation, allowing a variety of preprocessing operations.

7.3
Vector and Matrix Multiplication

7.3.1 Review of Standard Algorithms

We begin by reviewing the well-known methods for multiplying matrices and vectors, noting the number of operations done by these methods, and giving the known lower bounds for the number of multiplications and divisions (*/'s). Throughout this section we use capital letters for the names of vectors and matrices and the corresponding small letters for their components. The components are real numbers.

Let $V = (v_1, v_2, \ldots, v_n)$ and $W = (w_1, w_2, \ldots, w_n)$ be two n-vectors, that is, vectors with n components in each. The dot product of V and W, denoted $V \cdot W$, is defined as $V \cdot W = \sum_{i=1}^{n} v_i w_i$. The computation of $V \cdot W$ implied by the definition requires n multiplications and $n-1$ additions. It has been shown that, even if one of the vectors is known in advance and some preprocessing of its components is permitted, at least n */'s are required in the worst case. Thus the straightforward computation of dot products is optimal.

Let A be an $m \times n$ matrix and let V be an n-vector. Let W be the product AV. By definition, the ith component of W is the dot product of the ith row of A with V. That is, for $1 \le i \le m$, $w_i = \sum_{j=1}^{n} a_{ij} v_j$. The computation of AV implied by the definition requires mn multiplications. This is known to be optimal. The number of additions done is $m(n-1)$.

Let A be an $m \times n$ matrix, let B be an $n \times q$ matrix, and let C be the product of A and B. By definition, c_{ij} is the dot product of the ith row of A and the jth column of B. That is, for $1 \le i \le m$ and $1 \le j \le q$, $c_{ij} = \sum_{k=1}^{n} a_{ik} b_{kj}$. If the entries of C are computed by the usual matrix multiplication algorithm, i.e., as indicated by this formula, mnq multiplications and $m(n-1)q$ additions are done. Much to the surprise of people studying the problem, attempts to prove that mnq */'s are required for matrix multiplication were unsuccessful, and eventually algorithms that do fewer */'s were sought and found. Two of them are presented here.

7.3.2 Winograd's Matrix Multiplication

Suppose that the dot product of $V = (v_1, v_2, v_3, v_4)$ and $W = (w_1, w_2, w_3, w_4)$ is computed by the following formula:

$$V \cdot W = (v_1 + w_2)(v_2 + w_1) + (v_3 + w_4)(v_4 + w_3) - v_1 v_2 - v_3 v_4 - w_1 w_2 - w_3 w_4.$$

Observe that the last four multiplications involve only components of V or only components of W. Only two multiplications involve components of both vectors. (Also observe that the formula relies on the commutativity of multiplication; e.g., it uses the fact that $w_2 v_2 = v_2 w_2$. Hence it would not hold if multiplication of the components were not commutative; in particular, it would not hold if the components were matrices.)

Generalizing from the example, when n is even (say, $n = 2p$),

$$V \cdot W = \sum_{i=1}^{p} (v_{2i-1} + w_{2i})(v_{2i} + w_{2i-1}) - \sum_{i=1}^{p} v_{2i-1} v_{2i} - \sum_{i=1}^{p} w_{2i-1} w_{2i}. \qquad (7.1)$$

If n is odd, we let $p = \lfloor n/2 \rfloor$ and add the final term $v_n w_n$ to Eq. 7.1. In each summation, p, or $\lfloor n/2 \rfloor$, multiplications are done, so in all $3 \lfloor n/2 \rfloor$ multiplications are done. This is worse than the straightforward way of computing the dot product. Even if one of the vectors is known in advance and the second or third summation can be considered preprocessing, n multiplications would still be done. If *both* vectors are known in advance, then the whole computation could be thought of as preprocessing, thus eliminating the whole problem! So what has been gained by looking at a more complicated formula for the dot product?

Suppose we are to multiply the $m \times n$ matrix A by the $n \times q$ matrix B. Each row of A is involved in q dot products, one with each column of B, and each column of B is involved in m dot products, one with each row of A. Thus terms such as the last two summations in Eq. 7.1 can be computed once for each row of A and column of B and used many times.

Algorithm 7.3 Winograd's Matrix Multiplication

Input: A, B, m, n, and q, where A and B are $m \times n$ and $n \times q$ matrices, respectively.
Output: $C = AB$.

> **procedure** *Winograd* (A, B : *Matrix*; m, n, q: *integer*; **var** C: *Matrix*);
> **var**
> $\quad$ p, i, j : *integer*;
> $\quad$ *rowTerm, colmTerm*: *RealArray*;
> $\quad$ { These arrays are for the results of the "preprocessing" of the rows
> $\quad\quad$ of A and the columns of B. }
> **begin**
> $\quad$ $p := n$ div 2;
>
> $\quad$ { "Preprocess" rows of A. }
> $\quad$ **for** $i := 1$ **to** m **do**
> $\quad\quad$ $rowTerm[i] := \sum_{j=1}^{p} a_{i,2j-1} {*} a_{i,2j}$
> $\quad$ **end**;
>
> $\quad$ { "Preprocess" columns of B. }
> $\quad$ **for** $i := 1$ **to** q **do**
> $\quad\quad$ $colmTerm[i] := \sum_{j=1}^{p} b_{2j-1,i} {*} b_{2j,i}$
> $\quad$ **end**;
>
> $\quad$ { Compute entries of C. }
> $\quad$ **for** $i := 1$ **to** m **do**
> $\quad\quad$ **for** $j := 1$ **to** q **do**
> $\quad\quad\quad$ $c_{ij} := \sum_{k=1}^{p} (a_{i,2k-1} + b_{2k,j}) {*} (a_{i,2k} + b_{2k-1,j}) - rowTerm[i] - colmTerm[j]$
> $\quad\quad$ **end** { **for** j }
> $\quad$ **end** { **for** i };
>
> $\quad$ { If n is odd, a final term is added to each entry of C. }
> $\quad$ **if** *odd*(n) **then**
> $\quad\quad$ **for** $i := 1$ **to** m **do**
> $\quad\quad\quad$ **for** $j := 1$ **to** q **do** $c_{ij} := c_{ij} + a_{in} {*} b_{nj}$ **end**
> $\quad\quad$ **end** { **for** i }
> $\quad$ **end** { **if** n odd }
> **end** { *Winograd* }

Analysis

Assume that n is even. (The case of odd n is left as an exercise.) We count multiplications first. Processing rows of A does mp multiplications, processing columns of B does qp, and computing the entries of C does mqp. The total, since $p = n/2$, is $(mnq/2)+(n/2)(q+m)$. If A and B are square matrices, both $n{\times}n$, then Winograd's algorithm does $(n^3/2)+n^2$ multiplications instead of the usual n^3. (See Algorithm 1.2.) The difference is significant even for small n. Unfortunately, Winograd's algorithm does extra $\pm$'s. We count the $\pm$'s as follows:

> Processing rows of A: $m(p-1)$
> Processing columns of B: $q(p-1)$
> Computing elements of C: For each of the mq entries of C, there are:
> $2p$ (the two pluses in each term of the summation)
> $+ p-1$ (to add the terms in the summation)
> $+ 2$ (to subtract $rowTerm[i]$ and $colmTerm[j]$).

Thus to compute the elements of C the algorithm does $mq(3p+1)$ $\pm$'s, and the total, again assuming n is even, is $(3/2)(mnq)+(n/2)(m+q)+mq-m-q$. For square $n{\times}n$ matrices, where the comparison between algorithms is a little easier to see, Winograd's algorithm does $(3/2)n^3+2n^2-2n$ $\pm$'s instead of the usual n^3-n^2.

Observe that Winograd's algorithm contains fewer instructions that require incrementing and testing loop counters than does the usual method. On the other hand, Winograd's algorithm uses more complex subscripting and requires fetching matrix entries more often. These differences are explored in the exercises.

7.3.3   Lower Bounds for Matrix Multiplication

Winograd's algorithm shows that $m{\times}n$ and $n{\times}q$ matrices can be multiplied using fewer than mnq multiplications. How many $*/$'s are necessary? Is it in $\Theta(mnq)$, or can the cubic term be eliminated? The best-known lower bound is surprisingly low: mn, or n^2 for square matrices. We stated earlier that mn $*/$'s are necessary to multiply an $m{\times}n$ matrix by an n-vector. We would expect matrix multiplication to be at least as hard, hence to require at least as many $*/$'s, and Fig. 7.3 illustrates that this

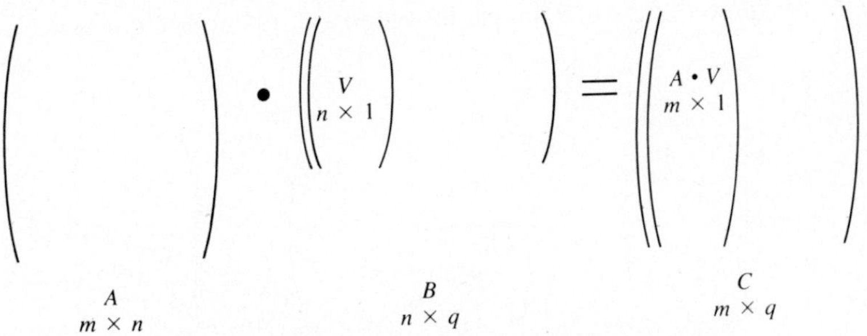

$$\left(\right) \cdot \left(\begin{matrix} V \\ n \times 1 \end{matrix} \right) = \left(\begin{matrix} A \cdot V \\ m \times 1 \end{matrix} \right)$$

$\begin{matrix} A \\ m \times n \end{matrix} \qquad\qquad \begin{matrix} B \\ n \times q \end{matrix} \qquad\qquad \begin{matrix} C \\ m \times q \end{matrix}$

Figure 7.3   Lower bound for matrix multiplication.

is true by showing that an algorithm to multiply matrices can be used to obtain a matrix vector product. (The two problems are the same, of course, if $q = 1$.) No known matrix multiplication algorithm does only mn */'s. However, there are algorithms that, for large matrices, do significantly fewer multiplications *and* ±'s than Winograd's.

7.3.4 Strassen's Matrix Multiplication

For the remainder of this section we assume that the matrices to be multiplied are $n \times n$ square matrices, A and B. The key to Strassen's algorithm is a method to multiply 2×2 matrices using seven multiplications instead of the usual eight. (Winograd's algorithm also uses eight.) For $n = 2$, first compute the following seven quantities, each of which requires exactly one multiplication:

$$
\begin{aligned}
x_1 &= (a_{11}+a_{22})*(b_{11}+b_{22}) & x_5 &= (a_{11}+a_{12})*b_{22} \\
x_2 &= (a_{21}+a_{22})*b_{11} & x_6 &= (a_{21}-a_{11})*(b_{11}+b_{12}) \\
x_3 &= a_{11}*(b_{12}-b_{22}) & x_7 &= (a_{12}-a_{22})*(b_{21}+b_{22}) \\
x_4 &= a_{22}*(b_{21}-b_{11})
\end{aligned}
\tag{7.2}
$$

Let $C = AB$. The entries of C are:

$$
\begin{aligned}
c_{11} &= a_{11}b_{11}+a_{12}b_{21} & c_{12} &= a_{11}b_{12}+a_{12}b_{22} \\
c_{21} &= a_{21}b_{11}+a_{22}b_{21} & c_{22} &= a_{21}b_{12}+a_{22}b_{22}
\end{aligned}
$$

They are computed as follows:

$$
\begin{aligned}
c_{11} &= x_1+x_4-x_5+x_7 & c_{12} &= x_3+x_5 \\
c_{21} &= x_2+x_4 & c_{22} &= x_1+x_3-x_2+x_6
\end{aligned}
\tag{7.3}
$$

Thus 2×2 matrices can be multiplied using seven multiplications and 18 additions. It is critical to Strassen's algorithm that commutativity of multiplication is not used in the formulas in Eq. 7.2, so that they can be applied to matrices whose components are also matrices. Let n be a power of two. Strassen's method consists of partitioning A and B each into four $n/2 \times n/2$ matrices as shown in Fig. 7.4 and multiplying them using the formulas in Eqs. 7.2 and 7.3; the formulas are used recursively to multiply the component matrices. Before considering extensions for the case when n is not a power of 2, we compute the number of multiplications and ±'s done.

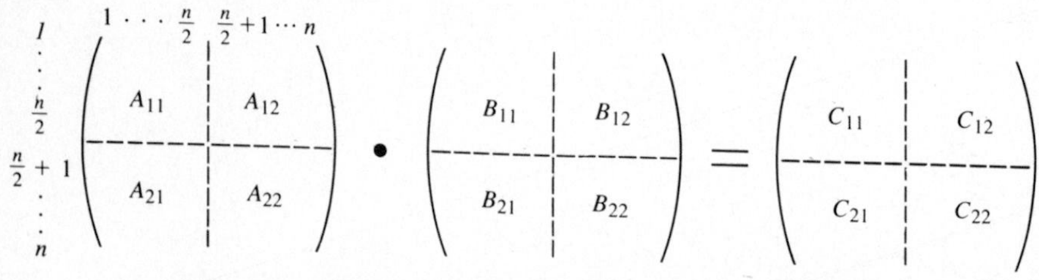

Figure 7.4 Partitioning for Strassen's matrix multiplication.

Suppose that $n = 2^k$ for some $k \geq 0$. Let $M(k)$ be the number of multiplications (of the underlying matrix components, i.e., real numbers) done by Strassen's method for $n \times n$ matrices. Then, since the formulas of Eq. 7.2 do seven multiplications of $2^{k-1} \times 2^{k-1}$ matrices,

$$M(0) = 1$$
$$M(k) = 7M(k-1) \qquad \text{for } k > 0.$$

This recurrence relation is very easy to solve. $M(k) = 7^k$, and $7^k = 7^{\lg n} = n^{\lg 7} \approx n^{2.81}$. Thus the number of multiplications is in $o(n^3)$.

Let $P(k)$ be the number of $\pm$'s done. Clearly $P(0) = 0$. There are 18 $\pm$'s in the formulas of Eqs. 7.2 and 7.3, so $P(1) = 18$. For $k \geq 1$, multiplying $2^k \times 2^k$ matrices involves 18 additions of $2^{k-1} \times 2^{k-1}$ matrices, plus all the $\pm$'s done by the seven matrix multiplications in Eqs. 7.2. So

$$P(0) = 0$$
$$P(k) = 18(2^{k-1})^2 + 7P(k-1) \qquad \text{for } k > 0.$$

Expanding the recurrence relation to see what the terms look like gives

$$P(k) = 18(2^{k-1})^2 + 7P(k-1) = 18(2^{k-1})^2 + 7 \cdot 18(2^{k-2})^2 + 7^2 P(k-2)$$
$$= 18(2^{k-1})^2 + 7 \cdot 18(2^{k-2})^2 + 7^2 \cdot 18(2^{k-3})^2 + 7^3 P(k-3).$$

Therefore

$$P(k) = \sum_{i=0}^{k-1} 7^i \, 18(2^{k-i-1})^2 = 18(2^k)^2 \sum_{i=0}^{k-1} \frac{7^i}{(2^{i+1})^2}$$

$$= \frac{9}{2}(2^k)^2 \sum_{i=0}^{k-1} \left[\frac{7}{4}\right]^i = \frac{9}{2}(2^k)^2 \left[\frac{\left[\frac{7}{4}\right]^k - 1}{\frac{7}{4} - 1} \right] = 6 \cdot 7^k - 6 \cdot 4^k$$

$$\approx 6n^{2.81} - 6n^2,$$

which is also in $o(n^3)$.

If n is not a power of two, some extension of Strassen's algorithm must be used and more work will be done. There are two simple approaches, both of which can be very slow. The first possibility is to add extra rows and columns of zeros to make the dimension a power of 2. The second is to use Strassen's formulas as long as the dimension of the matrices is even and then use the usual algorithm when the dimension is odd. Another, more complicated possibility is to modify the algorithm so that at each level of the recursion, if the matrices to be multiplied have odd dimension, one extra row and one extra column are added. Strassen described a fourth strategy, one that combines the advantages of the first two. The matrices are embedded in larger ones with dimension $2^k m$, where $k = \lfloor \lg n - 4 \rfloor$ and $m = \lfloor n/2^k \rfloor + 1$. Strassen's formulas are used recursively until the matrices to be multiplied are $m \times m$; then the usual method is applied. The total number of arithmetic operations done on the matrix entries will be less than $4.7n^{\lg 7}$.

Table 7.1
Comparison of matrix multiplication methods for $n \times n$ matrices.

	The usual algorithm	Winograd's algorithm	Strassen's algorithm
Multiplications	n^3	$\frac{1}{2}n^3 + n^2$	$7^k \approx n^{2.81}$, where $n = 2^k$
Additions/subtractions	$n^3 - n^2$	$\frac{3}{2}n^3 + 2n^2 - 2n$	$6 \cdot 7^k - 6 \cdot 4^k \approx 6n^{2.81} - 6n^2$, where $n = 2^k$
Total	$2n^3 - n^2$	$2n^3 + 3n^2 - 2n$	$4.7n^{\lg 7} \approx 4.7n^{2.81}$ (n need not be a power of 2)

Table 7.1 compares the numbers of arithmetic operations done by the three matrix multiplication methods for $n \times n$ matrices. For large n, Strassen's algorithm does fewer multiplications *and* fewer $\pm$'s than either of the other methods. In practice, however, it is not a very good algorithm. Because of its recursive nature, implementing this algorithm would require a lot of bookkeeping that would be very slow and/or complicated. The other, much simpler algorithms will be more efficient for moderate-size n.

The primary importance of Strassen's algorithm is that it broke the $\Theta(n^3)$ barrier for matrix multiplication *and* the $\Theta(n^3)$ barrier for a number of other matrix problems. These problems, which include matrix inversion, computing determinants, and solving systems of simultaneous linear equations, have well-known $\Theta(n^3)$ solutions, but since they can be reduced to matrix multiplication, they too can be solved in $O(n^{\lg 7})$ time. Strassen's result has been improved upon several times in recent years. There is now a matrix multiplication algorithm with running time in $O(n^{2.376})$. It still remains for very practical algorithms with complexity in $o(n^3)$ to be developed. The lower bound of n^2 multiplications has not been increased; whether or not matrix multiplication can be done in $\Theta(n^2)$ steps is still an open question.

*7.4
The Fast Fourier Transform and Convolution

7.4.1   Introduction

Let U and V be n-vectors with components indexed from 0 to $n-1$. The *convolution* of U and V, denoted $U \otimes V$, is, by definition, an n-vector W with components $w_i = \sum_{j=0}^{n-1} u_j v_{i-j}$, where $0 \le i \le n-1$ and the indexes on the right-hand side are taken modulo n. For example, for $n = 5$,

$$w_0 = u_0 v_0 + u_1 v_4 + u_2 v_3 + u_3 v_2 + u_4 v_1$$
$$w_1 = u_0 v_1 + u_1 v_0 + u_2 v_4 + u_3 v_3 + u_4 v_2$$
$$\vdots$$
$$w_4 = u_0 v_4 + u_1 v_3 + u_2 v_2 + u_3 v_1 + u_4 v_0$$

The problem of computing the convolution of two vectors arises naturally and frequently in probability problems, engineering, and other areas. Symbolic polynomial multiplication, which will be examined in this section, is a convolution computation. An operator called the discrete Fourier transform (which will be defined later) can be used to compute convolutions and has many other applications. It is used in interpolation problems, in solving partial differential equations, in circuit design, in crystallography, and, very extensively, in signal processing.

The discrete Fourier transform of an n-vector and the convolution of two n-vectors can each be computed in a straightforward way using n^2 multiplications and fewer than n^2 additions. We will present a Divide and Conquer algorithm to compute the discrete Fourier transform using $\Theta(n\lg n)$ arithmetic operations. This algorithm (which appears in the literature in many variations) is known as the fast Fourier transform, or FFT. We will then use the FFT to compute convolutions in $\Theta(n\lg n)$ time. This time-saving is very valuable in the applications.

Throughout this section all matrix, array, and vector indexes will begin at 0. The complex roots of unity and some of their elementary properties are used in the FFT. The basic definitions and required properties are reviewed in the appendix (Section 7.4.4). Readers who are unfamiliar with nth roots of unity should read the appendix before proceeding.

7.4.2 The Fast Fourier Transform

The discrete Fourier transform transforms an n-vector with real components into a complex n-vector. For $n \geq 1$, let ω be a primitive nth root of 1, and let F_n be the $n \times n$ matrix with entries $f_{ij} = \omega^{ij}$, where $0 \leq i, j \leq n-1$. The *discrete Fourier transform* of the n-vector $P = (p_0, p_1, \ldots, p_{n-1})$ is the product $F_n P$. The components of $F_n P$ are:

$$\omega^0 p_0 + \omega^0 p_1 + \cdots + \omega^0 p_{n-2} + \omega^0 p_{n-1}$$
$$\omega^0 p_0 + \omega p_1 + \cdots + \omega^{n-2} p_{n-2} + \omega^{n-1} p_{n-1}$$
$$\vdots$$
$$\omega^0 p_0 + \omega^i p_1 + \cdots + \omega^{i(n-2)} p_{n-2} + \omega^{i(n-1)} p_{n-1}$$
$$\vdots$$
$$\omega^0 p_0 + \omega^{n-1} p_1 + \cdots + \omega^{(n-1)(n-2)} p_{n-2} + \omega^{(n-1)(n-1)} p_{n-1}.$$

Rewritten in a slightly different form, the ith component is

$$p_{n-1}(\omega^i)^{n-1} + p_{n-2}(\omega^i)^{n-2} + \cdots + p_1 \omega^i + p_0.$$

Thus if we interpret the components of P as coefficients of the polynomial $p(x)=p_{n-1}x^{n-1}+p_{n-2}x^{n-2}+\cdots+p_1x+p_0$, then the ith component is $p(\omega^i)$ and computing the discrete Fourier transform of P means evaluating the polynomial $p(x)$ at $\omega^0, \omega, \omega^2, \ldots, \omega^{n-1}$, i.e., at each of the nth roots of 1. We will approach the problem from this point of view. We will develop a recursive Divide and Conquer algorithm first and then examine it closely to remove the recursion. We assume that $n=2^k$ for some $k\geq 0$. (Adjustments to the algorithm can be made if it is to be used for n not a power of 2.)

The strategy of Divide and Conquer is to divide the problem into smaller instances, solve those, and use the solutions to get the solution for the current instance. Here, to evaluate p at n points, we evaluate two smaller polynomials at a subset of the points and then combine the results appropriately. Recall that $\omega^{n/2}=-1$ and thus for $0\leq j\leq n/2-1$, $\omega^{(n/2)+j}=-\omega^j$. Group the terms of $p(x)$ with even powers and the terms with odd powers as follows:

$$p(x) = \sum_{i=0}^{n-1}p_ix^i = \sum_{i=0}^{n/2-1}p_{2i}x^{2i}+x\sum_{i=0}^{n/2-1}p_{2i+1}x^{2i}.$$

Define

$$p_{even}(x) = \sum_{i=0}^{n/2-1}p_{2i}x^i \quad \text{and} \quad p_{odd}(x) = \sum_{i=0}^{n/2-1}p_{2i+1}x^i.$$

Then

$$p(x) = p_{even}(x^2)+x*p_{odd}(x^2) \quad \text{and} \quad p(-x) = p_{even}(x^2)-x*p_{odd}(x^2). \tag{7.4}$$

Equation 7.4 shows that, to evaluate p at $1, \omega, \ldots, \omega^{(n/2)-1}, -1, -\omega, \ldots, -\omega^{(n/2)-1}$, it suffices to evaluate p_{even} and p_{odd} at $1, \omega^2, \ldots, (\omega^{(n/2)-1})^2$ and then do $n/2$ multiplications (for $x*p_{odd}(x^2)$) and n additions and subtractions. The polynomials p_{even} and p_{odd} can be evaluated recursively by the same scheme. That is, they are polynomials of degree $n/2-1$ and will be evaluated at the $n/2$th roots of unity: $1, \omega^2, \ldots, (\omega^{(n/2)-1})^2$. Clearly, when the polynomial to be evaluated is a constant, there is no work to be done.

The recursive algorithm follows.

Algorithm 7.4 The Fast Fourier Transform (Recursive Version)

Input: P, the vector $(p_0, p_1, \ldots, p_{2^k-1})$; k; and m. We assume the 2^kth roots of 1: $\omega^0, \omega, \ldots, \omega^{2^k-1}$, are stored in the global array *omega* in the order listed here. We use m to select roots from this array. *RecursiveFFT* would be called initially with $m=1$. In general, the set consisting of every mth entry, that is, $\omega^0, \omega^m, \omega^{2m}, \ldots,$ is the set of $2^k/m$th roots of 1.

Output: The discrete Fourier transform of P stored in the array *transform*.

```
procedure RecursiveFFT (P: RealArray; k, m: integer;
                        var transform: ComplexArray);
var
    evens, odds : ComplexArray;
    xPOdd: Complex;
    j: integer;
begin
    if k = 0 then
        transform[0] := p₀;  transform[1] := p₀;
    else
        { Evaluate p_even at the 2^(k-1)th roots of 1. }
        RecursiveFFT ((p₀, p₂, . . . , p₂ₖ₋₂); k−1; 2m; evens);
        { Evaluate p_odd at the 2^(k-1)th roots of 1. }
        RecursiveFFT ((p₁, p₃, . . . , p₂ₖ₋₁); k−1; 2m; odds);
        for j := 0 to 2^(k-1)−1 do { evaluate p(ωʲ) and p(ω^(2^(k-1)+j)) } }
            xPOdd := omega[mj]*odds[j];
            { Compute p(ωʲ) }
            transform[j] := evens[j]+xPOdd;
            { Compute p(ω^(2^(k-1)+j)) }
            transform[2^(k-1)+j] := evens[j]−xPOdd
        end { for }
    end { if }
end { RecursiveFFT }
```

The recursiveness of the algorithm makes it easy to find a recurrence relation for the number of operations done. We will count the arithmetic operations done on components of P and roots of 1. Let $M(k)$, $A(k)$, and $S(k)$ be the number of multiplications, additions, and subtractions, respectively, done by *RecursiveFFT* to compute the discrete Fourier transform of a 2^k-vector. The three operations are done, one each, in the body of the **for** loop, so $M(k) = A(k) = S(k)$. We solve for $M(k)$.

$$M(0) = 0$$

and

$$M(k) = 2^{k-1} + 2M(k-1),$$

where the first term in $M(k)$, i.e., 2^{k-1}, counts the multiplications in the **for** loop, and the second term, $2M(k-1)$, counts multiplications done by the recursive calls to *RecursiveFFT* to compute the values in the arrays *evens* and *odds*. It is easy to see that $M(k) = 2^{k-1}k$. Thus $M(k) = A(k) = S(k) = 2^{k-1}k$, or $(n/2)\lg n$. Since the operations are done on complex numbers, this result should be multiplied by a small constant to reflect the fact that each complex operation requires several ordinary ones.

Algorithm 7.4 would require a lot of extra time and space for the bookkeeping necessitated by the recursion. Yet the breakdown of the polynomial seems systematic enough that we should be able to obtain a scheme for carrying out the same

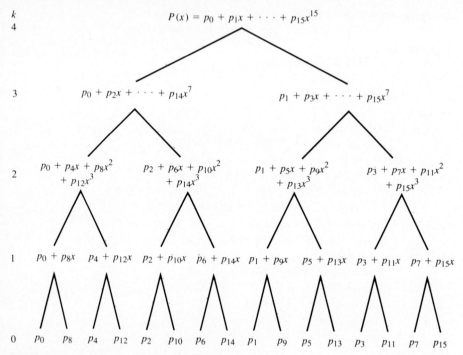

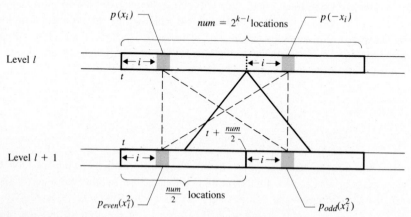

Figure 7.5 Polynomial evaluation at roots of unity. For a polynomial p at any internal node, the left child is p_{even} and the right child is p_{odd}.

Figure 7.6 Illustration for the *FFT*. At the node shown in level l, the polynomial p is to be evaluated at $x_0, x_1, \ldots, x_{2^{k-(l+1)}-1}, -x_0, -x_1, \ldots, -x_{2^{k-(l+1)}-1}$, where $p(x_i) = p_{even}(x_i^2) + x_i p_{odd}(x_i^2)$ and $p(-x_i) = p_{even}(x_i^2) - x_i p_{odd}(x_i^2)$. The diagram shows which values from the previous levels are used to compute $p(x_i)$ and $p(-x_i)$.

computation "from the bottom up" without using a recursive program. The example in the tree diagram of Fig. 7.5 should help suggest the pattern of the computation. The levels of the tree correspond to the levels of recursion, but the recursion could be eliminated if we start the computation at the leaves. The leaves are the components of the vector P permuted in a particular way. Determining the correct permutation, π_k, is the key to constructing an efficient implementation of the evaluation scheme. π_k will be given (and its correctness proved) after the presentation of the nonrecursive algorithm. The reader is invited to try to determine how to define π_k before proceeding.

Observe that, at each level of the tree, the same number of values is to be computed: 2^k, since at level l there are 2^l nodes, or polynomials, to be evaluated at 2^{k-l} roots of unity. Since the values computed at one level are needed only to compute two values in the next level, one array, *transform*, with 2^k entries suffices to store the results of the computations. Figure 7.6 illustrates how two values at a node at level l are computed using one value from each of its children. The diagram may help clarify the algorithm's indexing.

Algorithm 7.5   The Fast Fourier Transform

Input: The n-vector $P = (p_0, p_1, \ldots, p_{n-1})$, where $n = 2^k$ for some $k > 0$.

Output: *transform*, the discrete Fourier transform of P.

Comment: We assume that *omega* is an array containing the nth roots of 1: $\omega^0, \omega, \ldots, \omega^{n-1}$. The array *transform* is initialized to contain the values for level $k-1$, not the leaves, in the tree of Fig. 7.5. π_k is a permutation on $\{0, 1, \ldots, n-1\}$.

    procedure *FFT* (*P: RealArray; n: integer;* **var** *transform: ComplexArray*);
    var
        l: integer;   { the level number }
        num: integer;   { the number of values to be computed at each node at
                      level l }
        t: integer;   { the index in *transform* for the first of these
                      values for a particular node }
        j: integer;   { counts off the pairs of values to be computed for that
                      node }
        m: integer;   { used as in *RecursiveFFT* to pick out the correct entry
                      from *omega* }

    begin
        { To initialize *transform*, evaluate polynomials of degree $1 = 2^l - 1$
        at square roots of 1. }
        for $t := 0$ **to** $n-2$ **by** 2 **do**
            transform$[t] := p_{\pi_k(t)} + p_{\pi_k(t+1)}$;
            transform$[t+1] := p_{\pi_k(t)} - p_{\pi_k(t+1)}$
        end { **for** t };

```
{ The main computation }
m := n/2;  num := 2;
{ Begin triply nested loop }
for l := k−2 to 0 by −1 do
    m := m/2;  num := 2*num;
    for t := 0 to (2^l −1)num by num do
    for j := 0 to (num/2)−1 do
        xPOdd := omega[mj] * transform[t+num/2+j];
        transform[t+num/2+j] := transform[t+j]−xPOdd;
        transform[t+j] := transform[t+j]+xPOdd
    end { for j }
    end { for t }
    { end of body of outer for loop }
    end { for l }
end { FFT }
```

An analysis of the number of operations done by *FFT* gives a result only slightly different from that obtained for *RecursiveFFT*. The statements that do the bulk of the computation (one complex multiplication, one complex addition, and one complex subtraction) are in a triply nested **for** loop. It is easy to verify that $num = 2^{k-l}$, so the ranges of the loop indexes indicate that the number of each operation done in these statements is

$$\sum_{l=0}^{k-2} 2^l \frac{num}{2} = \sum_{l=0}^{k-2} 2^l 2^{k-l-1} = \sum_{l=0}^{k-2} 2^{k-1}$$

$$= (k-1)2^{k-1}, \text{ or } \frac{n}{2}\lg\frac{n}{2}.$$

The first **for** loop, initializing *transform*, does $n/2$ additions and $n/2$ subtractions, so the total is $(3/2)n\lg(n/2)+n$ complex arithmetic operations. We claim that the permutation π_k can be computed easily enough so that the running time of the FFT is in $\Theta(n\lg n)$.

Note that the FFT allows us to evaluate a polynomial of degree $n-1$ at n distinct points at a cost of only $(n/2)\lg(n/2)$ complex multiplications. The lower bound on polynomial evaluation given in Section 7.2 suggests that this is not possible for n arbitrary points. The speed of the FFT derives from its use of some of the properties of roots of unity.

Now, what is π_k? Let t be an integer between 0 and $n-1$, where $n = 2^k$. Then t can be represented in binary by $[b_0 b_1 ... b_{k-1}]$, where each b_j is 0 or 1. Let $rev_k(t)$ be the number represented by these bits in reverse order, i.e., by $[b_{k-1} ... b_1 b_0]$. We claim that $\pi_k(t) = rev_k(t)$. Lemma 7.3 describes the values computed by the FFT using $\pi_k = rev_k$. It is used in Theorem 7.4 to establish the correctness of the algorithm, thus also establishing the correctness of this choice of π_k. The proof of the lemma follows the theorem.

Lemma 7.3 Let π_k in Algorithm 7.5 be rev_k. The following statements hold for $l = k-1$ before the triply nested **for** loop is first entered and for each l such that $k-2 \geq l \geq 0$ at the end of each execution of the body of the outer **for** loop.

1. $m = 2^l$ and $num = 2^{k-l}$.

2. For $t = r2^{k-l}$ where $0 \leq r \leq 2^l - 1$, $transform[t], \ldots, transform[t+num-1]$ contain the values of $P^{t,l}$ evaluated at the 2^{k-l}th roots of 1, where $P^{t,l}$ is the polynomial of degree $2^{k-l}-1$ with coefficients $c_j^{t,l} = p_{2^l j + rev_k(t)}$ for $0 \leq j \leq 2^{k-l}-1$.

Theorem 7.4 Algorithm 7.5 computes the values of

$$p(x) = p_0 + p_1 x + \cdots + p_{n-2} x^{n-2} + p_{n-1} x^{n-1}$$

at the nth roots of 1 — i.e., it computes the discrete Fourier transform of $P = (p_0, p_1, \ldots, p_{n-1})$.

Proof. Let $l = 0$ in Lemma 7.3. Then the only value for t is 0, and the lemma says that $transform[0], \ldots, transform[2^k-1]$ contain the values of $P^{0,0}$ at the 2^kth roots of 1. The coefficients of $P^{0,0}$ are $c_j^{0,0} = p_{2^0 j + rev(0)} = p_j$ for $0 \leq j \leq 2^k-1$, so $P^{0,0}$ is the polynomial p. $\square$

Proof of Lemma 7.3. We prove the lemma by induction on l, where l ranges from $k-1$ down to 0. Let $l = k-1$ for the basis. Statement 1 is clearly true. Statement 2 says that t ranges from 0 to 2^k-2 (i.e., $n-2$) in steps of 2, and that, for each t, $transform[t]$ and $transform[t+1]$ contain $P^{t,k-1}(1)$ and $P^{t,k-1}(-1)$, where the coefficients of $P^{t,k-1}$ are $c_0^{t,k-1} = p_{2^{k-1}0+rev(t)} = p_{rev(t)}$ and $c_1^{t,k-1} = p_{2^{k-1}+rev(t)} = p_{rev(t+1)}$ (see Lemma 7.5). That is, $P^{t,k-1}(x) = p_{rev(t)} + p_{rev(t+1)}x$. This corresponds exactly to the initial values assigned to $transform$ in the first **for** loop.

Now suppose that $0 \leq l < k-1$ and that statements 1 and 2 hold for $l+1$. It follows easily that statement 1 holds for l. Note that ω^{2^l} and $\omega^{2^{l+1}}$ are primitive 2^{k-l}th and $2^{k-(l+1)}$th roots of 1, respectively. Using the induction assumption, we see that for $0 \leq i \leq num/2-1$, the body of the triply nested **for** loop computes

$$xPOdd = (\omega^{2^l})^i P^{t+num/2,l+1}((\omega^{2^{l+1}})^i),$$

$$transform[t+num/2+i] = P^{t,l+1}((\omega^{2^{l+1}})^i) - (\omega^{2^l})^i P^{t+num/2,l+1}((\omega^{2^{l+1}})^i),$$

$$transform[t+i] = P^{t,l+1}((\omega^{2^{l+1}})^i) + (\omega^{2^l})^i P^{t+num/2,l+1}((\omega^{2^{l+1}})^i).$$

Thus $P^{t,l}(x) = P^{t,l+1}(x^2) + xP^{t+num/2,l+1}(x^2)$ and $transform[0], \ldots, transform[t+num-1]$ contain $P^{t,l}$ evaluated at

$$(\omega^{2^l})^0, \omega^{2^l}, \ldots, (\omega^{2^l})^{2^{k-l-1}}, -(\omega^{2^l})^0, -\omega^{2^l}, \ldots, -(\omega^{2^l})^{2^{k-l-1}},$$

i.e., at the 2^{k-l}th roots of 1. The coefficients of $P^{t,l}$ are derived as follows for $0 \leq j \leq 2^{k-l}-1$:

$$c_j^{t,l} = c_{j/2}^{t,l+1} \quad \text{for even } j$$

$$c_j^{t,l} = c_{(j-1)/2}^{t+num/2,l+1} \quad \text{for odd } j.$$

Therefore, using the induction hypothesis, for even j,

$$c_j^{t,l} = c_{j/2}^{t,l+1} = P_{2^{l+1}(j/2)+rev(t)} = P_{2^l j+rev(t)},$$

as required. For odd j,

$$c_j^{t,l} = c_{(j-1)/2}^{t+num/2,l+1}$$

$$= P_{2^{l+1}(j-1)/2+rev(t+2^{k-(l+1)})}$$

$$= \text{(by Lemma 7.5)} \ P_{2^l(j-1)+rev(t)+2^l} = P_{2^l j+rev(t)},$$

also as required.

□

The proof used the following lemma, whose proof is left as an exercise.

Lemma 7.5   For $k, a, b > 0$, $b \leq k$, and $a + 2^{k-b} < 2^k$, if a is a multiple of 2^{k-b+1}, then $rev_k(a + 2^{k-b}) = rev_k(a) + 2^{b-1}$.

7.4.3 Convolution

To motivate the convolution computation we will examine the problem of symbolic polynomial multiplication. Suppose that we are given the coefficient vectors

$$P = (p_0, p_1, \ldots, p_{m-1})$$

and

$$Q = (q_0, q_1, \ldots, q_{m-1})$$

for the polynomials $p(x) = p_{m-1}x^{m-1} + p_{m-2}x^{m-2} + \cdots + p_1x + p_0$ and $q(x) = q_{m-1}x^{m-1} + q_{m-2}x^{m-2} + \cdots + q_1x + q_0$. The problem is to find the vector $R = (r_0, r_1, \ldots, r_{2m-1})$ of coefficients of the product polynomial $r(x) = p(x)q(x)$. The coefficients of r are given by the formula

$$r_i = \sum_{j=0}^{i} p_j q_{i-j} \quad \text{for } 0 \leq i \leq 2m-1$$

with p_k and q_k taken as 0 for $k > m-1$. (Note that $r_{2m-1} = 0$ since r has degree $2m-2$; it is included as a convenience.) R is very much like the convolution of P and Q. Let $\bar{P}$ and $\bar{Q}$ be the 2m-vectors obtained by adding m zeroes to P and Q, respectively. Then $R = \bar{P} \otimes \bar{Q}$. Thus our investigation of polynomial multiplication should lead to a convolution algorithm.

Consider the following outline for polynomial multiplication.

1.   Evaluate $p(x)$ and $q(x)$ at 2m points: $x_0, x_1, \ldots, x_{2m-1}$.

2.   Multiply pointwise to find the values of $r(x)$ at these 2m points; i.e., compute $r(x_i) = p(x_i)q(x_i)$ for $0 \leq i \leq 2m-1$.

3. Find the coefficients of the unique polynomial of degree $2m-2$ that passes through the points $\{(x_i, r(x_i)) \mid 0 \le i \le 2m-1\}$. (It is a well-known theorem that the coefficients of a polynomial of degree d can be determined if the values of the polynomial are known at $d+1$ points.)

If the points $x_0, \ldots, x_{2m-1}$ were chosen arbitrarily, the method outlined would require much more work than a straightforward computation of $\bar{P} \otimes \bar{Q}$, but the FFT can evaluate p and q at the $(2m)$th roots of 1 very efficiently (assume that m is a power of 2). So step 1 can be done in $\Theta(m \lg m)$ time. Step 2 requires only $2m$ multiplications. How do we carry out step 3?

Let ω be a primitive $(2m)$th root of 1, and for $0 \le j \le 2m-1$, let $w_j = r(\omega^j) = p(\omega^j) q(\omega^j)$. The coefficients of r can be found by solving the following set of simultaneous linear equations for $r_0, r_1, \ldots, r_{2m-1}$.

$$r_0 + r_1 \omega^0 + \cdots + r_{2m-2}(\omega^0)^{2m-2} + r_{2m-1}(\omega^0)^{2m-1} = w_0$$
$$r_0 + r_1 \omega + \cdots + r_{2m-2}(\omega)^{2m-2} + r_{2m-1}(\omega)^{2m-1} = w_1$$
$$\vdots \tag{7.5}$$
$$r_0 + r_1 \omega^{2m-1} + \cdots + r_{2m-2}(\omega^{2m-1})^{2m-2} + r_{2m-1}(\omega^{2m-1})^{2m-1} = w_{2m-1}$$

If r had been evaluated at $2m$ arbitrary points, an $\Theta(m^3)$ algorithm such as Gaussian elimination might be used to solve the equations. Again, we take advantage of the fact that the points are roots of unity to obtain an $\Theta(m \lg m)$ algorithm. The formulas of Eq. 7.5 can be written as a matrix equation $F_{2m} R = W$, where W is the vector $(w_0, w_1, \ldots, w_{2m-1})$. Thus

$$\bar{P} \otimes \bar{Q} = R = F_{2m}^{-1} W = F_{2m}^{-1}(F_{2m} \bar{P} * F_{2m} \bar{Q}),$$

where $*$ denotes pointwise multiplication. Three problems remain: to show that F_n is in fact invertible for all $n > 0$, to show that the formula $U \otimes V = F_n^{-1}(F_n U * F_n V)$ holds for arbitrary n-vectors U and V, and to find an efficient way to compute the inverse transform. The formula for $U \otimes V$ does not follow immediately from the formula for R because $\bar{P}$ and $\bar{Q}$ have the property that half of their components are zero.

Lemma 7.6   For $n > 0$, F_n is invertible and the (i, j)th entry of its inverse is $(1/n) \omega^{-ij}$ for $0 \le i, j \le n-1$.

Proof.   Let $\tilde{F}_n$ be the matrix that the lemma claims is F_n^{-1}. We show that $F_n \tilde{F}_n = I$; $\tilde{F}_n F_n = I$ similarly.

$$(F_n \tilde{F}_n)_{ij} = \sum_{k=0}^{n-1} \omega^{ik} \frac{1}{n} \omega^{-kj} = \frac{1}{n} \sum_{k=0}^{n-1} (\omega^{i-j})^k.$$

For nondiagonal entries, i.e., for $i \ne j$, $(F_n \tilde{F}_n)_{ij} = 0$ by Property 7.1 for roots of unity since $0 < |i-j| < n$ (see the appendix at the end of this section). For diagonal entries, i.e., for $i = j$,

$$(F_n \tilde{F}_n)_{ij} = \frac{1}{n} \sum_{k=0}^{n-1} (\omega^0)^k = \frac{1}{n} \sum_{k=0}^{n-1} 1 = 1. \qquad \square$$

Theorem 7.7 Let U and V be n-vectors. Then $U \otimes V = F_n^{-1}(F_n U * F_n V)$, where $*$ denotes pointwise multiplication.

Proof. We show that $F_n(U \otimes V) = F_n U * F_n V$. For $0 \le i \le n-1$, the ith component of $F_n U * F_n V$ is

$$\left[\sum_{j=0}^{n-1} \omega^{ij} u_j \right] \left[\sum_{k=0}^{n-1} \omega^{ik} v_k \right] = \sum_{j=0}^{n-1} \sum_{k=0}^{n-1} u_j v_k \omega^{i(j+k)}.$$

The tth component of $U \otimes V$ is $\sum_{j=0}^{n-1} u_j v_{t-j}$, where subscripts are taken modulo n. Thus the ith component of $F_n(U \otimes V)$ is

$$\sum_{t=0}^{n-1} \left[\omega^{it} \sum_{j=0}^{n-1} u_j v_{t-j} \right] = \sum_{j=0}^{n-1} \sum_{t=0}^{n-1} u_j v_{t-j} \omega^{it}.$$

Let $k = t-j \pmod{n}$ in the inner summation. For each j, since t ranges from 0 to $n-1$, k will also range from 0 to $n-1$, although in a different order. Also, for any p, $\omega^p = \omega^{p \pmod{n}}$, so the ith component of $F_n(U \otimes V)$ is

$$\sum_{j=0}^{n-1} \sum_{k=0}^{n-1} u_j v_k \omega^{i(j+k)},$$

which is exactly the ith component of $F_n U * F_n V$. $\qquad \square$

Lemma 7.6 indicates that the matrix F_n^{-1} is not very much different from F_n. The entries of $n F_n^{-1}$ are ω^{-ij}. Its rows are the rows of F_n arranged in a different order. Specifically, since $\omega^{n-i} = \omega^{-i}$, for $1 \le i \le n-1$ the ith row of F_n is the $(n-i)$th row of $n F_n^{-1}$. Row 0 is the same for both matrices. Thus the inverse discrete Fourier transform of an n-vector A may be computed as follows.

Algorithm 7.6 Inverse Discrete Fourier Transform

Input: A, n, where A is an n-vector; n is a power of 2.
Output: The vector $B = (b_0, b_1, \ldots, b_{n-1})$, where $B = F_n^{-1} A$.

```
procedure InverseFT (A: ComplexArray; n: integer; var B: ComplexArray);
var
    i: integer;
    transform: ComplexArray;
begin
    FFT(A, n, transform);
    b₀ := transform[0]/n;
    for i := 1 to n−1 do
        bᵢ := transform[n−i]/n
    end { for }
end { InverseFT }
```

Analysis

The FFT does $(n/2)\lg(n/2)$ complex multiplications (and the same number of complex additions and subtractions), so Algorithm 7.6 does $(n/2)\lg(n/2)+n$ complex *'s, and both run in $\Theta(n\lg n)$ time. Computing the convolution of two n-vectors using the FFT takes $\Theta(n\lg n)$ time.

7.4.4 Appendix: Complex Numbers and Roots of Unity

C, the field of complex numbers, is obtained by joining i, the square root of -1, to the field of real numbers **R**. Thus $\mathbf{C} = \mathbf{R}(i) = \{a+bi \mid a, b \in \mathbf{R}\}$. If $z = a+bi$, a is called the real part of z and b the imaginary part. Let $z_1 = a_1 + b_1 i$ and $z_2 = a_2 + b_2 i$. Then by definition,

$$z_1 + z_2 = (a_1 + a_2) + (b_1 + b_2)i$$

and

$$z_1 z_2 = (a_1 a_2 - b_1 b_2) + (a_1 b_2 + b_1 a_2)i.$$

A complex number may be represented as a vector in a plane using the real and imaginary parts for the Cartesian coordinates. The geometric interpretation of multiplication of complex numbers is seen more easily by using polar coordinates, r and θ, where r is the length of the vector and θ is the angle (measured in radians) that it subtends with the horizontal, or real, axis. (See Fig. 7.7.) The product of two complex numbers (r_1, θ_1) and (r_2, θ_2) is $(r_1 r_2, \theta_1 + \theta_2)$. An example is given in Fig. 7.8.

C is an algebraically closed field, which means that every polynomial of degree n with coefficients in C has n roots. Therefore, $x^n - 1$ has n roots, which are called the *nth roots of unity*. The polar coordinates of 1 are $(1, 0)$. To find a root (r, θ) of $x^n - 1$ we solve the equation $(r^n, n\theta) = (1, 0)$. Since r is real and nonnegative, r must be 1, so all roots of unity are represented by vectors of unit length.

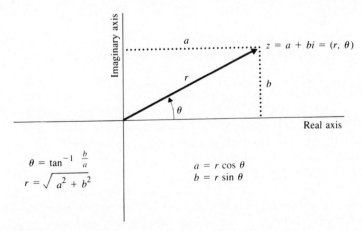

Figure 7.7 Cartesian and polar coordinates for complex numbers.

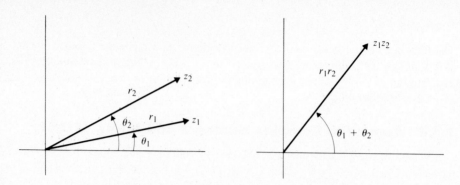

Figure 7.8   Multiplication of complex numbers.

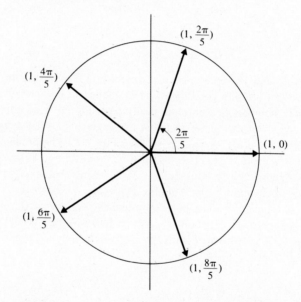

Figure 7.9   Fifth roots of unity (polar coordinates).

Since $n\theta = 0$, $\theta = 0$ so we have found that $(1,0)$ — i.e., 1 — is a solution, hardly a surprise. To find the other roots of unity, we use the fact that an angle of 0 radians is equivalent to an angle of $2\pi j$ radians for any integer j. The n distinct roots are $\{(1, 2\pi j/n) \mid 0 \leq j \leq n-1\}$. The vectors representing these numbers slice the unit circle into n equal pie slices as shown in Fig. 7.9.

If ω is an nth root of 1, then ω^k is also an nth root of 1, since $(\omega^k)^n = \omega^{nk} = (\omega^n)^k = 1^k = 1$. If ω is an nth root of 1 and $1, \omega, \omega^2, \ldots, \omega^{n-1}$ are all distinct, then ω is called a *primitive nth root of unity*. One primitive nth root of unity is $(1, 2\pi/n)$, or $\cos(2\pi/n) + i\sin(2\pi/n)$. The following properties are used in Section 7.4.

Property 7.1 For $n \geq 2$, the sum of all the nth roots of 1 is zero. Also, if ω is a primitive nth root of 1 and n does not divide c, then $\sum_{j=0}^{n-1}(\omega^c)^j = 0$.

Proof. Let ω be a primitive nth root of 1. Then $\omega^0, \omega, \omega^2, \ldots, \omega^{n-1}$ are all of the nth roots of 1. Their sum is

$$\sum_{j=0}^{n-1}\omega^j = \frac{\omega^n-1}{\omega-1} = \frac{1-1}{\omega-1} = 0.$$

The second statement is proved similarly.   □

Property 7.2 If n is even and ω is a primitive nth root of 1, then

1. ω^2 is a primitive $(n/2)$th root of 1, and

2. $\omega^{n/2} = -1$.

Proof. The proof is left as an exercise.

Exercises

Section 7.2: Evaluating Polynomial Functions

7.1. Any   polynomial $p(x)=a_n x^n + a_{n-1}x^{n-1} + \cdots + a_1 x + a_0$ may   be   factored: $a_n(x-r_1)(x-r_2)\cdots(x-r_n)$, where $r_1, \ldots, r_n$ are the roots of p. Could this factorization be used as the basis of an algorithm to evaluate $p(x)$? How, or why not?

7.2. We claimed that an algorithm to evaluate a polynomial of degree n must do at least n multiplications and/or divisions in the worst case. For special cases, we may get algorithms that do better. Devise a fast algorithm for evaluating each of the following polynomials. The inputs are n and x.

a) $p(x) = x^n + x^{n-1} + \cdots + x + 1$.

How many arithmetic operations does your algorithm do?

b) $p(x) = \sum_{k=0}^{n} \binom{n}{k} x^k$,

where $\binom{n}{k}$ are the binomial coefficients. How many arithmetic operations does your algorithm do?

7.3. Write   out   the   factorization   that   would   be   used   to   evaluate $p(x) = x^7 + 6x^6 - 7x^5 + 12x^4 + 2x^2 - 3x - 8$ by

a)   Horner's method.

b)   preprocessing coefficients.

7.4. What part(s) of the proofs of Theorem 7.1 and/or Lemma 7.2 would not work if division were permitted?

7.5. What modifications or additions must be made to the procedure for evaluating polynomials with preprocessing of coefficients so that it will work for nonmonic polynomials? How many multiplications and/or divisions are done by the extended algorithm?

7.6. Suppose that $A(1) = 1$ and for $k > 1$, $A(k) = 2A(k-1)+2$. Show that $A(k) = (3n-1)/2$, where $n = 2^k - 1$.

7.7. Using the terminology of the first paragraph of Section 7.2.3, what are I, I', A_I, and the algorithm that gives A_I from I for the problem of evaluating a polynomial with preprocessing of coefficients by the method described in that section?

Section 7.3: Vector and Matrix Multiplication

7.8. In Section 7.3.1, we stated that computing the dot product $U \cdot V$ of two n-vectors with real components requires at least n */'s. How many */'s are required if U always has integer components?

7.9. Compute the exact number of multiplications and additions done by Algorithm 7.3 when n is odd.

7.10. Let A and B be $n \times n$ matrices that are to be multiplied and suppose that a matrix entry must be fetched from storage each time it is used in the computation. How many times is each entry of A and B fetched to compute AB

   a)   by the usual algorithm?

   b)   by Winograd's algorithm (for n even)?

*7.11. Prove that Strassen's algorithm, using the fourth modification described toward the end of Section 7.3.4, does fewer than $4.7n^{\lg 7}$ arithmetic operations on the matrix entries, whether or not n is a power of 2.

Section 7.4: The Fast Fourier Transform and Convolution

7.12. a)   Why are the restrictions "$n \geq 2$" and "n does not divide c" needed in Property 7.1?

   b)   Prove Property 7.2.

7.13. Let $p(x) = p_7 x^7 + p_6 x^6 + \cdots + p_1 x + p_0$. Carry out the steps of the FFT on p to show how it evaluates p at the 8th roots of 1: $1, \omega, i, i\omega, -1, -\omega, -i, -i\omega$.

7.14. Suppose that you are given the real and imaginary parts of two complex numbers. Show that the real and imaginary parts of their product can be computed using only three multiplications.

7.15. Prove Lemma 7.5.

*7.16. Let $n = 2^k$ for some $k > 0$, let ω be a primitive nth root of 1, and let V be an n-vector. This problem describes the FFT (recursively) from the point of view of the matrix-vector product $F_n V$ rather than as polynomial evaluation. The reader should note the correspondence of various steps of this algorithm with steps of *RecursiveFFT*.

Let $\tilde{F}_n$ be the $n \times n$ matrix obtained from F_n by putting all the even-indexed columns before the odd-indexed columns, and let $\tilde{V}$ have all the even-indexed components of V precede all the odd-indexed components; i.e., for $0 \leq j \leq n/2 - 1$, $\tilde{f}_{ij} = \omega^{i2j}$, $\tilde{f}_{i,j+n/2} = \omega^{i(2j+1)}$, $\tilde{v}_j = v_{2j}$, and $\tilde{v}_{j+n/2} = v_{2j+1}$. Partition $\tilde{F}_n$ into four $(n/2) \times (n/2)$ matrices F_1, F_2, F_3, and F_4, and partition $\tilde{V}$ into two $(n/2)$-vectors V_1 and V_2 as shown in Fig. 7.10. Now

$$\tilde{F}_n \tilde{V} = \begin{bmatrix} F_1 V_1 + F_2 V_2 \\ F_3 V_1 + F_4 V_2 \end{bmatrix}$$

Prove the following statements.

   a) $F_n V = \tilde{F}_n \tilde{V}$.

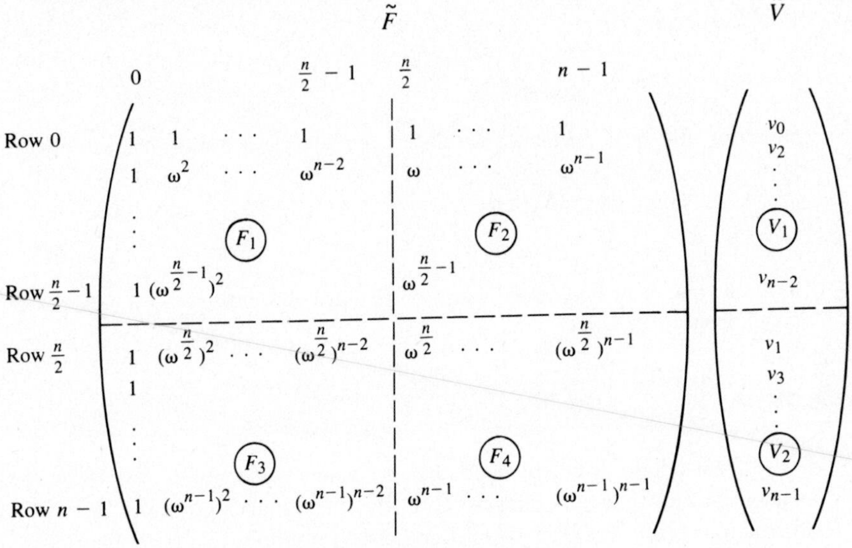

$$\tilde{F}$$

$$\tilde{V}$$

Figure 7.10

b) $F_1 = F_3$.

c) $F_4 = -F_2$.

d) $F_2 = DF_1$, where D is an $(n/2) \times (n/2)$ diagonal matrix with $d_{ii} = \omega^i$ for $0 \leq i \leq (n/2) - 1$.

e) F_1 has entries $f_{ij} = x^{ij}$, where x is a primitive $(n/2)$th root of 1. Thus

$$F_n V = \begin{vmatrix} F_1 V_1 + DF_1 V_2 \\ F_1 V_1 - DF_1 V_2 \end{vmatrix}$$

and the computation can be carried out by recursively computing the discrete Fourier transform of V_1 and V_2, both $(n/2)$-vectors.

f)   Derive recurrence relations for the number of multiplications, additions, and subtractions that would be done by the algorithm described here. Let $\bar{D} = (1, \omega, \ldots, \omega^{(n/2)-1})$. (Note that the product $D(F_1 V_2)$ can be computed as a component-wise product of $\bar{D}$ with $F_1 V_2$, requiring $n/2$ multiplications.) Compare your recurrence relations with those obtained for *RecursiveFFT*.

Additional Problems

7.17. Observe that the Fibonacci numbers satisfy the following matrix equation for $n \geq 2$.

$$\begin{bmatrix} F_n \\ F_{n-1} \end{bmatrix} = \begin{bmatrix} 1 & 1 \\ 1 & 0 \end{bmatrix} \begin{bmatrix} F_{n-1} \\ F_{n-2} \end{bmatrix}$$

Let

$$A = \begin{bmatrix} 1 & 1 \\ 1 & 0 \end{bmatrix}.$$

Then

$$\begin{bmatrix} F_n \\ F_{n-1} \end{bmatrix} = A \begin{bmatrix} F_{n-1} \\ F_{n-2} \end{bmatrix} = A^2 \begin{bmatrix} F_{n-2} \\ F_{n-3} \end{bmatrix} = \cdots = A^{n-1} \begin{bmatrix} F_1 \\ F_0 \end{bmatrix} = A^{n-1} \begin{bmatrix} 1 \\ 0 \end{bmatrix}.$$

How many arithmetic operations are done if F_n is computed using the following formula?

$$\begin{bmatrix} F_n \\ F_{n-1} \end{bmatrix} = A^{n-1} \begin{bmatrix} 1 \\ 0 \end{bmatrix}$$

How does this method compare with the recursive and iterative algorithms for computing Fibonacci numbers?

Programs

1. Write and debug efficient assembly language subroutines for Winograd's matrix multiplication algorithm and for the usual algorithm. How many instructions are executed by each program to multiply two $n \times n$ matrices? (If instruction timing tables for your computer are available, compute the actual time required by each.)

2. Implement the FFT (Algorithm 7.5). Make the computation of rev_k and the other bookkeeping as efficient as you can.

Notes and References

The lower bounds given in Sections 7.2 and 7.3 for polynomial evaluation, with or without preprocessing of coefficients, and for vector-matrix products are established in Reingold and Stocks (1972) and in Winograd (1970). Winograd's matrix multiplication algorithm also appears in the latter. Winograd's proofs use field theory. Reingold and Stocks use simpler arguments such as that in the proof of Theorem 7.1.

Strassen's matrix multiplication algorithm was presented in Strassen (1969), a short paper that gives no indication of how he discovered his formulas. The $O(n^{2.376})$ method is in Coppersmith and Winograd (1987). Several matrix problems that can be reduced to multiplication and therefore have $O(n^{2.376})$ solutions are described in Aho, Hopcroft, and Ullman (1974).

Versions of the fast Fourier transform are presented in Cooley and Tukey (1965) and in Aho, Hopcroft, and Ullman (1974). Gentleman and Sande (1966) discuss applications of the FFT and some special implementation situations, e.g., when not all the data fit in main memory. Brigham (1974) is a book on the FFT. Aho, Hopcroft, and Ullman (1974) present an application of the FFT to integer multiplication. (The string of digits $d_n d_{n-1} \ldots d_1 d_0$ representing an integer in base b is a polynomial $\sum_{i=0}^{n} d_i b^i$.) There are many other references on the FFT since it is used widely.

8

Transitive Closure, Boolean Matrices, and Equivalence Relations

8.1
The Transitive Closure of a
Binary Relation

8.1.1   Definitions and Background

Let S be a finite set with elements $s_1, s_2, \ldots, s_n$. A *binary relation* on S is a subset, say A, of $S \times S$. If $(s_i, s_j) \in A$, we say that s_i is A-related to s_j and use the notation $s_i A s_j$. The relation A can be represented by an $n \times n$ matrix with entries

$$a_{ij} = \begin{cases} 1 & \text{if } s_i A s_j \\ 0 & \text{otherwise.} \end{cases}$$

The adjacency relation on the set of vertices of a graph or digraph, used extensively in Chapter 4, is an important example of a relation. Other common examples are equivalence relations and partial orders. We will use the same (capital) letter to denote a relation and its matrix representation (which assumes a particular ordering on the elements of the underlying set), and the corresponding lowercase letters for the matrix entries. Unless otherwise stated, we assume that the set in question is $S = \{s_1, \ldots, s_n\}$.

A relation A on S is *transitive* if and only if for all i, j, and k between 1 and n, $s_i A s_j$ and $s_j A s_k$ implies $s_i A s_k$. Equivalence relations and partial orders are transitive relations; usually the adjacency relation for a graph or digraph is not. The *transitive closure* of a relation A on S is the relation R defined by $s_i R s_j$ if and only if for some $m \geq 2$, there exist $s_{k_1}, s_{k_2}, \ldots, s_{k_m}$ in S such that $s_i = s_{k_1}$, $s_{k_t} A s_{k_{t+1}}$ for $1 \leq t \leq m-1$ and $s_{k_m} = s_j$. The transitive closure of a transitive relation A is the relation A itself. The transitive closure of the adjacency relation for a graph or digraph with vertex set V is the *reachability relation* defined by $v R w$ if and only if there is a path (of length > 0) from v to w; i.e., if w is reachable from v by following edges.

In Sections 8.1 through 8.4 we study a variety of methods for finding the transitive closure of a relation. The application to graphs and digraphs is useful, and, in fact, if we extend the definition of a digraph to allow a vertex to be adjacent to itself, any binary relation A on a finite set S can be interpreted as the digraph $G = (S, A)$. Thus, with the extended definition assumed throughout these sections, the problems of finding the reachability relation, or matrix, for a digraph and finding the transitive closure of a binary relation are equivalent. We will use the terminology of whichever problem seems to motivate the particular algorithm studied. The form in which the input is given, however, will depend on how the problem arises in a particular application.

We assume throughout that $|S| = n$ and $|A| = m$.

8.1.2   Finding the Reachability Matrix by
Depth-first Search

A fairly obvious way to construct R, the reachability matrix for a digraph $G = (S, A)$, is to do a depth-first search (see Section 4.4) from each vertex to find all other vertices that can be reached from it. Initially R would be the zero matrix. Visiting, or processing, a vertex s_j encountered in the depth-first search from s_i would consist of assigning 1 to r_{ij}. Thus each depth-first search fills one row of R. This may seem overly simpleminded and inefficient since during a depth-first search from, say, s_i, entries may be made in rows other than the ith row; specifically, when a vertex s_j is encountered, r_{kj} may be assigned 1 for all k such that s_k is on the path from s_i to s_j. These vertices s_k may be found on the stack. How significant is this modification? Does it eliminate the need to do a depth-first search from s_k? How does it affect the amount of work done in the worst case?

Since depth-first search was illustrated by many examples in Chapter 4, we will not work out the details of an algorithm here, but simply comment on the amount of work done. If the adjacency list structure described in Chapter 4 is used for G and a depth-first search is done for each vertex, the worst-case running time will be in $\Theta(nm)$. Inserting 1's in more than one row of R during each depth-first search, as just suggested, can improve the algorithm's behavior for many graphs, but the worst case will still be in $\Theta(nm)$. (See Exercise 8.2.)

In Chapter 4 we defined the condensation of a digraph. Informally, the condensation is the digraph obtained by collapsing each strongly connected component to a single point; it is acyclic. We mentioned that some problems can be simplified by working with the condensation instead of the original digraph. The reachability relation for a digraph $G = (S, A)$ can be computed as follows:

1. Find the strong components of G (in $\Theta(n+m)$ time). Let G' be the condensation of G.

2. Find the reachability relation for G'. (Any of the methods presented in this chapter can be used, but note that the depth-first search method can be simplified somewhat if the input is known to be acyclic.)

3. Expand the reachability relation for G' by replacing each vertex of G' by all the vertices in G that were collapsed to it ($O(n^2)$ time).

The amount of work done at step 2 and hence by this method as a whole depends on the particular digraph G. If G has several large components, reduction to G' may save a lot of time.

Efficient depth-first search uses linked adjacency lists. In the next section we present a fairly simple $\Theta(n^3)$ algorithm for finding the reachability matrix without using any extra data structures.

8.2
Warshall's Algorithm

8.2.1 Warshall's Algorithm for Transitive Closure

If we interpret a binary relation A on a finite set S as a digraph, then finding elements of R, the transitive closure of the relation, corresponds to inserting edges in the digraph. In particular, for any pair of edges $s_i s_k$ and $s_k s_j$ inserted so far, we add the edge $s_i s_j$. That is, we can conclude that $s_i R s_j$ if we already know that, for some k, $s_i R s_k$ and $s_k R s_j$. Thus it should be easy to see that the following algorithm computes R.

Algorithm 8.1   Transitive Closure

Input: A and n, where A is an $n \times n$ matrix that represents a binary relation.

Output: R, the matrix for the transitive closure of A.

```
procedure TransitiveClosure (A: Matrix; n: integer; var R: Matrix);
var
    i, j, k : integer;
begin
    R := A;  { i.e., copy A into R }
    repeat
        for i := 1 to n do
        for j := 1 to n do
        for k := 1 to n do
            if r_ik = 1 and r_kj = 1 then r_ij := 1 end { if }
        end end end { triple for loop }
    until no entry of R changed during one complete pass
end { TransitiveClosure }
```

Figure 8.1 illustrates that the **repeat** loop could not be omitted. When a particular s_i and s_j are first considered, there may be no s_k that joins them. Later in the processing, because of the insertion of other edges, we may be able to insert $s_i s_j$ and hence must reconsider it. The complexity of Algorithm 8.1 is proportional to n^3 times the number of repetitions of the triple **for** loop. Investigating this number is left as an exercise, since we will revise the algorithm to reduce the amount of work done.

Suppose that we refer to the work done in the **if** statement as processing the triple (k, i, j). In Fig. 8.1, if the triple (k', i, k) were processed before (k, i, j), then none would have to be considered twice. Is there some order that eliminates the need for processing any triple more than once? Or, no matter what order we try, can we find an example in which repetition is required? We suggest that the reader try to answer these questions before proceeding.

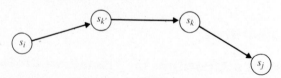

Figure 8.1   Suppose that $i < k'$ and that $j < k$. Then (k, i, j) and (k', i, j) are processed before (k', i, k) and (k, k', j), so (k, i, j) or (k', i, j) must be reprocessed.

Warshall's algorithm is simply an algorithm that processes the triples in the correct order — specifically, with k varying in the outermost loop. A proof of correctness follows the algorithm.

Algorithm 8.2   Warshall's Transitive Closure Algorithm

Input: A and n, where A is an $n \times n$ matrix that represents a binary relation.

Output: R, the $n \times n$ matrix for the transitive closure of A.

```
procedure WarshallTransitiveClosure (A: Matrix; n: integer; var R: Matrix);
var
    i, j, k : integer;
begin
    R := A;  { i.e., copy A into R }
    for k := 1 to n do
    for i := 1 to n do
    for j := 1 to n do
        if r_ik = 1 and r_kj = 1 then r_ij := 1 end { if }
    end end end { triple for loop }
end { WarshallTransitiveClosure }
```

Clearly the total number of triples processed is n^3. Initializing R takes $\Theta(n^2)$ time, so the number of matrix entries examined and/or changed for any input is in $\Theta(n^3)$.

Theorem 8.1   When Algorithm 8.2 terminates, R is the matrix representing the transitive closure of A.

Proof.   For each k, let $r_{ij}^{(k)}$ be the value of r_{ij} after the body of the "**for** k" loop is executed for the kth time. Let $S = \{s_1, s_2, \ldots, s_n\}$ and for $0 \le k \le n$, define $S_k = \{s_1, \ldots, s_k\}$. We claim that $r_{ij}^{(k)} = 1$ if and only if there is a path (of length > 0) from s_i to s_j using only vertices in S_k (besides s_i and s_j, which are at the ends of the path). This characterization of $r_{ij}^{(k)}$, which clearly proves the theorem, will be established by induction on the triples (k, i, j) in lexicographical order, the same order in which the $r_{ij}^{(k)}$ are computed.

For the basis of the induction we establish the claim for $k = 0$ and all i and j, i.e., for when the loop has not yet been executed but the initialization, $R := A$, has been done. Then $r_{ij}^{(0)} = a_{ij} = 1$ if and only if there is an edge from s_i to s_j, which is equivalent to the characterization of $r_{ij}^{(0)}$ since $S_0 = \emptyset$. Now suppose that $k, i, j > 0$ and the characterization holds for all triples that precede (k, i, j). Observe that when the triple (k, i, j) is processed, the superscripts on r_{ik} and r_{kj} may be (k) or $(k-1)$ depending on the relative order of the indexes. If $k \ge j$ then $r_{ik} = r_{ik}^{(k-1)}$ since r_{ik} has not yet been changed during the kth pass. The triple for $r_{ik}^{(k-1)}$, i.e., $(k-1, i, k)$, precedes (k, i, j). Similarly, if $k \ge i$, then $r_{kj} = r_{kj}^{(k-1)}$ and its triple, $(k-1, k, j)$, precedes (k, i, j). If $k < j$, then $r_{ik} = r_{ik}^{(k)}$, and its triple, (k, i, k), precedes (k, i, j); similarly for $k < i$. So the induction hypothesis may be applied to r_{ik} and r_{kj} in the **if** statement. Therefore $r_{ij}^{(k)} = 1$ if and only if $r_{ij}^{(k-1)} = 1$ (and hence is not changed in the **if** statement) or both $r_{ik}^{(p)} = 1$ and $r_{kj}^{(p')} = 1$ where $p, p' \in \{k{-}1, k\}$; that is, if and only if there is a path from s_i to s_j using only vertices in S_{k-1} or S_k (depending on p and p'). This, finally, is equivalent to there being a path from s_i to s_j using only vertices in S_k. This establishes the induction hypothesis for (k, i, j) and proves the theorem. $\square$

8.2.2 An Extension of Warshall's Algorithm

For those who have read Section 6.4, Warshall's algorithm should look somewhat familiar. There we used a dynamic programming approach to solve the following problem:

Given a weighted graph or digraph $G = (V, E, W)$ with $V = \{v_1, \ldots, v_n\}$, represented by the weight matrix with entries

$$w_{ij} = \begin{cases} W(v_i v_j) & \text{if } v_i v_j \in E \\ \infty & \text{if } i \neq j \text{ and } v_i v_j \notin E \\ 0 & \text{if } i = j, \end{cases}$$

compute the $n \times n$ matrix D defined by $d_{ij} = $ the distance from v_i to v_j. (The distance is the weight of a minimum weight path.)

The dynamic programming solution to the all-pairs distances problem is

Algorithm 8.3 (Algorithm 6.5) Distances in a Graph or Digraph

Input: W, the weight matrix for a graph or digraph with vertices $v_1, \ldots, v_n$; and n.
Output: D, an $n \times n$ matrix such that $d_{ij} = $ the distance from v_i to v_j.

```
procedure Distances (W: Matrix; n: integer; var D: Matrix);
var
    i, j, k : integer;
begin
    D := W;  { i.e., copy W into D }
    for k := 1 to n do
    for i := 1 to n do
    for j := 1 to n do
        d_ij := min(d_ij, d_ik + d_kj)
    end  end  end  { for loops }
end { Distances }
```

The all-pairs distances problem is more general than the problem of finding R, the reachability matrix; R can be obtained from D simply by changing all entries less than ∞ to 1's and all ∞'s to 0's. Thus the dynamic programming solution for computing D is a generalization of Warshall's algorithm. For D, processing the triple (k, i, j) means computing $d_{ij} := \min(d_{ij}, d_{ik} + d_{kj})$. Here too, the order in which triples are processed is critical to getting the correct result without repeated processing.

8.2.3   Warshall's Algorithm for Bit Matrices

If the matrices A and R are stored with one entry per bit, then Warshall's algorithm has the following fast implementation using the logical **or** (or Boolean sum, or union) instruction available on most large general-purpose computers. The **or** instruction will be denoted by $\vee$ and is defined bitwise by: $1 \vee 0 = 0 \vee 1 = 1 \vee 1 = 1$ and $0 \vee 0 = 0$.

Algorithm 8.4   Warshall's Algorithm for Bit Matrices

Input: A and n as in Algorithm 8.2, but A is a bit matrix.

Output: R, the transitive closure of A, also as a bit matrix.

Notation: Let R_i be the ith row of R for $1 \le i \le n$.

> **procedure** *WarshallBitMatrices* (*A: BitMatrix; n: integer;* **var** *R: BitMatrix*);
> **var**
>   *i, k : integer*;
> **begin**
> $R := A$; { i.e., copy A into R }
>   **for** $k := 1$ **to** n **do**
>   **for** $i := 1$ **to** n **do**
>   **if** $r_{ik} = 1$ **then** $R_i := R_i \vee R_k$ **end** { if }
>   **end end** { for loops }
> **end** { WarshallBitMatrices }

At most n^2 logical **or**'s are done on rows of R. However, a row may not fit in one memory word and more than one **or** instruction may be needed in the **if** statement. (On some computers one machine instruction will compute the Boolean sum of two long bit strings, say, up to 256 bytes — i.e., 2048 bits — though the time required to execute the instruction depends on the length of the operands.) The number of **or**'s required for each row is $\lceil n/c \rceil$, where c is the word size (or the size of the operand of the Boolean **or** instruction), so Algorithm 8.4 does $\lceil n^3/c \rceil$ Boolean **or** instructions in the worst case. The complexity is in $\Theta(n^3)$, but the constant multiple of n^3 is small.

8.3
Computing Transitive Closure by Matrix Operations

Suppose that A is the matrix for a binary relation on $S = \{s_1, \ldots, s_n\}$ and that we interpret A as the adjacency relation on the digraph (S, A). Then $a_{ij} = 1$ if and only if there is a path of length one from s_i to s_j since a path of length one is an edge. Suppose we define matrices $A^{(p)}$ by

$$a_{ij}^{(p)} = \begin{cases} 1 & \text{if there is a path of length } p \text{ from } s_i \text{ to } s_j \\ 0 & \text{otherwise.} \end{cases}$$

Then $A^{(0)} = I$, the identity matrix, and $A^{(1)} = A$. How can we compute $A^{(2)}$? By definition, $a_{ij}^{(2)} = 1$ if and only if there is a path of length two from s_i to s_j, hence if and only if there is a vertex s_k such that $a_{ik} = 1$ and $a_{kj} = 1$. Thus

$$a_{ij}^{(2)} = \bigvee_{k=1}^{n} (a_{ik} \wedge a_{kj}),$$

where $\vee$ is the Boolean **or** and $\wedge$ is the Boolean **and**, or product: $1 \wedge 1 = 1$ and $1 \wedge 0 = 0 \wedge 1 = 0 \wedge 0 = 0$. The formula for $a_{ij}^{(2)}$ is the formula for an entry in the product of A with itself *as a Boolean matrix*. The product $C = AB$ of $n \times n$ Boolean matrices A and B (matrices whose entries are all 1's and 0's) is by definition the Boolean matrix with entries

$$c_{ij} = \bigvee_{k=1}^{n} (a_{ik} \wedge b_{kj}) \quad \text{for } 1 \leq i, j \leq n.$$

For the remainder of this section all matrix products will be Boolean, so we may denote $A^{(2)}$ as A^2. The *Boolean matrix sum* of A and B, $D = A + B$, is defined by $d_{ij} = a_{ij} \vee b_{ij}$ for $1 \leq i, j \leq n$.

We have concluded that A^2 indicates which vertices are connected by paths of length 2. It is easy to generalize and prove the following lemma by induction on p. The proof is left as an exercise.

Lemma 8.2 Let A be the adjacency matrix for a digraph with vertices $\{s_1, \ldots, s_n\}$. Denote elements of A^p, for $p \geq 0$, by $a_{ij}^{(p)}$. Then $a_{ij}^{(p)} = 1$ if and only if there is a path of length p from s_i to s_j.

The entries of R are defined by $r_{ij} = 1$ if and only if there is a path of *any* length (greater than 0) from s_i to s_j. Observe that for any p and p', the (i, j)th entry of the matrix $A^p + A^{p'}$ is 1 if and only if there is a path of length p or a path of length p' from s_i to s_j. Thus $R = \sum_{p=1}^{\infty} A^p$. ($R$ is often denoted A^+ since it is the Boolean sum of the powers of A.) This formula for R is not useful for computation since it indicates a computation of infinitely many matrices, but the problem is easily handled by the next lemma, the proof of which is also left to the reader.

Lemma 8.3 In a digraph with n vertices, if there is a path from vertex v to vertex w, then there is a path from v to w of length at most n.

Thus $R = \sum_{p=1}^{s} A^p$ for any $s \geq n$. The amount of work needed to compute R by this formula is still large, since s powers of the matrix A are computed. We will carry out some algebraic manipulations on the formula for R to put it in a form that suggests a more efficient computation. Some of the following properties of Boolean matrix operations will be useful. Assume that A, B, and C are $n \times n$.

$$A+B = B+A \qquad\qquad A+A = A$$
$$A+(B+C) = (A+B)+C \qquad A(BC) = (AB)C$$
$$A(B+C) = (AB)+(AC) \qquad (B+C)A = (BA)+(CA)$$
$$IA = AI = A, \text{ where } I \text{ is the } n\times n \text{ identity matrix.}$$

Now let $s \geq n$. Then

$$R = \sum_{p=1}^{s} A^p = A + A^2 + \cdots + A^s = A(I + A + \cdots + A^{s-1}).$$

Lemma 8.4 $I + A + A^2 + \cdots + A^{s-1} = (I+A)^{s-1}$ for $s \geq 1$.

Proof, by induction on s.   For $s = 1$, both sides are equal to I. Assume that $s > 1$ and that the equality holds for $s-1$, i.e., $I + A + \cdots + A^{s-2} = (I+A)^{s-2}$. Then

$$(I+A)^{s-1} = (I+A)^{s-2}(I+A) = (I + A + \cdots + A^{s-2})(I+A)$$
$$= I + A + A + A^2 + A^2 + \cdots + A^{s-2} + A^{s-2} + A^{s-1}$$
$$= I + A + A^2 + \cdots + A^{s-1}. \qquad\qquad\qquad \square$$

Theorem 8.5   Let A be an $n\times n$ Boolean matrix representing a binary relation. Then R, the matrix for the transitive closure of A, is $A(I+A)^{s-1}$ for any $s \geq n$.

How much work is needed to compute R using the formula of Theorem 8.5? Computing $I+A$ requires inserting 1's in the diagonal of A, $\Theta(n)$ operations. Since s may be any integer at least as large as n, we choose s so that $s-1$ is a power of 2, in particular $2^{\lceil \lg(n-1) \rceil}$. Then $(I+A)^{s-1}$ can be computed by doing $\lceil \lg(n-1) \rceil$ matrix multiplications, and R can be computed by a total of $\lceil \lg(n-1) \rceil + 1$ matrix multiplications. Boolean matrix products can be computed as indicated by the definition in $\Theta(n^3)$ time, or by using an $o(n^3)$ matrix multiplication algorithm (e.g., Strassen's), and then replacing all positive entries by 1's. Thus R can be computed in $o(n^3 \lg n)$ time. In fact, it can be computed (asymptotically) faster. Note that in the transitive closure of a relation, $r_{ii} = 1$ if and only if there is a path of length greater than zero from s_i to itself. The reflexive transitive closure, $A^* = I + A^+$ (recall that A^+ is R), allows paths of length zero. It is known that the complexity of computing A^* and the complexity of Boolean matrix multiplication are of the *same* order. (See the notes and references at the end of the chapter.) Since A^+ can be computed from A^* by one more matrix multiplication, the complexity of computing transitive closure and the complexity of Boolean matrix multiplication are also of the same order. None of the algorithms we have examined for transitive closure is as fast as matrix multiplication. The algorithm for transitive closure that is of the same order as matrix multiplication is recursive and involves repeated partitioning of matrices. It is similar in style to Strassen's algorithm, and it has a lot of overhead.

Multiplying Boolean matrices is a more specialized problem than multiplying matrices with real entries, and it is worth seeking specialized algorithms. In the next section we develop a fast Boolean matrix multiplication algorithm for bit matrices.

8.4
Multiplying Bit Matrices —
Kronrod's Algorithm

8.4.1   Introduction

Let A and B be $n \times n$ Boolean matrices stored with one entry per bit. Using the logical **or** instruction, their product $C = AB$ can be computed as follows, where C_i and B_k are the ith row of C and kth row of B, respectively, and initially all entries of C are 0:

> **for** $i := 1$ **to** n **do**
> **for** $k := 1$ **to** n **do**
>   **if** $a_{ik} = 1$ **then** $C_i := C_i$ **or** B_k **end** { if }
> **end   end**

(Compare this to Algorithm 8.3; notice how similar they are, although they compute different things.) We may think of the logical **or** operation as performing a union of sets. That is, if we view, say, A_i (the ith row of A) as the set $\{k \mid a_{ik} = 1\}$ (a subset of $\{1, 2, \ldots, n\}$), and similarly view rows of B and C, then $C_i = \bigcup_{k \in A_i} B_k$. The algorithm above does at most n^2 row unions (each of which may require several logical **or** machine instructions). We will derive an algorithm that does fewer row unions. This algorithm is sometimes referred to as the Four Russians' algorithm, although apparently it is the work of M. A. Kronrod, one of the four.

8.4.2   Kronrod's Algorithm

Certain groups of rows of B may appear in the unions for several different rows of C. For example, suppose that A is as shown in Fig. 8.2. Then $B_1 \cup B_3 \cup B_4$ is contained in rows 1, 3, and 7 of the product, and nine unions are done where fewer would suffice. How can some or all of the duplicated work be reduced? The approach that suggests itself is to first compute a lot of unions of small numbers of

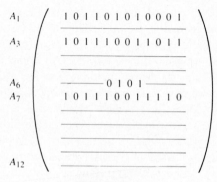

Figure 8.2

rows of B (such as $B_1 \cup B_3 \cup B_4$), and then to combine these unions appropriately to obtain the rows of the product. Several questions come to mind immediately: How many and which rows of B should be combined in the first step? How can these unions be stored so that they can be accessed efficiently during the second step? How much additional storage is needed? Will any time really be saved in the worst case? If so, how much?

The answers to most of the questions depend on the answer to the first. We adopt a straightforward strategy: Divide up the rows of B into several groups of t rows each and compute all possible unions within each group. We will ignore all implementation details until we see whether, with an appropriate choice for t, this strategy can produce an algorithm that does fewer than n^2 row unions. The rows of B are grouped as follows:

Group 1: $B_1, \ldots, B_t$
Group 2: $B_{t+1}, \ldots, B_{2t}$
$\vdots$

Group $\lceil n/t \rceil$: $B_p, \ldots, B_n$, where $p = (\lceil n/t \rceil - 1)t + 1$.

Suppose, for example, that matrix A in Fig. 8.2 is to be multiplied with a 12×12 matrix B; let $t = 4$. Unions of all combinations of rows B_1, B_2, B_3, and B_4 would be computed once. If done in the right order (first all combinations of two rows, then three, and lastly all four), all the unions can be obtained by doing 11 row union operations. The same would be done for the groups $B_5, \ldots, B_8$ and $B_9, \ldots, B_{12}$. Then, to obtain the first row of AB, only two more row union operations would be needed to compute $(B_1 \cup B_3 \cup B_4) \cup (B_6 \cup B_8) \cup (B_{12})$. $B_1 \cup B_3 \cup B_4$ is used again in the third and seventh rows of the product, and $B_6 \cup B_8$ is used again in the sixth row.

We will roughly estimate the total number of unions done as a function of t, and then see if we can choose a value for t that gives a total lower than n^2. For each group of rows (except perhaps the last), there are 2^t sets of rows to be combined. No unions are needed to compute the empty set or the sets consisting of only one row. Since each union of rows within a group can be computed by combining sets already computed with one more row, a total of $2^t - 1 - t$ union operations are done for each group. There are $\lceil n/t \rceil$ groups, so roughly $\lceil n/t \rceil(2^t - 1 - t)$ unions are done in the first phase of the proposed algorithm. Now any desired union of rows from B can be obtained by computing the union of at most one combination for each group. Therefore computing each row of the product matrix requires at most $\lceil n/t \rceil - 1$ additional unions, or at most $n(\lceil n/t \rceil - 1)$ additional unions for all n rows. The total number of unions done by this method is at most $\lceil n/t \rceil(2^t - 1 - t) + n(\lceil n/t \rceil - 1)$. To simplify our work, we approximate and consider only the high-order terms: $(n 2^t)/t + (n^2/t)$. If $t = 1$ or $t = n$, this expression is in $\Theta(n^2)$ or $\Theta(2^n)$, respectively. Suppose that we try to minimize $(n 2^t)/t + (n^2/t)$ under the assumption that the first term is of higher order than the second. We would want to make t as small as possible but if $t < \lg n$, the first term would no longer dominate. Similarly, if we assume that the second term is of higher order, we would want t to be as large as possible, but it cannot be larger than $\lg n$. This by no means rigorous argument suggests that we try $t = \lg n$ or, more precisely, $\lceil \lg n \rceil$ since t is an integer. The number of unions done for $t = \lceil \lg n \rceil$ is roughly $2n^2/\lceil \lg n \rceil$, which is of lower order than n^2. Thus this approach is worth pursuing with $t = \lceil \lg n \rceil$. We will now work out some of the implementation details and determine how much extra space is needed.

For each group of rows of B there are 2^t, or $2^{\lceil \lg n \rceil}$, sets to be stored. They are stored in the array *unions* according to the following scheme. Interpret the indexes for *unions* as t-bit binary numbers $b_1 b_2 \ldots b_t$. The bits of an index i indicate which rows of B are included in the union stored in *unions*$[i]$; B_j is included if and only if bit b_j in i is 1. Thus the first group of unions is stored as follows:

i	Contents of *unions*$[i]$
$00\ldots00$	$\varnothing$
$00\ldots01$	B_t
$00\ldots10$	B_{t-1}
$00\ldots11$	$B_{t-1} \cup B_t$
$\vdots$	$\vdots$
$11\ldots11$	$B_1 \cup B_2 \cup \cdots \cup B_t$

Exactly 2^t cells (each the size of one row) are used to store the unions for the first group of rows. We may suppose for now that the unions for the other groups are stored in successive blocks of cells, offset by the appropriate multiple of 2^t from the beginning of the array; later we will show how to make do with only 2^t (roughly n) cells, instead of using $2^t \lceil n/t \rceil$, or roughly $n^2/\lg n$.

This storage setup was devised to make it easy to find the unions needed for a given row of the product. Recall that the ith row of the product is $\bigcup_{k \in A_i} B_k$. Suppose that we break up each row of A into segments of t entries each, and let A_{ij} be the jth segment of t entries in the ith row. (See Fig. 8.3.) Interpreted as a binary number, A_{ij} is the correct index (minus the appropriate multiple of 2^t, in particular $(j-1)2^t$) in the array *unions* for the union of rows of B from the jth group. At this point, our algorithm looks like this:

Compute and store in *unions* unions of all combinations of
    rows of B within each group of $t = \lceil \lg n \rceil$ successive rows.
for $i := 1$ **to** n **do** { i indexes rows of A and C. }
    for $j := 1$ **to** $\lceil n/t \rceil$ **do** { j indexes groups of rows of B. }
        $C_i := C_i \cup$ *unions*$[(j-1)2^t + A_{ij}]$
end   **end**

The amount of space used to store the unions can be cut down merely by changing the order in which the work is done. In its present form the algorithm computes one complete row of C before going on to the next, so all groups of unions must be available. If instead it works with one group at a time, selecting the union

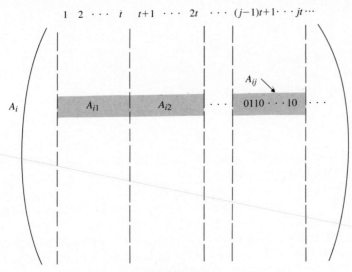

Figure 8.3

needed from that group for each row of C, succeeding groups of unions could use the same memory locations.

The last two details to work out are an efficient scheme for computing the unions within each group and a way of handling the case when the last group has fewer than t rows. We leave the latter problem to the reader. The former is taken care of easily in the final form of the algorithm.

Algorithm 8.5 Bit Matrix Multiplication — Kronrod's Algorithm

Input: A, B, and n, where A and B are $n \times n$ bit matrices.

Output: C, the Boolean matrix product.

Comment: B_i and C_i are the ith rows of B and C. A_{ij} denotes the jth segment of t bits in the ith row of A. As written, the algorithm assumes that t divides n.

```
procedure KronrodBitMatrices (A, B : BitMatrix; n: integer; var C : BitMatrix);
var
    t, i, j, k : integer;

begin
    C := 0;           { i.e., all zeros }
    t := ⌈lg n⌉;  unions[0] := 0;
    for j := 1 to ⌈n/t⌉ do
        {Compute all unions within the jth group of rows of B.}
        for k := 0 to t−1 do
        for i := 0 to 2^k −1 do
            unions[i+2^k] := unions[i] ∪ B_{t−k}
        end end { for k and i };
```

```
        { Select the appropriate union for each row of C. }
        for i := 1 to n do
            Cᵢ := Cᵢ ∪ unions[Aᵢⱼ]
        end { for i }
    end { for j }
end { KronrodBitMatrices }
```

Analysis

Note that $2^t - 1$ union operations are done to get all unions within a group (in the **for** k, **for** i loop) because single rows of B are computed as $\varnothing \cup B_k$. The extra t unions could be eliminated by starting i at 1, but the sets $B_1, \ldots, B_t$ must be put in the *unions* array anyway; not much time would be saved by having another loop to insert them without doing unions. Algorithm 8.5 does a total of $(n/t)(2^t - 1 + n)$, or $\Theta(n^2/\lg n)$, row unions. Each requires $\lceil n/c \rceil$ Boolean **or** instructions (where c is the word size, or the size of the operand of the **or** instruction), so the running time is in $\Theta(n^3/\lg n)$, but is a fairly small multiple of $n^3/\lg n$. The running time does not depend on the particular input; the same operations are done for all inputs of size n.

The formula derived in Section 8.3 for the matrix of the transitive closure of a relation (Theorem 8.5) uses approximately $\lg n$ Boolean matrix multiplications. Thus using Kronrod's algorithm, the transitive closure can be computed with only $\Theta(n^2)$ row unions.

Note that both Warshall's algorithm for transitive closure (Section 8.2) and Kronrod's Boolean matrix multiplication algorithm save time or space by doing their computations in a particular order. In both cases the natural, or more usual, order in which one would think of doing the work is less efficient.

*8.4.3   Lower Bound

Is Kronrod's algorithm optimal? If we consider the time it takes to do the row unions, then it is not, because it takes $\Theta(n^3 \lg n)$ time, and $n^{2.81}$ (the order of Strassen's algorithm) is in $o(n^3 \lg n)$. The various algorithms for Boolean matrix multiplication assume different representations for the matrices (bit matrices versus one entry per word) and do different kinds of operations (for example, Boolean operations on words, arithmetic operations if Strassen's method is used, or row unions as in Kronrod's algorithm). If we restrict our attention to the class of algorithms that compute rows of the product by forming unions of rows of the second factor matrix, then we can show that, within this class, Kronrod's algorithm is nearly optimal for large matrices; the number of unions done by an optimal algorithm would also be in $\Theta(n^2/\lg n)$. The result we prove is not really strong; it says that the number of row unions required is at least a very tiny bit less than $n^2/4\lg n$ for *sufficiently large* n. The n may have to be much larger than the dimension of any matrices to be multiplied in practice. One of the reasons for including the proof of

the theorem is that it illustrates a "counting argument," a useful approach for establishing lower bounds that involves counting all possible algorithms (ignoring differences not relevant to the sequence of basic operations — in this case, row unions
done by the algorithms).

The phrase "for sufficiently large n" means for all n larger than some particular
integer. A synonymous phrase is "almost everywhere," abbreviated "a.e."

To derive the lower bound we will use an abstracted model of algorithms (as
we did with decision trees for sorting and straight-line programs for polynomial
evaluation). Let P be an algorithm that computes $C = AB$ by forming unions of rows
of B (and possibly copying rows) and that can do no other operations on B. For a
particular input, A and B, we can make an indexed list of the union operations done
by P, denoting such an operation by $union(r, s)$, where r and s may be a row of B or
the result of a previous $union$ specified by its index in the list.

The sequence of $unions$ done is not sufficient to describe the result produced
by the algorithm; we must know which unions computed in the sequence are to be
rows of the product, and which rows in the product they are. Suppose that A and B
are $n \times n$, and let $steps$ be the number of steps in the list of $union$ operations. Then
the additional information needed can be provided by an n-vector $V = (j_1, \ldots, j_n)$,
where $-n \leq j_i \leq steps$ and j_i describes the ith row of the product as follows: If $j_i > 0$,
the ith row is the result of the j_ith $union$ operation; if $j_i = 0$, the ith row is all zeros
(the empty set); and if $j_i < 0$, the ith row is the $|j_i|$th row of B.

Example 8.1   If

$$A = \begin{bmatrix} 1 & 1 & 0 & 1 \\ 1 & 0 & 1 & 1 \\ 1 & 1 & 1 & 1 \\ 0 & 0 & 0 & 1 \end{bmatrix},$$

an algorithm might carry out the following sequence of $union$ operations:

1. $union($row 1, row 4$)$
2. $union(1,$ row 2$)$
3. $union(1,$ row 3$)$
4. $union(2, 3)$

The vector V for this example is $(2, 3, 4, -4)$.   ∎

Theorem 8.6   Let ε be a real number between 0 and 1. Then, for sufficiently large
n, any algorithm that does Boolean matrix multiplication using row unions must do
at least $n^2/[(4+\varepsilon)\lg n]$ union operations to multiply $n \times n$ matrices in the worst case.

Proof.   Let P be an algorithm that computes $C = AB$, and suppose that P is at least
as efficient as Kronrod's algorithm for sufficiently large n; i.e., for all but perhaps
finitely many values of n, P does at most $2n^2/\lg n$ row unions. Let $F(n)$ be the

number of unions done by P to multiply an arbitrary $n \times n$ matrix A and the identity matrix I_n in the worst case. The number of unions done by P in the worst case for all inputs is at least $F(n)$, and any lower bound derived for $F(n)$ is a lower bound for any algorithm in the class under consideration. We will show that, for $0 < \varepsilon < 1$, $F(n) \geq n^2/[(4+\varepsilon)\lg n]$ a.e.

Let S_n be the set of all valid sequences of $F(n)$ union operations. (A sequence is valid if, for each i, the ith operation refers to rows of B between 1 and n and/or to the results of operations with indexes between 1 and $i-1$.) Let V_n be the set of all n vectors with integer entries between $-n$ and $F(n)$. The operations done by P and its output for a given input A are described by an element of $S_n \times V_n$. If P does fewer than $F(n)$ unions for a particular A, S_n contains a sequence that does the work of P and is then padded out to length $F(n)$ with repetitions of, say, $union(1,1)$. We will derive an upper bound and a lower bound on $|S_n \times V_n|$ and use the resulting inequality to get a lower bound for $F(n)$.

Since each *union* has two operands, each of which is a row of B or an index between 1 and $F(n)$, there are $(n+F(n))^2$ choices for each *union* operation. Therefore $|S_n| \leq (n+F(n))^{2F(n)}$. $|V_n| = (n+1+F(n))^n$, so $|S_n \times V_n| \leq (n+1+F(n))^{2F(n)+n}$. To get a lower bound on $|S_n \times V_n|$, observe that $S_n \times V_n$ contains a distinct element for each $n \times n$ matrix A since $A_1 I_n \neq A_2 I_n$ if $A_1 \neq A_2$. Thus $|S_n \times V_n| \geq 2^{n^2}$ since there are 2^{n^2} $n \times n$ Boolean matrices. So

$$2^{n^2} \leq |S_n \times V_n| \leq (n+1+F(n))^{2F(n)+n}$$

or

$$n^2 \leq (2F(n)+n)\lg(n+1+F(n)) \text{ for all } n > 0. \tag{8.1}$$

We observe that $F(n) > n^{3/2}$ a.e., because if not, Equation 8.1 would imply that

$$n^2 \leq (2n^{3/2}+n)\lg(n+1+n^{3/2}) \leq 3n^{3/2}\lg(2n^{3/2}) \leq \frac{9}{2}n^{3/2}\lg(2n)$$

for infinitely many n, i.e., that n^2 is in $O(n^{3/2}\lg n)$, and this is not true. Since $F(n) > n^{3/2}$, for any $\varepsilon' > 0$, $2F(n)+n < (2+\varepsilon')F(n)$ a.e. Also, $F(n) \leq 2n^2/\lg n$ (by choice of P), so $n+1+F(n) < 2n^2$ a.e. Substituting these inequalities in Eq. 8.1 gives

$$n^2 \leq (2+\varepsilon')F(n)\lg(2n^2) = (2+\varepsilon')F(n)(1+2\lg n) \text{ a.e.}$$

For any ε'', $1+2\lg n \leq (2+\varepsilon'')\lg n$ a.e., so

$$n^2 \leq (2+\varepsilon')(2+\varepsilon'')F(n)\lg n \text{ a.e.}$$

Suppose that $0 < \varepsilon < 1$ is given. Let $\varepsilon' = \varepsilon'' = \varepsilon/6$. Then from the previous inequality we have

$$n^2 \leq (4+\varepsilon)F(n)\lg n, \quad \text{or} \quad \frac{n^2}{(4+\varepsilon)\lg n} \leq F(n) \text{ a.e.} \qquad \square$$

8.5
Dynamic Equivalence Relations and Union-Find Programs

8.5.1 Dynamic Equivalence Relations

An *equivalence relation* R on a finite set S is a binary relation on S that is reflexive, symmetric, and transitive. That is, for all s, t, and u in S, it satisfies these properties: sRs; if sRt then tRs; and, if sRt and tRu, then sRu. The equivalence class of an element s in S is the subset of S that contains all elements equivalent to s. The equivalence classes form a partition of S, i.e., they are disjoint and their union is S. The symbol $\equiv$ will be used from now on to denote an equivalence relation. The problem in this section is to represent, modify, and answer certain questions about an equivalence relation that changes with time. The applications include another minimum spanning tree algorithm.

The equivalence relation is initially the equality relation. The problem is to process a sequence of instructions of the following two types, where s_i and s_j are elements of S:

1.   IS $s_i \equiv s_j$?
2.   MAKE $s_i \equiv s_j$ (where $s_i \equiv s_j$ is not already true).

Question 1 is answered "yes" or "no." The correct answer depends on the instructions of the second type that have been received already; the answer is yes if and only if the instruction "MAKE $s_i \equiv s_j$" has already appeared or $s_i \equiv s_j$ can be derived by applying the reflexive, symmetric, and transitive properties to pairs that were explicitly made equivalent by the second type of instruction. The response to the latter (i.e., the MAKE instructions) is to modify the data structure that represents the equivalence relation so that later instructions of the first type will be answered correctly.

Consider the following example, in which $S = \{1, 2, 3, 4, 5\}$. The sequence of instructions is in the left-hand column. The right-hand column shows the response — either a yes or a no answer, or the set of equivalence classes for the relation as defined at the time.

Equivalence classes to start: $\{1\}, \{2\}, \{3\}, \{4\}, \{5\}$

IS 2 $\equiv$ 4?	No
IS 3 $\equiv$ 5?	No
MAKE 3 $\equiv$ 5.	$\{1\}, \{2\}, \{3,5\}, \{4\}$
MAKE 2 $\equiv$ 5.	$\{1\}, \{2,3,5\}, \{4\}$
IS 2 $\equiv$ 3?	Yes
MAKE 4 $\equiv$ 1.	$\{1,4\}, \{2,3,5\}$
IS 2 $\equiv$ 4?	No

8.5.2 Application: Kruskal's Minimum Spanning Tree Algorithm

Let $G = (V, E, W)$ be a weighted graph. In Section 4.2 we studied the Prim/Dijkstra algorithm to find a minimum spanning tree for G. The algorithm started at an arbitrary vertex and then branched out from it by "greedily" choosing edges with low weight. At any time, the edges chosen formed a tree. Here we examine an algorithm that uses a greedier strategy. At each step it chooses the lowest-weighted remaining edge from anywhere in the graph, discarding any edge that would form a cycle with those already chosen. At any time the edges chosen so far will form a forest but not necessarily one tree. The correctness of the strategy is stated formally in the following theorem, the proof of which is left as an exercise.

Theorem 8.7   Let $G = (V, E, W)$ be a weighted graph and let $E' \subseteq E$. If E' is contained in a minimum spanning tree for G and if e is an edge of minimum weight in $E-E'$ such that $E' \cup \{e\}$ has no cycles, then $E' \cup \{e\}$ is contained in a minimum spanning tree for G.

The algorithm starts with $E' = \varnothing$ and continues adding edges to E' until $|E'| = |V|-1$, or until it determines that G does not have a spanning tree. The only unresolved problem is how to determine whether an edge will form a cycle with others already in E'. The following lemma provides the criterion. The proof is easy and is left to the reader.

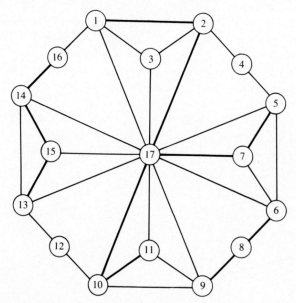

Figure 8.4   The darkened edges are in the subgraph G'. The equivalence classes are $\{1,2,5,7,10,11,17\}$, $\{6,8,9\}$, $\{13,14,15,16\}$, $\{3\}$, $\{4\}$, and $\{12\}$.

Lemma 8.8 Suppose that G' is a subgraph of G and that G' is contained in a spanning tree for G (hence G' is a forest). Let $e = vw$ be an edge of G that is not in G'. There is a cycle consisting of e and edges in G' if and only if v and w are in the same connected component of G'.

We define a relation, $\equiv$, on the vertices in a subgraph G' by $v \equiv w$ if and only if v and w are in the same connected component of G'. It is easy to check that $\equiv$ is an equivalence relation. (See Fig. 8.4 for an example.) Thus, by Lemma 8.8, an edge vw is chosen if and only if it is not true that $v \equiv w$. Each time an edge is chosen, the subgraph G' and the equivalence relation $\equiv$ change; each new edge causes two connected components, or two equivalence classes, to be merged into one.

Algorithm 8.6 Kruskal's Minimum Spanning Tree Algorithm

Input: $G = (V, E, W)$, a weighted graph, with $|V| = n, |E| = m$.

Output: T, a subset of E that forms a minimum spanning tree for G, or a minimum spanning forest if G is not connected.

```
procedure KruskalMST (G: Graph; n: integer; var T: EdgeArray);
var
    L : EdgeArray;
    count: integer;
begin
    Sort the edges in nondecreasing order by weight and put them
        in a list L;
    count := 0;
    { Initially, each vertex is a separate equivalence class. }
    while count < n-1 and L not empty do
        Let vw be the next edge in L; remove it from L;
        if not v ≡ w then
            Add vw to T;
            MAKE v ≡ w;
            count := count+1
        end { if }
    end { while }
end { KruskalMST }
```

Before analyzing this algorithm we need to investigate data structures and algorithms for dynamic equivalence relations.

8.5.3 Some Obvious Implementations

To compare various implementation strategies we will count operations of various kinds done by each strategy to process a sequence of n MAKE and/or IS instructions. We start by examining two fairly obvious data structures to represent the relation: matrices and arrays.

A matrix representation of an equivalence relation uses $|S|^2$ cells (or roughly $|S|^2/2$ if the symmetry is used). For an IS instruction only one entry need be examined; however, a MAKE instruction would involve copying information from several rows. A sequence of n MAKE instructions (hence, a worst-case sequence of MAKEs and ISs) would require at least $n|S|$ operations.

The amount of space used can be reduced to $|S|$ by using an array, say, $class$, where $class[i]$ is a label or name for the equivalence class containing s_i. An instruction IS $s_i \equiv s_j$? requires looking up and comparing $class[i]$ and $class[j]$. For MAKE $s_i \equiv s_j$, each entry is examined to see whether it equals $class[i]$, and if so is assigned $class[j]$. Again, for a sequence of n MAKEs (hence, for a worst-case sequence) at least $n|S|$ operations are done.

Both methods have inefficient aspects — the copying in the first and the search (for elements in $class[i]$) in the latter. Better solutions use links to avoid the extra work.

8.5.4 Union-Find Programs

The effect of a MAKE instruction is to form the union of two subsets of S. An IS can be answered easily if we have a way to find out of which set a given element is a member. Thus we turn our attention to *Union* and *Find* operators and a particular data structure in which they can be implemented easily. Each equivalence class, or subset, is represented by a tree. Each root will be used as a label or identifier for its tree. The instruction $r := Find(v)$ finds and assigns to r the root of the tree containing v. The arguments of *Union* must be roots; $Union(t, u)$ attaches the trees with roots t and u ($t \neq u$). To make it easy to implement *Find* and *Union*, the links in the trees point from each node to its parent (with null pointers in the roots). The pointers can be stored in an array with $|S|$ entries.

Find and *Union* would be used as follows to implement the equivalence instructions:

IS $s_i \equiv s_j$	MAKE $s_i \equiv s_j$
$t := Find(s_i)$	$t := Find(s_i)$
$u := Find(s_j)$	$u := Find(s_j)$
Is $t = u$?	$Union(t, u)$

A sequence of n *Union* and/or *Find* operations interspersed in any order will be considered an input, or *Union-Find* program, of size n. We take the number of operations on tree links (e.g., traversal, change, comparison) as the measure of work done. (It will be clear that the total number of operations is proportional to the number of link operations.) The obvious implementation for $Find(t)$ is simply to follow pointers from the node for t until the root is reached, and for $Union(t, u)$ to make the link in the node for t point to the node for u. Each *Union* does one link operation, and each *Find* does $l+1$, where l is the level of the argument of the *Find* in its tree. The program in Fig. 8.5, which does $n/2+(n/2)(n/2+1)$ link operations,

1.   *Union* (1, 2)

2.   *Union* (2, 3)

$\vdots$

$\frac{n}{2}.$ *Union* (n/2, n/2 + 1)

$\frac{n}{2} + 1.$ *Find* (1)

$\vdots$

n.   *Find* (1)

Figure 8.5   A *Union-Find* program with $S = \{1, 2, \dots, n\}$, *n* even.

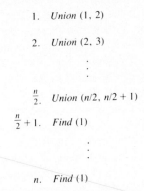
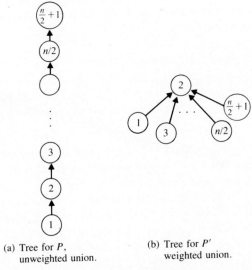

(a) Tree for *P*,
unweighted union.

(b) Tree for *P'*
weighted union.

Figure 8.6   Trees obtained using unweighted union and weighted union.

demonstrates that, using these methods, the worst-case time for a *Union-Find* program is in $\Omega(n^2)$. It is not hard to show that no such program does more than n^2 link operations, so the worst case is in $\Theta(n^2)$. This is not generally better than the methods described earlier. We will improve the implementation of the *Union* and *Find* instructions.

The cost of the program in Fig. 8.5 is high because the tree constructed by the *Union* instructions (see Fig. 8.6a) has large depth. It could be reduced by a more careful implementation of *Union* aimed at keeping the trees short. Let *W-Union* (for "weighted union") be the strategy that makes the root of the tree with fewer nodes point to the root of the other tree (and, say, makes the first argument point to the second if the trees have the same number of nodes). (Exercise 8.14 examines the possibility of using the depth rather than the number of nodes as the "weight" of each tree.) To distinguish between the two implementations of the *Union* operation, we will call the first one *UW-Union*, for unweighted union. For *W-Union* the number of nodes in each tree can be stored in the root; it can be distinguished from a pointer by the use of the sign bit. *W-Union* must compare the numbers of nodes, compute and store the size of the new tree, and set a pointer, so the cost of a *W-Union* is 3, still a small constant. Now if we go back to the program in Fig. 8.5 (call it *P*) to see how much work it requires using *W-Union*, we find that *P* is no longer a valid program because not all the arguments of *Union* in instructions 3 through $n/2$ will be roots of trees. We can expand *P* to the program *P'* by replacing each instruction of the form $Union(i, j)$ by the three instructions $t := Find(i)$, $u := Find(j)$, $Union(t, u)$. Then, using *W-Union*, *P'* requires only $4n-1$ operations! Figure 8.6 shows the trees constructed for *P* and *P'* using *UW-Union* and *W-Union*, respectively. We cannot conclude that *W-Union* allows linear time implementations in all cases; *P'* is not a worst-case program for *W-Union*. The following lemma helps us obtain an upper bound on the worst case.

Lemma 8.9   If $Union(t, u)$ is implemented by *W-Union* — i.e., so that the tree with root u is attached as a subtree of t if and only if the number of nodes in the tree with root u is smaller, and the tree with root t is attached as a subtree of u otherwise — then, after any sequence of *Union* instructions, any tree that has k nodes will have depth at most $\lfloor \lg k \rfloor$.

Proof, by induction on k.   Let $k = 1$; a tree with one node has depth 0, which is $\lfloor \lg 1 \rfloor$. Now suppose that $k > 1$ and that any tree constructed by a sequence of *Union* instructions and containing k' nodes, for $k' < k$, has depth at most $\lfloor \lg k' \rfloor$. Consider the tree T in Fig. 8.7, which has k nodes and depth d, and was constructed from the trees T_1 and T_2 by a *Union* instruction. Suppose, as indicated in the figure, that u, the root of T_2, was made to point to t, the root of T_1. Let k_1, d_1, k_2, and d_2 be the number of nodes in and the depths of T_1 and T_2, respectively. By the induction assumption $d_1 \le \lfloor \lg k_1 \rfloor$ and $d_2 \le \lfloor \lg k_2 \rfloor$. Either $d = d_1$ (if $d_1 > d_2$) or $d = d_2+1$ (if $d_1 \le d_2$). In the first case,

$$d = d_1 \le \lfloor \lg k_1 \rfloor \le \lfloor \lg(k_1+k_2) \rfloor = \lfloor \lg k \rfloor.$$

The second case uses the fact that, since *W-Union* was used, $k_2 \le k_1$. Thus

$$d = d_2+1 \le \lfloor \lg k_2 \rfloor +1 = \lfloor \lg 2k_2 \rfloor \le \lfloor \lg(k_1+k_2) \rfloor = \lfloor \lg k \rfloor.$$

So in both cases $d \le \lfloor \lg k \rfloor$.

$\square$

Theorem 8.10   A *Union-Find* program of size n does $\Theta(n \lg n)$ link operations in the worst case if the weighted union and the straightforward *Find* are used.

Proof.   A program that does at most n weighted union instructions can build a tree with at most $n+1$ nodes. Hence by the lemma, each tree has depth at most

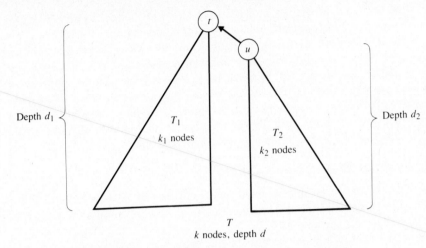

Figure 8.7 An example for the proof of Lemma 8.9.

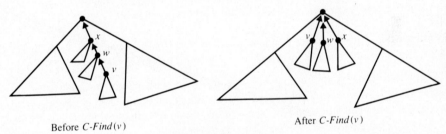

Before *C-Find* (v)   After *C-Find* (v)

Figure 8.8 *C-Find.*

$\lfloor \lg(n+1) \rfloor$, so the cost of each *Find* is at most $\lfloor \lg(n+1) \rfloor + 1$. The total number of link operations is therefore less than $3n + n(\lfloor \lg(n+1) \rfloor + 1)$, which is in $\Theta(n\lg n + 4n) = \Theta(n\lg n)$. Showing, by example, that programs requiring $\Theta(n\lg n)$ steps can be constructed is left to the reader (Exercise 8.15).   □

Implementation of the *Find* operation can also be modified to improve the speed of a *Union-Find* program by performing a process called *path compression*. Given argument v, *C-Find* (for "compressing-find") follows pointers from the node for v to the root, and then resets the pointers in all the nodes on the path just traversed so that they all point to the root. *C-Find* is illustrated in Fig. 8.8. *C-Find* does twice as much work as the straightforward *Find* for a given node in a given tree, but the use of *C-Find* keeps the trees very short so that, overall, the work is reduced. It can be shown (see the notes and references at the end of the chapter) that, using find-with-path-compression and the unweighted union, the worst-case running time for programs of length n is in $O(n\lg n)$. Exercises 8.17 through 8.21 show that there is in fact a program that requires $\Theta(n\lg n)$ steps. Thus using either the improved implementation of *Union* or the improved implementation of *Find* lowers the worst-case complexity of a program from $\Theta(n^2)$ to $\Theta(n\lg n)$. The next step is to combine the two improvements, hoping for a further reduction.

Are *C-Find* and *W-Union* compatible? *C-Find* changes the structure of the tree it acts on but does not alter the number of nodes in that tree; it may, however, change the depth. Recall that it might have seemed more natural for *W-Union* to compare the depths of the trees it was joining rather than the number of nodes in each since the point was to keep the trees short. It would be difficult to update the depth of a tree correctly after *C-Find* modified it; the number of nodes was used specifically to make *W-Union* and *C-Find* compatible.

*Analysis of *W-Union* and *C-Find*

We will now derive an upper bound on the number of link operations done by a *Union-Find* program using *W-Union* and *C-Find*. Several lemmas are needed to get the desired result: Theorem 8.14.

The *height* of a node v in a tree is the depth (i.e., height) of the subtree rooted at v. Suppose that P is a *Union-Find* program of length n. Let F be the forest constructed by the sequence of *Union* instructions in P assuming that *W-Union* is used and that the *Find*'s are ignored. We derive a few properties of F.

Lemma 8.11   In the forest F constructed by the *Union* instructions in program P using *W-Union*, there are at most $n/2^{h-1}$ nodes with height h, for $h>0$.

Proof. It follows from Lemma 8.9 that any tree with height h constructed by a sequence of *W-Union*'s has at least 2^h nodes. Each subtree in F (i.e., a node and all its descendants) was at one time a separate tree, so any subtree in F rooted at a node of height h has at least 2^h nodes. The set S on which the program P acts may be large, but P can affect at most $2n$ elements; the others will have height 0. For $h>0$, since the subtrees with root at height h are disjoint, there can be at most $2n/2^h$ of them, hence at most $2n/2^h = n/2^{h-1}$ nodes of height h.   □

Lemma 8.12   No node has height in F greater than $\lfloor \lg(n+1) \rfloor$.

Proof. Use Lemma 8.9 and the observation that n *Union* instructions can build a tree with at most $n+1$ nodes.   □

Lemmas 8.11 and 8.12 describe properties of the forest F constructed by the *Union* instructions of a *Union-Find* program. If the *Find* instructions are executed as they occur using *C-Find*, a different forest results, and the heights of the various nodes will be different from their heights in F. To avoid possible confusion about which height is meant, we define the *rank* of a node to be its height in F. Thus the word "height" in Lemmas 8.11 and 8.12 may be replaced by the word "rank."

Lemma 8.13   At any time during execution of a *Union-Find* program P, the ranks of the nodes on a path from a leaf to a root of a tree form a strictly increasing sequence. When a *C-Find* changes the pointer in a node, the new parent has higher rank than the old parent of that node.

Proof.   Certainly in F the ranks form an increasing sequence on a path from leaf to root. If during execution of P a node v becomes a child of a node w, v must be a descendant of w in F, and hence the rank of v is lower than the rank of w. If v is made a child of w by a *C-Find*, then w was an ancestor of the previous parent of v; hence the second statement of the lemma follows.   □

In Theorem 8.14 we will establish an upper bound of $O(nG(n))$ on the running time of a *Union-Find* program using *W-Union* and *C-Find*, where G is a function that grows extremely slowly. $G(n)$ is in $O(\lg^p(n))$ for any $p \geq 1$. To define G we first define the function H as follows:

$$H(0) = 1$$

and

$$H(i) = 2^{H(i-1)} \text{ for } i > 0.$$

For example,

$$H(5) = 2^{2^{2^{2^2}}}.$$

$G(j)$ is defined as the least i such that $H(i) \geq j$; i.e., informally, $G(j)$ is the number of 2's that must be "piled up" to reach or exceed j. Some values of H and G are shown in Table 8.1. For any conceivable input that might be used, $G(n) \leq 5$.

Theorem 8.14   The number of link operations done by a *Union-Find* program of length n implemented with *W-Union* and *C-Find* is in $O(nG(n))$.

Proof.   The *Union* instructions in the program do at most $3n$ link operations in total. The work done by the *C-Find*'s is counted in a tricky way. Some of the work done by each *C-Find* will be assigned, or charged, to the *Find* itself and some will be charged to the nodes on which it acts. The nodes are grouped so that those with rank h are in group $G(h)$. (For example, nodes with rank 0 or 1 are in group 0 and nodes with ranks 5 through 16 are in group 3.) Suppose that when *Find(v)* is executed, w is a node on the path from v to the root. If w is the root or if w and its parent (before being changed by the *C-Find*) are in different groups, one link operation is charged to the *Find* instruction itself. If w and its parent are in the same group, one is charged to w. (The total charged will be multiplied by 2 later because the path is traversed twice by the *C-Find*.) Since the indexes of the groups containing the nodes on the path form a nondecreasing sequence, the number of operations charged to a

Table 8.1
The functions H and G.

i	0	1	2	3	4	5	6...16	17...65536
$H(i)$	1	2	4	16	65536	2^{65536}		
$G(i)$	0	0	1	2	2	3	3...3 4	... 4

single *Find* cannot exceed the number of distinct groups, which is $1+G(\lfloor \lg(n+1) \rfloor) \le G(n+1) \le G(n)+1$. Thus the number of operations charged to all *Find*'s is at most $n(G(n)+1)$.

Next we count the operations charged to nodes. A node w is charged if it is on the path traversed by a *C-Find* and is in the same group as its parent. If w is not the child of a root, it is relinked and its new parent has higher rank than its old parent. Once a new parent of w is in a different group, w will never be charged again. Thus the total charges to w that occur while w is not the child of a root is at most the number of ranks in its group. The ranks in group i are $H(i-1)+1$, $H(i-1)+2,\ldots,$ $H(i)$. (See Table 8.1.) The sum of charges for all w is at most

$$\sum_{i=0}^{G(n)} (\text{number of ranks in group } i) \cdot (\text{number of nodes in group } i). \qquad (8.2)$$

There are at most $n/2^{h-1}$ nodes of rank h, so the number of nodes in group i is

$$\sum_{h=H(i-1)+1}^{H(i)} \frac{n}{2^{h-1}} = \frac{n}{2^{H(i-1)}} \sum_{j=0}^{H(i)-H(i-1)-1} \frac{1}{2^j} \le \frac{n}{2^{H(i-1)}} 2 = \frac{2n}{H(i)}.$$

Thus the summation in Eq. 8.2 is

$$\sum_{i=0}^{G(n)} [H(i)-H(i-1)] \frac{2n}{H(i)} = 2n \sum_{i=0}^{G(n)} \left[1 - \frac{H(i-1)}{H(i)}\right] \le 2n(G(n)+1).$$

The only operations not yet counted are those that may be charged to a node while it is the child of a root. There is at most one such operation per *Find*, so there are at most n altogether. The total number of link operations done by the *Union*'s and *Find*'s is at most $3n+2[n(G(n)+1)+2n(G(n)+1)+n] \in \Theta(nG(n))$. $\square$

Since $G(n)$ grows so slowly and the estimates made in the proof of the theorem are fairly loose, it is natural to wonder whether we can in fact prove a stronger theorem, i.e., that the running time of *Union-Find* programs of size n implemented with *W-Union* and *C-Find* is in $\Theta(n)$. It has been shown that this is not true (see the notes and references at the end of the chapter); for any constant c, there are programs of size n that require more than cn operations using these (and a variety of other) techniques. It is an open question whether there exist different techniques that implement *Union-Find* programs in linear time. Nevertheless, as Theorem 8.14 shows, the use of *C-Find* and *W-Union* results in a very efficient implementation of *Union-Find* programs. We will assume this implementation when discussing the following applications. The algorithms for *W-Union* and *C-Find* are very easy to write out; we leave them to the reader.

8.5.5 Back to Equivalence Programs and Kruskal's Algorithm

We began by attempting to find a good way to represent a dynamic equivalence relation so that instructions of the forms MAKE $s_i \equiv s_j$ and IS $s_i \equiv s_j$? can be handled efficiently. We define an equivalence program of length n to be a sequence of n such instructions interspersed in any order. Since, as we observed earlier, each MAKE or IS instruction can be implemented by three instructions from {*Union, Find*, comparison}, an equivalence program of length n can be implemented in $O(nG(n))$ time.

Now we can analyze Kruskal's minimum spanning tree algorithm, which is reproduced here for convenience.

Algorithm 8.6 Kruskal's Minimum Spanning Tree Algorithm

Input: $G = (V, E, W)$, a weighted graph, with $|V| = n, |E| = m$.

Output: T, a subset of E that forms a minimum spanning tree for G, or a minimum spanning forest if G is not connected.

```
procedure KruskalMST (G: Graph; n: integer; var T: EdgeArray);
var
    L : EdgeArray;
    count: integer;
begin
    Sort the edges in nondecreasing order by weight and put them
        in a list L;
    count := 0;
    { Initially, each vertex is a separate equivalence class. }
    while count < n−1 and L not empty do
        Let vw be the next edge in L; remove it from L;
        if not v ≡ w then
            Add vw to T;
            MAKE v ≡ w;
            count := count+1
        end { if }
    end { while }
end { KruskalMST }
```

The edges of the graph can be sorted with $\Theta(m\lg m)$ comparisons. The condition $v \equiv w$ in the **if** statement may be tested as many as m times. At most $\min(n-1, m)$ edges are added to T. The total number of operations done in these lines therefore is bounded by a multiple of the number of operations done to execute an equivalence program of size m, i.e., it is in $O(mG(m))$. Thus the worst-case running time of the algorithm is in $\Theta(m\lg m)$. Algorithm 4.2 is in $\Theta(n^2)$ in the worst case. Which is better depends on the relative sizes of n and m. For sparse graphs, Kruskal's algorithm is faster; for dense graphs, the Prim/Dijkstra algorithm is better. If the edges of G were already sorted, Kruskal's algorithm would run in $O(mG(m))$ time, which is quite good.

8.5.6 Other Applications

References on all of the applications described briefly here can be found in the notes and references at the end of the chapter.

The *Union* and *Find* operators can be used to implement a sequence of two other types of instructions that act on the same kind of tree structures: *Link(r, v)*, which makes the tree rooted at *r* a subtree of *v*, and *Depth(v)*, which determines the current depth of *v*. A sequence of *n* such instructions can be implemented in $O(nG(n))$ time.

The study of equivalence programs was motivated by the problem of processing EQUIVALENCE declarations in FORTRAN and other programming languages. An EQUIVALENCE declaration indicates that two or more variables or array entries are to share the same storage locations. The problem is to correctly assign storage addresses to all variables and arrays. The declaration

EQUIVALENCE $(A, B(3)), (B(4), C(2)), (X, Y, Z), (J(1), K), (B(1), X), (J(4), L, M)$

indicates that *A* and *B(3)* share the same location, that *B(4)* and *C(2)* share the same location, and so forth. (FORTRAN uses parentheses, not square brackets, for array indexes.) The complete storage layout indicated by this EQUIVALENCE statement is shown in Fig. 8.9, which assumes for simplicity that each array has five entries.

If there were no arrays, the problem of processing EQUIVALENCE declarations would be essentially the same as the problem of processing an equivalence program. The inclusion of arrays requires some extra bookkeeping and introduces the possibility of an unacceptable declaration. For example,

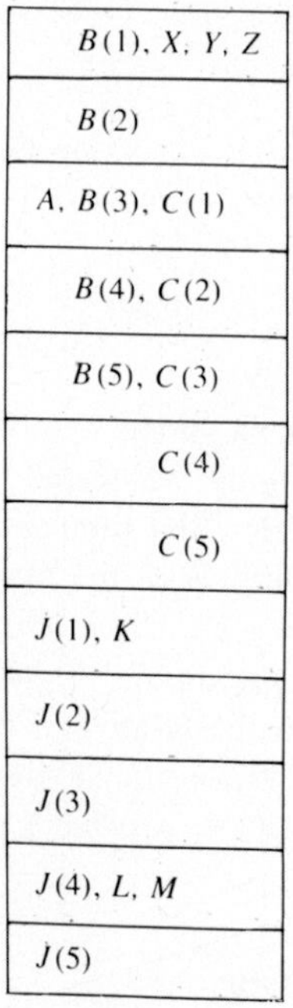

$B(1), X, Y, Z$
$B(2)$
$A, B(3), C(1)$
$B(4), C(2)$
$B(5), C(3)$
$C(4)$
$C(5)$
$J(1), K$
$J(2)$
$J(3)$
$J(4), L, M$
$J(5)$

Figure 8.9 Storage arrangement for EQUIVALENCE $(A, B(3)), (B(4), C(2)), (X, Y, Z),$ $(J(1), K), (B(1), X), (J(4), L, M)$.

EQUIVALENCE $(A(1), B(1)), (A(2), B(3))$

could not be allowed since the elements of each array must occupy consecutive memory locations. The *Link* and *Depth* instructions mentioned previously can be used in processing EQUIVALENCE declarations.

Union and *Find* are only two of many possible operations on collections of subsets. Some others are *Insert*, which inserts a new member in a set; *Delete*, which removes an item from a set; *Min*, which finds the smallest item in a set; *Intersect*, which produces a third set from two given sets; and *Member*, which indicates whether or not a specified element is in a particular set. Techniques and data structures for efficiently processing "programs" consisting of sequences of two or three types of such instructions have been studied. In some cases, the *Union* and *Find* techniques can be used to implement such programs of size n in $O(nG(n))$ time.

Exercises

Section 8.1: The Transitive Closure of a Binary Relation

8.1. a)   Let $G = (V, E)$ be a graph (not directed) and let R be a relation on V defined by vRw if and only if $v = w$ or there exists a path from v to w. Show that R is an equivalence relation. Would R be an equivalence relation if the case $v = w$ were omitted from its definition? Why?

   b)   What are the equivalence classes of this relation?

   c)   Show that the reachability matrix R for a graph (undirected) with n vertices can be constructed in $O(n^2)$ time.

8.2. a)   Try to write an algorithm using depth-first search to construct R, the reachability matrix for a digraph, given A, the adjacency matrix. (Assume that a vertex may be adjacent to itself.) The algorithm should use the suggestion in Section 8.1.2 that entries of R in several rows be computed during one depth-first search. Use whatever other tricks you can think of to design an efficient algorithm.

   b)   What is the order of the worst-case running time of your algorithm?

   c)   Test your algorithm on the digraph in Fig. 8.10. If it does not work correctly, modify it so that it does and redo (b).

Section 8.2: Warshall's Algorithm

*8.3.   Construct the worst example you can for Algorithm 8.1, that is, an example for which the triple **for** loop is repeated many times. How many times will the loop be repeated in your example?

8.4.   Construct an example of a binary relation matrix A for which Algorithm 8.2 would not work correctly if k were varied in the innermost loop.

Section 8.3: Computing Transitive Closure by Matrix Operations

8.5.   Prove Lemma 8.2.

8.6.   Prove Lemma 8.3.

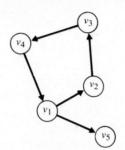

Figure 8.10

8.7. Show that A^+, the transitive closure of the Boolean matrix A, can be computed with one matrix multiplication if the reflexive transitive closure, A^*, is known.

Section 8.4: Multiplying Bit Matrices — Kronrod's Algorithm

8.8. Prove that if A and B are $n \times n$ Boolean matrices with rows interpreted as subsets of $\{1, 2, \ldots, n\}$ as described in Section 8.4.1, then if $C = AB$, the ith row of C is $\bigcup_{k \in A_i} B_k$, where A_i is the ith row of A and B_k is the kth row of B.

Section 8.5: Dynamic Equivalence Relations and Union-Find Programs

8.9. Prove Theorem 8.7.

8.10. Prove Lemma 8.8.

8.11. Find the minimum spanning tree for the graph in Fig. 8.11 that would be found by Kruskal's algorithm (Algorithm 8.6), assuming that the edges are sorted as shown.

8.12. Write algorithms for processing a sequence of MAKE and IS instructions using a matrix to represent the equivalence relation. How many matrix entries are examined or changed in the worst case when processing a list of n instructions?

8.13. Prove that a *Union-Find* program of size n does at most n^2 link operations if implemented with the unweighted union and straightforward find.

8.14. The weighted union, *W-Union*, uses the number of nodes in a tree as its weight. Let *WD-Union* be an implementation that uses the depth of a tree as its weight and makes the tree with the smaller depth a subtree of the other.

a) Write out an algorithm for *WD-Union*.

b) Either prove that the trees constructed for all *Union-Find* programs are the same regardless of whether *W-Union* or *WD-Union* is used, or exhibit a program for which they differ. (For both implementations, if the trees are of equal sizes, make the first argument point to the second.)

c) What is the worst-case complexity of *Union-Find* programs using the straightforward *Find* (without path compression) and *WD-Union*?

8.15. Exhibit a *Union-Find* program of size n that requires $\Theta(n \lg n)$ time if the straightforward *Find* (without path compression) and the weighted union (*W-Union*) are used.

8.16. Let $S = \{1, 2, \ldots, 9\}$ and assume that weighted union and find-with-path-compression are used. (If the sizes of the trees rooted at t and u are equal, *Union*(t,u) makes u the

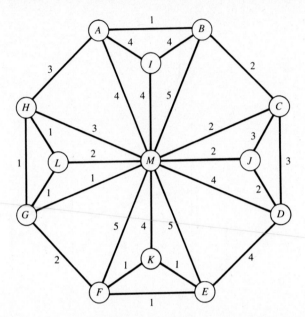

Figure 8.11 Sorted edges: *AB, EF, EK, FK, GH, GL, GM, HL, BC, CM, DJ, FG, JM, LM, AH, CD, CJ, HM, AI, AM, BI, DE, DM, IM, KM, BM, EM,* and *FM.*

root of the new tree.) Draw the trees after the last *Union* and after each *Find* in the following program.

>*Union*(1, 2)
>*Union*(3, 4)
>*Union*(2, 4)
>*Union*(6, 7)
>*Union*(8, 9)
>*Union*(7, 9)
>*Union*(4, 9)
>*Find*(1)
>*Find*(4)
>*Find*(6)
>*Find*(1)

8.17. S_k trees are defined as follows: S_0 is a tree with one node. For $k>0$, an S_k tree is obtained from two disjoint S_{k-1} trees by attaching the root of one to the root of the other. See Fig. 8.12 for examples.

Prove that if T is an S_k tree, T has 2^k vertices, depth k, and a unique vertex at level k. The node at level k is called the *handle* of the S_k tree.

8.18. Using the definitions and results of Exercise 8.17, prove the following characterization of an S_k tree: Let T be an S_k tree with handle v. There are disjoint trees $T_0, T_1, \ldots, T_{k-1}$, not containing v, with roots $r_0, r_1, \ldots, r_{k-1}$, respectively, such that

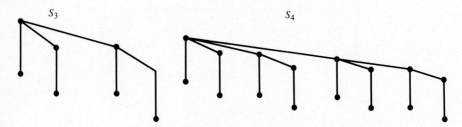

Figure 8.12 S_k trees.

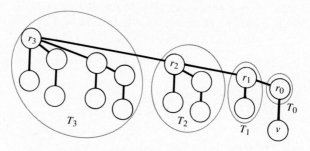

Figure 8.13 Decomposition of S_4 for Exercise 8.18.

1. T_i is an S_i tree, $0 \le i \le k-1$, and

2. T results from attaching v to r_0, and r_i to r_{i+1}, for $0 \le i < k-1$.

This decomposition of an S_4 tree is illustrated in Fig. 8.13.

8.19. Using the definitions and results of Exercise 8.17, prove the following characterization of an S_k tree: Let T be an S_k tree with root r and handle v. There are disjoint trees $T_0', T_1', \ldots, T_{k-1}'$ not containing r, with roots $r_0', r_1', \ldots, r_{k-1}'$, respectively, such that

    1. T_i' is an S_i tree, $0 \le i \le k-1$,

    2. T is obtained by attaching each r_i' to r for $0 \le i \le k-1$, and

    3. v is the handle of T_{k-1}'.

This decomposition of an S_4 tree is illustrated in Fig 8.14.

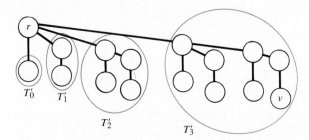

Figure 8.14   Decomposition of S_4 for Exercise 8.19.

*8.20.   An *embedding* of a tree T in a tree T' is a one-to-one function $f:T \rightarrow T'$ (i.e., from the vertices of T to the vertices of T') such that for all v and w in T, v is the parent of w if and only if $f(v)$ is the parent of $f(w)$. An embedding f is an *initial embedding* if it maps the root of T to the root of T'; it is a *proper embedding* otherwise. Using the results of Exercises 8.17 through 8.19, show that, if T is an S_k tree with handle v, and f is a proper embedding of T in a tree U, then there is an S_k tree T' initially embedded in U', the tree that results from doing a *C-Find* on $f(v)$ in U.

*8.21.   Show that a *Union-Find* program of length n can be constructed so that, if *C-Find* and *UW-Union* are used to implement it, $\Theta(n \lg n)$ operations are done. (Hint: Read Exercises 8.17 through 8.20.)

*8.22.   We stated that some *Union-Find* programs take more than linear time even when weighted union and find-with-path-compression are used. Show that in a program of size n, if all the *Union*s occur before the *Find*s, then the total number of operations is in $O(n)$.

8.23.   Design an algorithm to process EQUIVALENCE declarations and assign memory addresses to all arrays and variables in the declarations. Assume that a DIMENSION statement gives the dimensions of all the arrays. Does your algorithm detect invalid EQUIVALENCEs?

Additional Problems

8.24.   A *triangle* in a graph is a cycle of length 3. Outline an algorithm that uses the adjacency matrix of a graph to determine whether it has a triangle. How many operations on matrix entries are done by your algorithm?

8.25.   Most of the time in Kruskal's minimum spanning tree algorithm (Algorithm 8.6) is spent sorting the edges. Suppose the edges were kept in a min-heap instead of a completely sorted list. How would that affect the worst-case behavior of the algorithm? In what cases, if any, would using a heap be an advantage?

8.26.   Suppose that you need to determine whether a large graph (with n vertices and m edges, m quite a bit larger than n) is connected. The input will consist of n and m and a sequence of edges (pairs of vertices). You do not have enough space to store the whole graph; you can use cn units of space, where c is a small constant, but you cannot use space proportional to m. Thus you can process each edge when you read it, but you cannot save edges. Describe an algorithm to solve the problem. How much time does your algorithm take in the worst case?

Programs

1. Write a program to multiply two bit matrices using Kronrod's algorithm (Algorithm 8.5). Allow for n to be larger than the number of bits per word. How much space is used?

2. Write a program to implement *Union-Find* programs using the weighted union and find-with-path-compression.

Notes and References

Warshall's algorithm (Algorithms 8.2 and 8.3) is presented in Warshall (1962). Proofs of the correctness of Algorithm 8.2 (Theorem 8.1) and Algorithm 8.4 can be found there and in Wegner (1974). The proof that computing the reflexive transitive closure can be done as fast as Boolean matrix multiplication is from I. Munro and appears in Aho, Hopcroft, and Ullman (1974). Kronrod's algorithm (Algorithm 8.5) is from Arlazarov, Dinic, Kronrod, and Faradzev (1970) (where implementation is not discussed). The proof of Theorem 8.6, the lower bound on Boolean matrix multiplication by row unions, is from Angluin (1976). This result and generalizations of Kronrod's algorithm appear in Savage (1974).

Van Leeuwen and Tarjan (1983) describes and analyzes a large number of techniques for implementing *Union-Find*, or equivalence, programs. Galler and Fischer (1964) introduced the use of tree structures for the problem of processing EQUIVALENCE declarations. Knuth (1968) describes the equivalence problem and some suggestions for a solution (see Section 2.3.3, Exercise 11). Fischer (1972) proves that, using the unweighted union and find-with-path-compression, there are programs that do $\Theta(n \lg n)$ link operations. Exercises 8.17 through 8.21 develop Fischer's proof. The upper bound of $O(n \lg n)$ was proved by M. Paterson (unpublished). Hopcroft and Ullman (1973) proved Theorem 8.14 — i.e., that, when *C-Find* and *W-Union* are used, a program of size n does $O(nG(n))$ operations. Tarjan (1975) establishes a lower bound for the worst-case behavior of *C-Find* and *W-Union*; he shows that it is not linear.

Kruskal's strategy for finding minimum spanning trees is from J.B. Kruskal (1956). The implementation using equivalence programs is attributed by Hopcroft and Ullman to M. D. McIlroy and R. Morris. Much of the material in this section plus additional applications and extensions appear in Aho, Hopcroft, and Ullman (1974).

9

NP-Complete Problems

9.1
P and *NP*

9.1.1   Some Sample Problems

In the previous chapters we have studied a variety of problems and algorithms. Some of the algorithms are straightforward, while others are complicated and tricky, but virtually all have complexity in $O(n^3)$, where n is the appropriately defined input size. For the purposes of this chapter, we will accept all the algorithms studied so far as having fairly low time requirements. Take another look at Table 1.1, which shows that algorithms whose complexity is described by simple polynomial functions can be run for fairly large inputs in a reasonable amount of time. The last column in the table shows that if the complexity is 2^n, the algorithm is useless except for very small inputs. In this chapter we are concerned with problems whose complexity may be described by exponential functions, problems for which the best-known algorithms would require many years or centuries of computer time for moderately large inputs. We will present definitions aimed at distinguishing between the *tractable* (i.e., "not-so-hard") problems we have encountered already and *intractable* (i.e., "hard," or very time-consuming) ones. We will study a class of important problems with an irksome property — we do not even know whether they can be solved efficiently. No reasonably fast algorithms for these problems have been found, but no one has been able to prove that the problems require a lot of time. Because many of these problems are optimization problems that arise frequently in applications, the lack of efficient algorithms is of real importance.

Before the formal definitions and theorems, we describe several problems that will be used as examples throughout this chapter. Most of these problems occur naturally as optimization problems (they are called *combinatorial optimization problems*), but they can also be formulated as *decision problems*, that is, problems for which the output is a simple *yes* or *no* answer for each input.

Graph Coloring

A *coloring* of a graph $G = (V, E)$ is a mapping $C : V \to S$, where S is a finite set (of "colors"), such that if $vw \in E$ then $C(v) \neq C(w)$; in other words, adjacent vertices are not assigned the same color. The *chromatic number* of G, denoted $\chi(G)$, is the smallest number of colors needed to color G, that is, the smallest k such that there exists a coloring C for G and $|C(V)| = k$.

Optimization problem: Given G, determine $\chi(G)$ (and produce an optimal coloring, i.e., one that uses only $\chi(G)$ colors).

Decision problem: Given G and a positive integer k, is there is a coloring of G using at most k colors? (If so, G is said to be *k-colorable*.)

The graph coloring problem is an abstraction of certain types of scheduling problems. For example, suppose that the final exams at a university are to be scheduled during one week, with three exam times each day for a total of 15 time slots. The exams for some courses, say Calculus I and Physics I, must be at different times because many students are in both classes. Let V be the set of courses, and let E be the pairs of courses whose exams must not be at the same time. Then the exams can be scheduled in the 15 time slots without conflicts if and only if the graph $G = (V, E)$ can be colored with 15 colors.

Job Scheduling with Penalties

Suppose that n jobs $J_1, \ldots, J_n$ are to be executed one at a time. We are given execution times $t_1, \ldots, t_n$, deadlines $d_1, \ldots, d_n$ (measured from the starting time for the first job executed), and penalties for missing the deadlines $p_1, \ldots, p_n$. Assume that the execution times, deadlines, and penalties are all positive integers. A schedule for the jobs is a permutation π of $\{1, 2, \ldots, n\}$, where $J_{\pi(1)}$ is the job done first, $J_{\pi(2)}$ is the job done next, and so forth. The total penalty for a particular schedule is

$$P_\pi = \sum_{j=1}^{n} \left[\textbf{if } t_{\pi(1)} + \cdots + t_{\pi(j)} > d_{\pi(j)} \textbf{ then } p_{\pi(j)} \textbf{ else } 0 \right].$$

Optimization problem: Determine the minimum possible penalty (and find an optimal schedule, i.e., one that minimizes the total penalty).

Decision problem: Given, in addition to the inputs described, a nonnegative integer k, is there a schedule with $P_\pi \leq k$?

Bin Packing

Suppose we have an unlimited number of bins each of capacity 1, and n objects with sizes $s_1, \ldots, s_n$, where $0 < s_i \leq 1$.

Optimization problem: Determine the smallest number of bins into which the objects can be packed (and find an optimal packing).

Decision problem: Given, in addition to the inputs described, an integer k, do the objects fit in k bins?

Applications of bin packing include packing data in computer memories (e.g., files on disk tracks, program segments into memory pages, and fields of a few bits each into memory words) and filling orders for a product (e.g., fabric or lumber) to be cut from large, standard-size pieces.

Knapsack

Suppose we have a knapsack of capacity C (a positive integer) and n objects with sizes $s_1, \ldots, s_n$ and "profits" $p_1, \ldots, p_n$ (where $s_1, \ldots, s_n$ and $p_1, \ldots, p_n$ are positive integers).

Optimization problem: Find the largest total profit of any subset of the objects that fits in the knapsack (and find a subset that achieves the maximum profit).

Decision problem: Given k, is there a subset of the objects that fits in the knapsack and has a total profit at least k?

The knapsack problem has a variety of applications in economic planning and loading, or packing, problems. For example, it could describe a problem of making investment decisions where the "size" of an investment is the amount of money required, C is the total amount one has to invest, and the "profit" of an investment is the expected return. In an application of a more complicated version of the problem, the objects are tasks or experiments that various organizations want to have performed on a space flight. In addition to its size (the volume of the equipment needed), each task may have a power requirement and a requirement for a certain amount of crew time. The space, power, and time available on the flight are all limited. Each task has some value, or profit. Which feasible subset of the tasks has the largest total value?

The next problem is a simpler version of the knapsack problem.

Subset Sum

The input is a positive integer C and n objects whose sizes are positive integers $s_1, \ldots, s_n$.

Optimization problem: Among subsets of the objects with sum at most C, what is the largest subset sum?

Decision problem: Is there a subset of the objects whose sizes add up to exactly C?

CNF-satisfiability

A logical (or Boolean) variable is a variable that may be assigned the value *true* or *false*. If v is a logical variable, then $\bar{v}$, the negation of v, has the value *true* if and only if v has the value *false*. A *literal* is a logical variable or the negation of a logical variable. A *clause* is a sequence of literals separated by the Boolean **or** operator ($\vee$). A logical expression in *conjunctive normal form* (*CNF*) is a sequence of clauses separated by the Boolean **and** operator ($\wedge$). An example of a logical expression in CNF is

$$(p \vee q \vee s) \wedge (\bar{q} \vee r) \wedge (\bar{p} \vee r) \wedge (\bar{r} \vee s) \wedge (\bar{p} \vee \bar{s} \vee \bar{q}),$$

where p, q, r, and s are logical variables.

Decision problem: Is there a truth assignment, i.e., a way to assign the values *true* and *false*, for the variables in the expression so that the expression has value *true*?

The CNF-satisfiability problem has applications in computerized theorem proving. It played a central role in developing the ideas in this chapter.

Hamilton Paths and Hamilton Circuits

A Hamilton path (Hamilton circuit, or cycle) in a graph or digraph is a path (cycle) that passes through every vertex exactly once. (*Circuit* is another term for *cycle*, and Hamilton cycles are most commonly called Hamilton circuits.)

Decision problem: Does a given graph or digraph have a Hamilton path (circuit)?

A related optimization problem is the minimum tour problem.

Minimum Tour (Traveling Salesperson Problem)

Optimization Problem: Given a weighted graph, find a minimum weighted Hamilton circuit.

This problem is widely known as the traveling salesperson problem; the salesperson wants to minimize total traveling while visiting all the cities in a territory. Other applications include routing trucks for garbage pickup and package delivery.

Decision Problem: Given a weighted graph and an integer k, is there a Hamilton circuit with total weight at most k?

The usefulness and apparent simplicity of these problems may intrigue the reader, who is invited to try to devise algorithms for some of them before proceeding.

9.1.2   The Class *P*

None of the algorithms known for the problems just described is guaranteed to run in a reasonable amount of time. We will not rigorously define "reasonable," but we will define a class *P* of problems that *includes* those with reasonably efficient algorithms. An algorithm is said to be *polynomial bounded* if its worst-case complexity is bounded by a polynomial function of the input size, i.e., if there is a polynomial p such that for each input of size n the algorithm terminates after at most $p(n)$ steps. A problem is said to be polynomial bounded if there is a polynomial bounded algorithm for it. All of the problems and algorithms studied in Chapters 1 through 8 are polynomial bounded.

Definition   *P* is the class of decision problems that are polynomial bounded.

P is defined only for decision problems, but the reader usually will not go wrong by thinking of the kinds of problems studied earlier in this book as being in *P*.

It may seem rather extravagant to use existence of a polynomial time bound as the criterion for defining the class of more or less reasonable problems — polynomials can be quite large. There are, however, a number of good reasons for this choice. First, while it is not true that *every* problem in *P* has an acceptably efficient algorithm, we can certainly say that if a problem is *not* in *P*, it will be extremely expensive and probably impossible to solve in practice. All of the problems described at the beginning of this section are probably not in *P*; there are no algorithms for them that are known to be polynomial bounded, and most researchers in the field believe that no such algorithms exist. Thus while the definition of *P* may be too broad to provide a criterion for problems with truly reasonable time requirements, it provides a useful criterion — not being in *P* — for problems that are intractable.

A second reason for using a polynomial bound to define *P* is that polynomials have nice "closure" properties. An algorithm for a complex problem may be obtained by combining several algorithms for simpler problems. Some of the simpler algorithms may work on the output or intermediate results of others. The complexity of the composite algorithm may be bounded by addition, multiplication, and composition of the complexities of its component algorithms. Since polynomials are closed under these operations, any algorithm built from several polynomial-bounded algorithms in various natural ways will also be polynomial bounded. No smaller class of functions that are useful complexity bounds has these closure properties.

A third reason for using a polynomial bound is that it makes *P* independent of the particular formal model of computation used. A number of formal models (formal definitions of algorithms) are used to prove rigorous theorems about the complexity of algorithms and problems. The models differ in the kinds of operations permitted, the memory resources available, and the costs assigned to different operations. A problem that requires $\Theta(f(n))$ steps on one model may require more than $\Theta(f(n))$ steps on another, but for virtually all the realistic models, if a problem is polynomial bounded for one, then it is polynomial bounded for the others.

9.1.3 The Class *NP*

Many decision problems (including all our sample problems) are phrased as existence questions: Does there exist a k-coloring of the graph *G*? Does there exist a truth assignment that makes a given logical expression true? For a given input, a "solution" is an object (e.g., a graph coloring or a truth assignment) that satisfies the criteria in the problem and hence justifies a *yes* answer (e.g., the graph coloring uses at most k colors; the truth assignment makes the CNF expression true). A "proposed solution" is simply an object of the appropriate kind — it may or may not satisfy the criteria. Loosely speaking, *NP* is the class of decision problems for which a given proposed solution for a given input can be checked quickly (in polynomial time) to see if it really is a solution, i.e., if it satisfies all the requirements of the problem. A proposed solution may be described by a string of symbols from some finite set, for example, the set of characters on the keyboard of a computer terminal, We simply need some conventions for describing graphs, sets, functions, etc. using these symbols. The size of a string is the number of symbols in it. Checking a proposed solution includes checking that the string makes sense (that is, has the correct syntax) as a description of the required kind of object, as well as checking that it satisfies the criteria of the problem. Thus any string of characters can be thought of as a proposed solution.

Decision problems may arise where there is no natural interpretation for "solutions" and "proposed solutions." A decision problem is, abstractly, some function from a set of inputs to the set {yes, no}. A formal definition of NP considers all decision problems. The definition uses nondeterministic algorithms, which we define next. Although such algorithms are not realistic, useful algorithms in practice, they are useful for classifying problems.

A *nondeterministic algorithm* has two phases:

1. The nondeterministic phase. Some completely arbitrary string of characters, s, is written beginning at some designated place in memory. Each time the algorithm is run, the string written may differ. (This string may be thought of as a guess at a solution for the problem, so this phase may be called the guessing phase, but s could just as well be gibberish.)

2. The deterministic phase. A deterministic (i.e., ordinary) algorithm begins execution. In addition to the decision problem's input, the algorithm may read s, or it may ignore s. Eventually it halts with an output of *yes* or *no* — or it may go into an infinite loop and never halt. (Think of this as the checking phase — the deterministic algorithm is checking s to see whether it is a solution for the decision problem's input.)

The number of steps carried out during one execution of a nondeterministic algorithm is defined as the sum of the steps in the two phases; that is, the number of steps taken to write s (simply the number of characters in s) plus the number of steps executed by the deterministic second phase.

Normally, each time we run an algorithm with the same input, we get the same output. This does not happen with nondeterministic algorithms; for a particular input x, the output from one run may differ from the output of another because it may depend on s. So what is the "answer" computed by a nondeterministic algorithm, say A, for a particular input x? A's answer for x is defined to be *yes* if and only if there is *some* execution of A that can give a *yes* output. Using our informal notion of s as a proposed solution, A's answer for x is *yes* if and only if there is some proposed solution that "works."

For example, suppose that the problem is to determine whether a graph is k-colorable. The first phase of a nondeterministic algorithm will write some string s that we interpret as a proposed coloring. A proposed coloring could be described by a list of colors $c_1, c_2, \ldots, c_n$ to be assigned to the vertices $v_1, v_2, \ldots, v_n$, respectively. To verify that the coloring is valid, the second phase can check that there are indeed n colors listed, scan the list of edges in the graph (or scan an adjacency matrix) and check that the two vertices incident with one edge have different colors, and finally, count the distinct colors used and check that there are at most k.

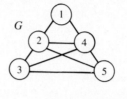

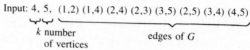

Figure 9.1 Input for nondeterministic graph coloring (Example 9.1).

Example 9.1 Nondeterministic graph coloring

Let $k = 4$ and let the graph G in Fig. 9.1 be the input. For simplicity, we will denote colors by letters B (blue), R (red), G (green), O (orange), and Y (yellow). (This is not a satisfactory notation in general because large graphs may need more than 26 colors.) Here is a list of a few possible strings s and the output that would result.

s	Output	Reason
RGRBG	*no*	v_2 and v_5, both green, are adjacent
RGRB	*no*	Not all vertices are colored
RBYGO	*no*	More than k colors
RGRBY	*yes*	A valid 4-coloring
R%*,G@	*no*	Bad syntax

Since there is one possible computation of the algorithm that produces a *yes* output, the answer for the input G is *yes*. ∎

A nondeterministic algorithm is said to be polynomial bounded if there is a polynomial p such that for each input of size n for which the answer is *yes*, there is some execution of the algorithm that produces a *yes* output in at most $p(n)$ steps.

Definition *NP* is the class of decision problems for which there is a polynomial-bounded nondeterministic algorithm. (The name *NP* comes from "*n*ondeterministic *p*olynomial bounded.")

Theorem 9.1 Graph coloring, the Hamilton path and circuit problems, job scheduling with penalties, bin packing, the subset sum problem, the knapsack problem, CNF-satisfiability, and the traveling salesperson problem are all in *NP*.

The proofs are straightforward and are left for the exercises. The work described earlier to check a possible graph coloring, for example, can be done easily in polynomial time.

Theorem 9.2 $P \subseteq NP$.

Proof. An ordinary (deterministic) algorithm for a decision problem is a special case of a nondeterministic algorithm. In other words, if *A* is a deterministic algorithm for a decision problem, let *A* be the second phase of a nondeterministic algorithm. *A* simply ignores whatever was written by the first phase and proceeds with its usual computation. A nondeterministic algorithm can do zero steps in the first phase (writing the null string), so if *A* runs in polynomial time, the nondeterministic algorithm with *A* as its second phase can also run in polynomial time and will give *A*'s correct *yes* or *no* answer.   □

The big question is: Does $P = NP$ or is *P* a proper subset of *NP*? In other words, is nondeterminism more powerful than determinism in the sense that some problems can be solved in polynomial time with a nondeterministic "guesser" that cannot be solved in polynomial time by an ordinary algorithm? If a problem is in *NP*, with polynomial time bound, say *p*, we can (deterministically) give the proper answer (*yes* or *no*) if we check all strings of length at most $p(n)$ — that is, run the second phase of the nondeterministic algorithm on each possible string, one at a time. The number of steps needed to check each string is at most $p(n)$. The trouble is that there are too many strings to check. If our character set contains *c* characters, then there are $c^{p(n)}$ strings of length $p(n)$. The number of strings is exponential, not polynomial, in *n*. Of course there is another way to solve problems: Use some properties of the objects involved and some cleverness to devise an algorithm that does not have to examine all possible solutions. When sorting, for example, we do not check each of the *n*! permutations of the given *n* keys to see which one puts the keys in order. The difficulty with the problems discussed in this chapter, however, is that this approach has not yielded efficient algorithms; all the known algorithms either examine all possibilities or, if they use tricks to reduce the work, the tricks are not good enough to give polynomial-bounded algorithms.

It is believed that *NP* is a much larger set than *P*, but there is not one single problem in *NP* for which it has been proved that the problem is not in *P*. No polynomial-bounded algorithms are known for many problems in *NP* (including all the sample problems in Section 9.1.1), but no larger-than-polynomial lower bounds have been proved for these problems. Thus the question we just asked, Does $P = NP$?, is still open.

9.1.4   The Size of the Input

Consider the following problem.

Given a positive integer *n*, are there integers $j, k > 1$ such that $n = jk$? (That is, is *n* nonprime?)

Is this problem in *P*? Consider the following algorithm, which looks for a factor of *n*.

```
found := false;
j := 2;
while not found and j < n do
    if n mod j = 0 then
        found := true else j := j+1
    end { if }
end { while }
```

The loop body is executed fewer than n times, so the running time of the algorithm is in $\Theta(n)$. Yet the problem of determining whether an integer is prime is *not* known to be in P, and, in fact, finding factors of large integers is the basis for various encryption algorithms exactly because it is considered a hard problem. What is the resolution of the apparent paradox?

The input for the prime-testing algorithm is the integer n, but what is the *size* of n? Until now, we have used any convenient and reasonable measure of input size; it was not important to count individual characters or bits. When our measure of the size of an input may make the difference of whether an algorithm is polynomial or exponential, we have to be more careful. The size of an input is the number of characters it takes to write the input. If $n = 150$, for example, we write three digits, not 150 digits. Thus an integer n written in decimal notation has size roughly $\log_{10} n$. If we choose to think of the binary representation used inside a computer then the size of n is roughly $\lg n$. These representations differ by a constant factor; that is, $\log_2 n = \log_2 10 \log_{10} n$, so which we use is not critical. The point, however, is that if the input size s is $\log_{10} n$ and the running time of an algorithm is n, then the running time of the algorithm is an exponential function of the input size ($n = 10^s$). Thus our algorithm for determining if n is prime is not in P. There is no algorithm presently known for prime testing in polynomial time.

In the problems we considered earlier in this book, the variable we used to describe the input size corresponded (more or less) to the amount of data in the input. For example, we used n as the input size when sorting a list of n keys. Each of the keys would be represented in, say, binary, but since there are n keys, there are at least n symbols in the input. So if the complexity of an algorithm is bounded by a polynomial in n, it is bounded by a polynomial in the exact size of the input. If each of two measures of input size, say n and $n \lg(\text{maximum key})$, which counts individual bits, is bounded by a polynomial function of the other, then determining whether the problem is in P will not depend on the specific measure used. So usually we do not have to be entirely precise about the input size. We must be careful, however, when the running time of an algorithm is expressed as a polynomial function of one of the input *values*, as is the case in the prime-testing problem.

A few of the sample problems described earlier have dynamic programming solutions that appear to be polynomial bounded at first glance, but, like our prime-testing program, are not. For example, recall the subset sum problem: Is there a subset of the n objects with sizes $s_1, s_2, \ldots, s_n$ that adds up to exactly C? This can be solved using an $n \times C$ table, with only a few operations needed to compute each table entry. (For $1 \leq i \leq n$ and $1 \leq j \leq C$, let *table*$[i,j]$ = *true* if and only if there is some subset of the first i objects whose sizes add up to exactly j. We leave it as an exercise to determine how to compute each entry from previously computed entries.) Similar dynamic programming solutions exist for various versions of the knapsack problem. The dynamic programming solution for the subset sum problem runs in $\Theta(nC)$ time. Since there are n objects in the input, the term n is no problem, but the number C is exponentially bigger (in general) than the input, because the datum C in the input would be represented in $\lg C$ bits. Thus the dynamic programming solution is not a polynomially bounded algorithm. Of course, if C is not too large, it may be useful in practice.

9.2
NP-Complete Problems

9.2.1 Polynomial Reductions

NP-complete is the term used to describe decision problems that are the hardest ones in *NP* in the sense that, if there were a polynomial-bounded algorithm for an *NP*-complete problem, then there would be a polynomial-bounded algorithm for each problem in *NP*.

Some of the sample problems described in Section 9.1.1 may seem easier than others and, in fact, the worst-case complexities of the algorithms that have been devised and analyzed for them do differ (they are fast-growing functions such as $2^{\sqrt{n}}$, 2^n, $(n/2)^{n/2}$, and $n!$), but, surprisingly, they are all equivalent in the sense that if any one is in P, they all are. They are all *NP*-complete.

The formal definition of *NP*-complete uses reductions, or transformations, of one problem to another. Suppose that we want to solve a problem Π_1 and that we already have an algorithm for another problem Π_2. Suppose that we also have a function T that takes an input x for Π_1 and produces $T(x)$, an input for Π_2 such that the correct answer for Π_1 on x is *yes* if and only if the correct answer for Π_2 on $T(x)$ is *yes*. Then, by composing T and the algorithm for Π_2, we have an algorithm for Π_1. See Fig. 9.2.

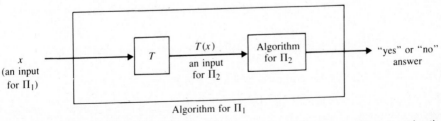

Figure 9.2 Reduction of problem Π_1 to problem Π_2. Π_2's answer for $T(x)$ must be the same as Π_1's answer for x.

Example 9.2 A simple reduction

Let the problem Π_1 be: Given n Boolean variables, does at least one of them have the value *true*? (In other words, this is a decision-problem version of computing Boolean *or*.) Let Π_2 be: Given n integers, is the maximum of the integers positive? Let $T(x_1, x_2, \ldots, x_n) = y_1, y_2, \ldots, y_n$ where $y_i = 1$ if $x_i = true$, and $y_i = 0$ if $x_i = false$. Clearly an algorithm to solve Π_2, when applied to $y_1, y_2, \ldots, y_n$, solves Π_1 for $x_1, x_2, \ldots, x_n$. ■

Definition Let T be a function from the input set for a decision problem Π_1 into the input set for a decision problem Π_2. T is a *polynomial reduction* (also called a *polynomial transformation*) from Π_1 to Π_2 if

1. T can be computed in polynomial-bounded time, and
2. For every input x for Π_1, the correct answer for Π_2 on $T(x)$ is the same as the correct answer for Π_1 on x.

Definition Π_1 is *polynomially reducible* (also called *polynomially transformable*) to Π_2 if there exists a polynomial transformation from Π_1 to Π_2. (We usually simply say that Π_1 is *reducible* to Π_2; the polynomial bound is understood.) The notation $\Pi_1 \propto \Pi_2$ is used to indicate that Π_1 is reducible to Π_2.

The point of the reducibility is that Π_2 is at least as "hard" to solve as Π_1. This is made more precise in the following theorem.

Theorem 9.3 If $\Pi_1 \propto \Pi_2$ and Π_2 is in P, then Π_1 is in P.

Proof. Let p be a polynomial bound on the computation of T, and let q be a polynomial bound on an algorithm for Π_2. Let x be an input for Π_1 of size n. Then the size of $T(x)$ is at most $p(n)$ (since, at worst, a program for T writes a symbol at each step). When the algorithm for Π_2 is given $T(x)$, it does at most $q(p(n))$ steps. So the total amount of work to transform x to $T(x)$ and then use the Π_2 algorithm to get the correct answer for Π_1 on x is $p(n)+q(p(n))$, a polynomial in n. □

Now we can give the formal definition of *NP*-complete.

Definition A problem Π is *NP-complete* if it is in *NP* and for every other problem Π' in *NP*, $\Pi' \propto \Pi$.

The following theorem follows easily from the definition and Theorem 9.3.

Theorem 9.4 If any *NP*-complete problem is in P, then $P = NP$.

This theorem indicates, on the one hand, how valuable it would be to find a polynomial-bounded algorithm for any *NP*-complete problem and, on the other, how unlikely it is that such an algorithm exists because there are so many problems in *NP* for which polynomial-bounded algorithms have been sought without success.

The first major theorem demonstrating that a specific problem is *NP*-complete is the following. (The proof of this and other theorems stated here without proof can be found in the references.)

Theorem 9.5   (Cook's theorem) The CNF-satisfiability problem is *NP*-complete.

The proof must show that any problem Π in *NP* is reducible to CNF-satisfiability. To do so, it gives an algorithm to construct a CNF logical expression for an input x for Π such that the expression, informally speaking, describes the computation of a nondeterministic algorithm for Π acting on x. The expression, which is very long but constructed in time bounded by a polynomial function of the length of x, will be satisfiable if and only if the computation produces a *yes* answer.

Once Theorem 9.5 was proved and the importance of *NP*-complete problems was recognized, many of the problems for which polynomial-bounded algorithms were being sought were shown to be *NP*-complete. In fact, the list of *NP*-complete problems grew to many hundreds in the 1970s.

Theorem 9.6   Graph coloring, the Hamilton path and circuit problems, job scheduling with penalties, bin packing, the subset sum problem, the knapsack problem, and the traveling salesperson problem are all *NP*-complete.

To prove that a problem in *NP* is *NP*-complete, it suffices to prove that some other *NP*-complete problem is polynomially reducible to it since the reducibility relation is transitive. Hence a theorem like Theorem 9.6 is proved by establishing chains of transformations beginning with the satisfiability problem. We will do a few as examples.

Students often become confused about the direction of the reduction needed to prove that a problem is *NP*-complete, so we emphasize: To show that the problem Π is *NP*-complete, choose some known *NP*-complete problem Π' and reduce Π' to Π, not the other way around. The logic is as follows:

Since Π' is *NP*-complete, all problems in $NP \propto \Pi'$.
Show $\Pi' \propto \Pi$.
Then all problems in $NP \propto \Pi$.
Therefore, Π is *NP*-complete.

Theorem 9.7   The directed Hamilton circuit problem is reducible to the undirected Hamilton circuit problem. (Thus if we know that the directed Hamilton circuit problem is *NP*-complete, we can conclude that the undirected Hamilton circuit problem is also *NP*-complete.)

Proof. Let $G = (V, E)$ be a directed graph with n vertices. G is transformed into the undirected graph $G' = (V', E')$, where $V' = \{v^i \mid v \in V, \ i = 1, 2, 3\}$ and $E' = \{v^1v^2, v^2v^3 \mid v \in V\} \cup \{v^3w^1 \mid vw \in E\}$. In other words, each vertex of G is expanded to three vertices connected by two edges, and an edge vw in E becomes an edge from the third vertex for v to the first for w. See Fig. 9.3 for an illustration.

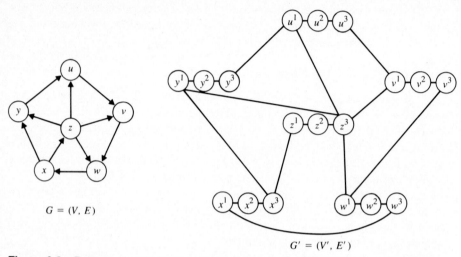

Figure 9.3 Reduction of the directed Hamilton circuit problem to the undirected Hamilton circuit problem.

The transformation is straightforward, and G' can certainly be constructed in polynomial-bounded time. If $|V| = n$ and $|E| = m$, then G' has $3n$ vertices and $2n+m$ edges.

Now suppose that G has a (directed) Hamilton circuit $v_1, v_2, \ldots, v_n$. (That is, $v_1, v_2, \ldots, v_n$ are distinct, and there are edges $v_i v_{i+1}$, for $1 \leq i < n$, and $v_n v_1$.) Then $v_1^1, v_1^2, v_1^3, v_2^1, v_2^2, v_2^3, \ldots, v_n^1, v_n^2, v_n^3$ is an undirected Hamilton circuit for G'. On the other hand, if G' has an undirected Hamilton circuit, the three vertices, say $v^1, v^2,$ and v^3, that correspond to one vertex from G must be traversed consecutively in the order v^1, v^2, v^3 or v^3, v^2, v^1 since v^2 cannot be reached from any other vertex in G'. Since the other edges in G' connect vertices with superscripts 1 and 3, if for any one triple the order of the superscripts is $1, 2, 3$, then the order is $1, 2, 3$ for all triples. Otherwise, it is $3, 2, 1$ for all triples. Since G' is undirected, we may assume that its Hamilton circuit is $v_{i_1}^1, v_{i_1}^2, v_{i_1}^3, \ldots, v_{i_n}^1, v_{i_n}^2, v_{i_n}^3$. Then $v_{i_1}, v_{i_2}, \ldots, v_{i_n}$ is a directed Hamilton circuit for G. Thus G has a directed Hamilton circuit if and only if G' has an undirected Hamilton circuit. $\qquad \square$

It is, of course, much easier to see that the G' defined in the proof is the proper transformation to use than it is to think up the correct G' in the first place, so we make a few observations to indicate how G' was chosen. To ensure that a circuit in G' corresponds to a circuit in G, we must simulate the directedness of the edges of G. This aim suggests giving G' two vertices, say v^1 and v^3, for each v in G with the interpretation that v^1 is used for edges in G whose head is v and v^3 is used for edges whose tail is v. Then wherever v^1 and v^3 appear consecutively in a circuit for G' they can be replaced by v to get a circuit for G, and vice versa. Unfortunately, there is nothing about G' that forces v^1 and v^3 to appear consecutively in all of its circuits; thus G' could have a Hamilton circuit that does not correspond to one in G (see Exercise 9.9). The third vertex, v^2, which can be reached only from v^1 and v^3, is introduced to force the vertices that correspond to v to appear together in any circuit in G'.

Theorem 9.8   The subset sum problem is reducible to the job scheduling problem.

Proof.   Let $s_1, \ldots, s_n$, C be an input I for the subset sum problem (which asks whether there is a subset of the objects that adds up to exactly C). If $\sum_{i=1}^{n} s_i < C$, then the output for I is *no*, and I can be transformed to any job scheduling input with a *no* output, e.g., $t_i = 2$, $d_i = p_i = 1$, and $k = 0$. If $\sum_{i=1}^{n} s_i \geq C$, then I is transformed into the following input: $t_i = p_i = s_i$ and $d_i = C$ for $1 \leq i \leq n$, and $k = \sum_{i=1}^{n} s_i - C$. Clearly the transformation itself takes little time.

Now suppose that the subset sum input produces a *yes* answer; i.e., there is a subset J of $N = \{1, 2, \ldots, n\}$ such that $\sum_{i \in J} s_i = C$. Then let π be any permutation of N that causes all jobs with indexes in J to be done before any jobs with indexes in $N-J$. The first $|J|$ jobs are completed by their deadline since $\sum_{i \in J} t_i = \sum_{i \in J} s_i = C$, and C is the deadline for all jobs. The penalty for the remaining jobs is

$$\sum_{i=|J|+1}^{n} p_{\pi(i)} = \sum_{i=|J|+1}^{n} s_{\pi(i)} = \sum_{i=1}^{n} s_i - \sum_{i \in J} s_i = \sum_{i=1}^{n} s_i - C = k.$$

Thus the jobs can be done with total penalty $\leq k$.

Conversely, let π be any schedule for the jobs with total penalty $\leq k$. Let m be largest such that

$$\sum_{i=1}^{m} t_{\pi(i)} \leq C; \tag{9.1}$$

i.e., m is the number of jobs completed by the deadline C. The penalty, then, is

$$\sum_{i=m+1}^{n} p_{\pi(i)} \leq k = \sum_{i=1}^{n} s_i - C. \tag{9.2}$$

See Fig. 9.4 for an illustration. Since $t_i = p_i = s_i$ for $1 \leq i \leq n$, we must have

$$\sum_{i=1}^{m} t_{\pi(i)} + \sum_{i=m+1}^{n} p_{\pi(i)} = \sum_{i=1}^{n} s_i,$$

and this can happen only if the inequalities in Eqs. 9.1 and 9.2 are equalities (i.e., if the shaded areas in Fig. 9.4 are zero). Thus $\sum_{i=1}^{m} t_{\pi(i)} = C$, so the objects with indexes $\pi(1), \ldots, \pi(m)$ are a solution to the subset sum problem. $\square$

There are similar problems in the exercises.

9.2.2   What Makes a Problem Hard?

If the set of inputs for an *NP*-complete problem is restricted in some way, the problem may be in *P*; in fact, it may have a very fast solution. On the other hand, of

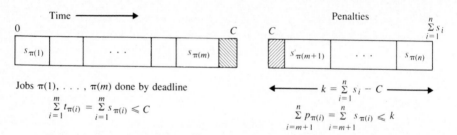

Figure 9.4 A satisfactory job schedule solves the subset sum problem.

course, even with restrictions, the problem may still be *NP*-complete. Knowing the effect on complexity of restricting the set of inputs for a problem is important because, in many applications, the inputs that actually occur have special properties that might allow a polynomial-bounded solution. Unfortunately, the results are discouraging; even with quite strong restrictions on the inputs, many *NP*-complete problems are still *NP*-complete.

For graph problems we can restrict $\Delta(G)$, the maximum degree of the vertices in the input graph G. (The degree of a vertex is the number of edges incident with it.) It is easy to test most graph properties on graphs with $\Delta \leq 2$. For such graphs, the Hamilton circuit problem and the k-colorability problem can be solved in polynomial time. However, even for graphs with $\Delta \leq 3$, the Hamilton circuit problem is *NP*-complete. For graphs with $\Delta \leq 4$, k-colorability is *NP*-complete. Thus it is not the possibility that vertices may have high degree that makes these problems hard.

A *planar* graph or digraph is one that can be drawn in a plane such that no two edges intersect. Planar graphs occur in many applications, so it is well worth knowing how hard various problems are if the inputs are restricted to such graphs. (Determining whether an arbitrary graph or digraph is planar is an important problem in itself; fortunately, it is known to be in *P*. The best algorithms for testing planarity are complicated but run in linear time.) The directed Hamilton path problem is *NP*-complete even when restricted to planar digraphs. The *vertex cover problem* asks, for a given graph and positive integer k, whether the graph has a subset of k vertices such that each edge is incident with one of the vertices in the subset. The vertex cover problem is *NP*-complete, and it is still *NP*-complete when restricted to planar graphs. Planarity simplifies another important problem, though. A *k-clique* in a graph is a subgraph consisting of k mutually adjacent vertices (i.e., a complete graph on k vertices.) The problem of determining whether a graph has a k-clique is *NP*-complete, but for planar graphs it is in *P* because a planar graph cannot have a clique with more than four vertices. (The clique problem is also in *P* for graphs with bounded degrees.)

The 3-CNF satisfiability problem is the CNF-satisfiability problem restricted to expressions with exactly three literals per clause. It is *NP*-complete. If there are at most two literals per clause, satisfiability can be checked in polynomial-bounded time. However, consider this variation of the problem: Given a CNF expression with at most two literals per clause, and given an integer k, is there a truth assignment for the variables that satisfies at least k clauses? This problem is *NP*-complete.

Another interesting phenomenon, illustrated by some of the following examples, is that two problems that seem to differ only slightly in their statement may differ very much in complexity; one may be in P while the other is *NP*-complete.

Although the vertex cover problem is *NP*-complete, its dual, the edge cover problem — Is there a set of k edges such that each vertex is incident with one of them? — is in P.

In Chapter 4 we saw that efficient algorithms exist for finding the shortest path between two specified vertices in a graph. The longest path problem is *NP*-complete. (The decision problem formulation for these two problems includes an integer k as input and asks whether there is a path shorter than k, or a path longer than k, respectively.)

Determining if a graph is 2-colorable is easy; determining if it is 3-colorable is *NP*-complete. It is still *NP*-complete if the graphs are planar and the maximum degree is 4.

The *feedback edge set problem* is: Given a digraph $G = (V, E)$ and an integer k, is there a subset $E' \subseteq E$ such that $|E'| \leq k$ and E' contains an edge from every (directed) cycle in G? This problem is *NP*-complete; however, the same problem for undirected graphs is in P.

The problem of job scheduling with penalties is *NP*-complete, but if the penalties are omitted and we simply ask whether there is a schedule such that at most k jobs miss their deadlines, then the problem is in P.

These examples do not yield any nice generalizations about *why* a problem is *NP*-complete. There are still a great many open questions in this field, the main one being, of course, Does $P = NP$?

9.2.3 Optimization Problems and Decision Problems

In our descriptions of sample *NP*-complete problems in Section 9.1.1, we included two aspects of the optimization problems: We may ask for the optimal solution *value*, e.g., the chromatic number of a graph or the minimum number of bins into which a set of objects fit, or we may ask for an actual solution (a coloring of the graph, a packing of the objects) that achieves the optimal value. Thus we have three kinds of problems:

Decision problem: Is there a solution better than some given bound?
Optimal value: What is the value of a best possible solution?
Optimal solution: Find a solution that achieves the optimal value.

It is easy to see that these problems are listed in order of increasing difficulty. For example, if we have an optimal coloring for a graph, we need only count the colors to determine the graph's chromatic number, and if we know its chromatic number, it is trivial to determine whether the graph is k-colorable for any given k. In real applications we usually want an optimal (or nearly optimal) solution. It has been easier to work out the theory of *NP*-completeness for decision problems, and since the optimization problems are at least as hard to solve as the related decision problems, we have not lost anything essential by doing so. That is, our comments about the difficulty of *NP*-complete decision problems apply to the optimization problems associated with them, and we will informally refer to these optimization problems as being *NP*-complete also.

But suppose it turns out that $P = NP$. If we had polynomial time algorithms for the decision problems, could we then find the optimal solution value in polynomial time? In many cases it is easy to see that we could. Consider graph coloring. Suppose we have a polynomial time Boolean function subprogram *CanColor*(G, k) that has value *true* if and only if the graph G can be colored with k colors. Then we can write the following program:

```
k := 0;
repeat
    k := k+1;
    cando := CanColor(G, k)
until cando;
chromaticNumber := k
```

Since any graph with n vertices can be colored with n colors, we know there will be at most n iterations of the **repeat** loop, so if *CanColor* runs in polynomial time, so does the whole program.

The same technique will show that for some other problems as well, if we can solve the decision problem in polynomial time, we can find the optimal solution value in polynomial time. However, it is not always this simple. Consider the traveling salesperson problem. We are given a complete graph with an integer cost assigned to each edge, and we want to find the cost of a minimum tour, or Hamilton circuit. If *TSPbound*(G, k) is a function with value *true* if and only if there is a tour of cost at most k, then the following program finds the cost of a minimum tour:

```
k := 0;
repeat
    k := k+1;
    cando := TSPbound(G, k)
until cando;
minTourCost := k
```

How many iterations of the loop can there be? Let W be the maximum of the edge weights. Since there are n edges in a Hamilton circuit, the weight of a minimum tour is at most nW, and so there will be at most nW iterations. Unfortunately, as the discussion of input size in Section 9.1.4 indicates, this is not good enough to conclude that the program runs in polynomial time. We leave it for the exercises to show that this program can be modified to find the cost of a minimum tour in polynomial time, and, in fact, that an optimal tour can be found in polynomial time — both of course, under the assumption that the decision problem has a polynomial time solution. (See Exercise 9.44.)

9.3
Approximation Algorithms

Many hundreds of important applications problems are *NP*-complete. What can we do if we must solve one of these problems? There are several possible approaches. Even though no polynomial-bounded algorithm may exist, there may still be significant differences in the complexities of the known algorithms; we can try, as usual, to develop the most efficient one possible. We might concentrate on average rather than worst-case behavior and look for algorithms that are better than others by that criterion, or, more realistically, we might seek algorithms that seem to work well for the inputs that usually occur; this choice may depend more on empirical tests than on rigorous analysis. In this section we study a different approach for solving *NP*-complete optimization problems: the use of fast (i.e., polynomial-bounded) algorithms that are not guaranteed to give the best solution but will give one that is close to the optimal. Such algorithms are called approximation algorithms or heuristic algorithms. In many applications an approximate solution is good enough, especially when the time required to find an optimal solution is considered. You do not win by finding an optimal job schedule, for example, if the cost of the computer time needed to find it exceeds the worst penalty you might have paid.

The strategies, or heuristics, used by many of the approximation algorithms are simple and straightforward, yet for some problems they provide surprisingly good results. To make precise statements about the behavior of an approximation algorithm (how good its results are, not how much time it takes), we need several definitions. In the following paragraphs, assume that we are considering a particular optimization problem and a particular input I.

FS_I is the set of feasible solutions for I. A feasible solution is an object of the right type but not necessarily an optimal one. For example, if the problem is to find an optimal graph coloring for a given graph G, FS_G is the set of all colorings; that is, if $G = (V, E)$, $FS_G = \{C{:}V \rightarrow \{1, 2, \ldots, |V|\}$ such that $C(v) \neq C(w)$ if $vw \in E\}$. (The "colors" are integers.) For the bin packing problem and an input $I = \{s_1, \ldots, s_n\}$, FS_I is the set of all packings, that is, all partitions of the index set $\{1, 2, \ldots, n\}$ into disjoint subsets $T_1, \ldots, T_p$ (for some p) such that $\sum_{i \in T_j} s_i \leq 1$ for $1 \leq j \leq p$. The set of feasible solutions for an input to the job scheduling problem is the set of permutations of the n jobs.

The function $v_I : FS_I \rightarrow \mathbf{N}$ assigns a nonnegative integer value to each feasible solution. For graph colorings, $v_G(C) = |C(V)|$, i.e., the number of colors used by the coloring C. For bin packing, if $T_1, \ldots, T_p$ is a feasible partition of the objects for an input I, $v_I(T_1, \ldots, T_p) = p$, the number of bins used. For job scheduling $v_I(\pi) = P_\pi$, the penalty for the schedule π. It should be easy for the reader to identify the feasible solution sets and the solution value functions for other optimization problems.

Depending on the problem, we want to find a solution that either minimizes or maximizes v; let "best" be "min" or "max," respectively. Then $opt(I) = best\{v_I(x) \,|\, x \in FS_I\}$. An optimal solution is an x in FS_I such that $v_I(x) = opt(I)$.

An *approximation algorithm* for a problem is a polynomial time algorithm that, when given input I, outputs an element of FS_I. There are several ways to describe how good a solution the approximation algorithm provides. Usually it is most useful to look at the ratio between the value of the algorithm's output and the value of an optimal solution (though sometimes we might want to look at the absolute difference between the two). Let A be an approximation algorithm. We denote by $A(I)$ the feasible solution A chooses for I. We define

$$r_A(I) = \frac{v_I(A(I))}{opt(I)}$$

if the problem is a minimization one, and

$$r_A(I) = \frac{opt(I)}{v_I(A(I))}$$

if it is maximization. In both cases, $r_A(I) \geq 1$. To summarize the behavior of A we would like to consider the worst-case ratio. Again, there are several choices: We could consider the worst-case ratio for all inputs of a certain size, or for all inputs with a certain optimal solution value, or for all inputs. Different approaches are useful for different problems. We define the following functions:

$$R_A(m) = \max\{r_A(I) \mid I \text{ such that } opt(I) = m\}$$

$$S_A(n) = \max\{r_A(I) \mid I \text{ of size } n\}.$$

(Note that $R_A(m)$ may be infinite for some m.) For some *NP*-complete problems, the maximum ratio is not well defined; this can occur when the set of inputs being considered is infinite. Thus it is sometimes necessary to define R and S as follows:

$$R_A(m) = \inf\{r \mid r_A(I) \leq r \text{ for all } I \text{ with } opt(I) = m\}$$

$$S_A(n) = \inf\{r \mid r_A(I) \leq r \text{ for all } I \text{ of size } n\}.$$

For some problems there are approximation algorithms for which R and S are arbitrarily close to 1, for others they are bounded by small constants, and for still others no algorithms guaranteed to produce reasonably close solutions are known. For some problems it can be shown that finding a nearly optimal solution is as hard as finding an optimal solution. We will present some approximation algorithms in the next few sections.

9.4
Bin Packing

9.4.1　The *First Fit* Strategy

Let $S = (s_1, \ldots, s_n)$ where $0 < s_i \leq 1$ for $1 \leq i \leq n$. The problem is to pack $s_1, \ldots, s_n$ into as few bins as possible, where each bin has capacity 1. An optimal solution can be found by considering all the ways to partition a set of n items into n or fewer

$S = (0.8, 0.5, 0.4, 0.4, 0.3, 0.2, 0.2, 0.2)$

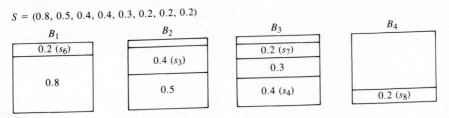

Figure 9.5   Example of nonincreasing first fit.

subsets, but the number of possible partitions is more than $(n/2)^{n/2}$. The approximation algorithm we present here uses a very simple strategy, called *first fit*; it has worst-case time complexity in $\Theta(n^2)$ and produces good solutions. The *first fit* strategy places an object in the first bin into which it fits. The following algorithm sorts the objects first so that they are considered in order of nonincreasing size. (This algorithm is called *FFD*, for *first fit decreasing*, by other authors. Since the sizes need not be distinct, we prefer the name *nonincreasing first fit*.) An example appears in Fig. 9.5.

Algorithm 9.1   Nonincreasing First Fit

Input: $S = (s_1, \ldots, s_n)$, where $0 < s_i \le 1$ for $1 \le i \le n$.

Output: An array *bin* where for $1 \le i \le$ n, *bin*[*i*] is the number of the bin into which s_i is placed.

> **procedure** *Niff* (*S: RealArray; n: integer;* **var** *bin: IntegerArray*);
> **var** *used*: **array**[1..*n*] **of** *real*;
>   { *used*[*j*] is the amount of space in bin *j* already used up. }
>   *i, j: integer*;
> **begin**
>   sort *S* into nonincreasing order;
>   { To keep the notation simple, we will refer to the objects
>   by their index in the sorted list, not the original list. }
>   **for** *j* := 1 **to** *n* **do** *used*[*j*] := 0 **end**;
>   **for** *i* := 1 **to** *n* **do**
>   { Look for a bin in which s_i fits. }
>   *j* := 1;
>   **while** *used*[*j*]+s_i > 1 **do** *j* := *j*+1 **end**;
>   *bin*[*i*] := *j*;
>   *used*[*j*] := *used*[*j*]+s_i
>   **end** { for }
> **end** { Niff }

The sort can be done in $\Theta(n \lg n)$ time, and *j* is incremented while searching for an appropriate bin at most $n(n-1)/2$ times. All of the other instructions are executed at most *n* times, so the worst-case time complexity is in $O(n^2)$.

Niff does not always give optimal packings; the packing in Fig. 9.5 is not optimal. Theorem 9.11, which gives upper bounds on the worst packings produced by *Niff*, is established via the next two lemmas. After the theorem we will mention some results about how well *Niff* does on the average.

Lemma 9.9   Let $S = (s_1, \ldots, s_n)$ be an input for the bin packing problem and let $opt(S)$ be the optimal (i.e., minimum) number of bins for S. All of the objects placed by *Niff* in extra bins (i.e., bins with index larger than $opt(S)$) have size at most 1/3.

Proof.   We assume that S is already sorted. Let i be the index of the first object placed by *Niff* in bin $opt(S)+1$. Since S is sorted in nonincreasing order, it suffices to show that $s_i \le 1/3$. We examine the contents of the bins at the time s_i is considered by *Niff*. Suppose that $s_i > 1/3$. Then $s_1, \ldots, s_{i-1} > 1/3$, so bins B_j for $1 \le j \le opt(S)$ contain at most two objects each. We claim that for some $k \ge 0$ the first k bins contain one object each and the remaining $opt(S)-k$ bins contain two each. Otherwise there would be bins B_p and $B_{p'}$ as in Fig. 9.6, with $p < p'$ such that B_p has two objects, say t and t' (with $t \ge t'$), and $B_{p'}$ only one, $\hat{t}$. Since the objects are considered in nonincreasing order, $t \ge \hat{t}$ and $t' \ge s_i$; so $1 \ge t+t' \ge \hat{t}+s_i$, and *Niff* would have put s_i in $B_{p'}$.

Thus the bins are filled by *Niff* as in Fig. 9.7. Since *Niff* did not put any of $s_{k+1}, \ldots, s_i$ into the first k bins, none of them can fit. Therefore, in an optimal solution there will be k bins that do not contain any of the objects $s_{k+1}, \ldots, s_i$; without loss of generality, we may assume that they are the first k bins. Then, in an optimal solution, although they may not be arranged exactly as in Fig. 9.7, $s_{k+1}, \ldots, s_{i-1}$ will be in bins $B_{k+1}, \ldots, B_{opt}$, and since these objects are all larger than 1/3, there will be two in each bin and $s_i > 1/3$ cannot fit. But an optimal solution must fit s_i into one of the first $opt(S)$ bins; therefore the assumption that $s_i > 1/3$ must be false. $\square$

Lemma 9.10   For any input $S = (s_1, \ldots, s_n)$ the number of objects placed by *Niff* in extra bins is at most $opt(S)-1$.

Proof.   Since all the objects fit into $opt(S)$ bins, $\sum_{i=1}^{n} s_i \le opt(S)$. Suppose that *Niff* puts $opt(S)$ objects with sizes $t_1, \ldots, t_{opt(S)}$ into extra bins, and let b_j be the final contents of bin B_j for $1 \le j \le opt(S)$. If $b_j + t_j \le 1$, *Niff* could have put t_j in B_j, so

$$\sum_{i=1}^{n} s_i \ge \sum_{j=1}^{opt(S)} b_i + \sum_{j=1}^{opt(S)} t_i = \sum_{j=1}^{opt(S)} (b_i + t_i) > opt(S),$$

which is impossible.

$\square$

Theorem 9.11 $R_{Niff}(m) \le 4/3 + 1/(3m)$. $S_{Niff}(n) \le 3/2$, and for infinitely many n, $S_{Niff} = 3/2$.

Proof.   Let $S = (s_1, \ldots, s_n)$ be an input with $opt(S) = m$. *Niff* puts at most $m-1$ objects, each of size at most 1/3, in extra bins, so *Niff* uses at most $m + \lceil (m-1)/3 \rceil$ bins. Thus

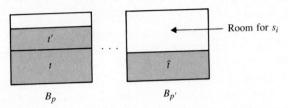

Figure 9.6   First illustration for the proof of Lemma 9.9.

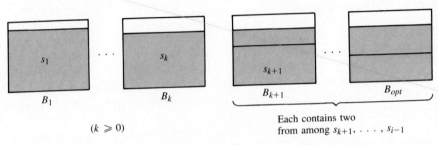

Figure 9.7   Second illustration for the proof of Lemma 9.9.

$$r_{Niff}(S) \le \frac{m + \lceil (m-1)/3 \rceil}{m} = 1 + \frac{m + \alpha}{3m} \quad \text{(where } -1 \le \alpha \le 1) \le \frac{4}{3} + \frac{1}{3m}.$$

Thus $R_{Niff} \le 4/3 + 1/(3m)$. For input size n, $r_{Niff}(S)$ is largest for $m = 2$ (if $m = 1$, *Niff* uses only one bin), so $S_{Niff}(n) \le 4/3 + 1/6 = 3/2$. Construction of a sequence of examples I_n for arbitrarily large n where $r_{Niff}(I_n) = 3/2$ is left as an exercise.   ☐

A stronger result than that stated in Theorem 9.11 is known: The number of extra bins used by *Niff* is bounded by $2opt/9 + 4$, i.e., about 22 percent of the optimal number. (That is, $R_{Niff}(m) \le 11/9 + 4/m$.) For arbitrarily large m, there are examples that show $R_{Niff}(m) \ge 11/9$, so we cannot improve the bound on the worst packings produced by *Niff*.

Niff usually does much better than these worst-case bounds would suggest. To determine the expected (average) number of extra bins used by *Niff* (i.e., the excess over the optimal number needed), extensive empirical studies have been done on large inputs. The data were generated randomly for various distributions. The alert reader may wonder how extensive studies of the number of extra bins used for large inputs could be done. Don't we have to know the number of optimal bins to determine the number of extra bins used by *Niff*? We are developing approximation algorithms because it takes too long to determine the number of optimal bins for large inputs! In fact, the empirical studies did not determine the optimal number of bins exactly; they estimated the number of extra bins by the amount of empty space in the packings produced by *Niff*. The empty space is the number of bins used by *Niff* minus $\sum_{i=1}^{n} s_i$. Clearly the number of extra bins used in a packing is bounded by the amount of empty space.

For inputs S with $n = 128,000$ and object sizes uniformly distributed between zero and one, *Niff* produced packings using roughly 64,000 bins. The strongest worst-case bound (mentioned previously) guarantees that the number of extra bins is at most $2opt(S)/9 + 4 \le 2 \cdot 64,000/9 + 4 \approx 14,200$. In fact, there were only about 100 units of empty space in the *Niff* packings. It has been shown that for n objects with sizes uniformly distributed between zero and one, the expected amount of empty space in packings by *Niff* is approximately $0.3\sqrt{n}$. Hence the expected number of extra bins is at most roughly $0.3\sqrt{n}$.

9.4.2 Other Heuristics

The first fit strategy can be used without sorting the objects. The results are not as good as for *Niff*, but it can be shown that the number of extra bins used by *first fit* is at most about 70 percent more than the optimal (and some examples are that bad). Empirical studies have shown that the expected behavior of *first fit* is not bad. For $n = 128,000$, for example, the number of extra bins used was no more than about 2 percent of the total number of bins used.

Another heuristic strategy is *best fit*: An object of size s is placed in a bin B_j, which is the fullest of those bins in which the object fits; i.e., $used[j]$ is maximum subject to the requirement $used[j] + s \le 1$. If the s_i are sorted in nonincreasing order, the best fit strategy works about as well as *Niff*. If the s_i are not sorted, the results can be worse but the number of bins would still be smaller than twice the optimal.

Another strategy, even simpler than first fit and best fit, gives an approximation algorithm that is faster and can be used in circumstances where the contents of all the bins cannot be stored but must be output as the packing progresses. The strategy is called *next fit*. The s_i are not sorted. One bin is filled at a time. Objects are put in the current bin until the next one does not fit; then a new bin is started and no more objects are packed in bins considered earlier.

Example 9.3 The *next fit* strategy

Let $S = (0.2, 0.2, 0.7, 0.8, 0.3, 0.6, 0.3, 0.2, 0.6)$. The objects are placed in six bins, as in Fig. 9.8, although they would fit in four. ∎

Clearly the next fit strategy can be implemented in a linear time algorithm. It may seem, however, that next fit will use a lot of extra bins. In fact, its worst-case behavior is worse than *Niff*, but the observation that the sum of the contents of any

$S = (0.2, 0.2, 0.7, 0.8, 0.3, 0.6, 0.3, 0.2, 0.6)$

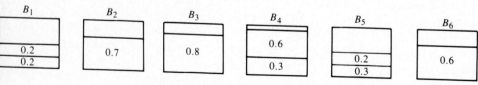

Figure 9.8 Example of next fit.

two consecutive bins must be greater than 1 allows us to conclude that $R_{nextfit}(m) < 2$.

For some bin packing strategies, if the s_i are bounded by some number less than one, better (i.e., lower) bounds on the ratio of actual to optimal output can be proved.

9.5
The Knapsack and Subset Sum Problems

An input for the knapsack problem consists of C and two vectors, $(s_1, \ldots, s_n)$ and $(p_1, \ldots, p_n)$. The problem is to find a subset T of the indexes $\{1, 2, \ldots, n\}$ that maximizes $\sum_{i \in T} p_i$ subject to the constraint $\sum_{i \in T} s_i \leq C$. Using the terminology and notation of Section 9.3, $FS_I = \{T \mid T \subseteq \{1, 2, \ldots, n\}$ and $\sum_{i \in T} s_i \leq C\}$; in other words, an approximation algorithm must output a set of objects that fits in the knapsack. The value of a feasible solution T, i.e., $v_I(T)$, is $\sum_{i \in T} p_i$, the total profit of the objects specified by T. (The subscript I will henceforth be omitted.) An optimal solution can be found by computing $v(T)$ for each $T \subseteq \{1, 2, \ldots, n\}$, but there are 2^n such subsets.

We will describe some approximation algorithms for a slightly simpler version of the problem in which the profit for each object is the same as its size. Thus the input is $(s_1, s_2, \ldots, s_n)$ and C, and we want to find a subset T of $\{1, 2, \ldots, n\}$ to maximize $\sum_{i \in T} s_i$ subject to the requirement that $\sum_{i \in T} s_i \leq C$. This is the optimization version of the subset sum problem. The algorithms can be extended to the general knapsack problem by starting with the list of objects in order by "profit density"; i.e., sorting so that $p_1/s_1 \geq p_2/s_2 \geq \cdots \geq p_n/s_n$. At a few places in the algorithms, references to sizes would have to be replaced by references to profits; these should be obvious. The theorems about the closeness of the approximations can be carried over to the general knapsack problem.

There is a very simple "greedy" strategy. Go through the list of objects and put each one in the knapsack if it fits. Let V be the sum of sizes of the objects chosen, and let $M = \max\{s_i \mid 1 \leq i \leq n\}$. If $V < M$, dump everything out of the knapsack and put in an object of value M. It is not hard to show that with this strategy the sum of the objects chosen will be at least half the optimal. ($R_{Greedy} \leq 2$.) We can do much better.

We will present a sequence of polynomial-bounded algorithms A_k for which the ratio of the optimal solution to the algorithm's output is $1 + 1/k$. Hence we can get as close to optimal as we choose. However, the amount of work done by A_k is in $O(kn^{k+1})$. Thus the closer the approximation, the higher the degree of the polynomial describing the time bound. Using the main idea in these algorithms along with an additional trick, it is possible to obtain a sequence of algorithms that achieve equally good results but that run in time $O(n + k^2 n)$. (See the notes and references at the end of this chapter.)

For $k \geq 0$ the algorithm A_k considers each subset T with at most k elements. If $\sum_{i \in T} s_i \leq C$, it goes through the remaining objects, $\{s_i \mid i \notin T\}$, and adds to the knapsack all that fit. The output is the set so obtained that gives the largest sum. An example follows the algorithm.

Algorithm 9.2   Approximation Algorithm A_k for the (Simplified) Knapsack Problem

Input: $s_1, s_2, \ldots, s_n$, and C, positive integers.

Output: *set*, a subset of $\{1, 2, \ldots, n\}$; and $maxSum = \sum_{i \in set} s_i$.

```
procedure Knapsack (S: IntegerArray; C: integer;
                var set: IndexSet; var maxSum: integer);
var
    T: IndexSet;
    j, sum: integer;
begin
    set := ∅; maxSum := 0;
    for each subset T of {1, 2, ..., n} with at most k elements do
        sum := ∑_{i ∈ T} s_i;
        if sum ≤ C then
            { Consider remaining objects. }
            for j := 1 to n do
                if j ∉ T and sum+s_j ≤ C then
                    sum := sum+s_j;
                    T := T ∪ {j}
                end { if }
            end { for j };
            { See if T fills the knapsack more than the best set found so far. }
            if maxSum < sum then
                maxSum := sum;
                set := T
            end { if }
        end { if sum ≤ C }
    end { for each subset }
end { Knapsack }
```

Note that tests "**if** $sum = C$" could be added in appropriate places to avoid considering extra elements and subsets once the knapsack has been filled to capacity. Such tests were omitted to keep the description of the strategy as clear as possible.

Example 9.4

Suppose that the input for the problem is $(54, 45, 43, 29, 23, 21, 14, 1)$ and that $C = 110$. Table 9.1 shows the subsets considered by A_0 and A_1. The optimal solution is $(43, 29, 23, 14, 1)$, which fills the knapsack completely. This solution would be found by A_2.   ∎

Table 9.1
Knapsack example.

	Subsets of size k	Objects added by inner **for** loop	*sum*
k=0	∅	54, 45, 1	100
	maxSum = 100	*set* = {54, 45, 1}	
k=1	54	45, 1	100
	45	54, 1	100
	43	54, 1	98
	29	54, 23, 1	107
	23	54, 29, 1	107
	21	54, 29, 1	105
	14	54, 29, 1	98
	1	54, 45	100
	maxSum = 107	*set* = {29, 54, 23, 1}	

Theorem 9.12   For $k>0$, algorithm A_k does $O(kn^{k+1})$ operations; A_0 does $\Theta(n)$. Hence $A_k \in P$ for $k \geq 0$.

Proof.   There are $\binom{n}{j}$ j-element subsets of $\{1, 2, \ldots, n\}$, so the outer loop is executed $\sum_{j=0}^{k}\binom{n}{j}$ times. Since $\binom{n}{j} \leq n^j$ and $\binom{n}{0} = 1$, $\sum_{j=0}^{k}\binom{n}{j} \leq kn^k+1$. The amount of work done in one pass through the loop is in $O(n)$, so for all passes it is in $O(kn^{k+1}+n)$. We leave it to the reader to show that the overhead for systematically generating one subset with at most k elements from the previous one can be done in $O(k)$ time. (See Exercise 9.31; this is not a trivial problem.) Thus the total work done is in $O(kn^{k+1}+n)$ and the theorem follows.   □

Theorem 9.13   For $k>0$, $R_{A_k}(m)$ and $S_{A_k}(n)$, the worst-case ratios of the optimal solution to the value found by A_k, are at most $1+1/k$ for all m and n.

Proof.   Fix k and let $s_1, \ldots, s_n$ and C be a particular input I with $opt(I) = m$. Suppose that an optimal solution is obtained by filling the knapsack with $s_{i_1}, s_{i_2}, \ldots, s_{i_p}$. If $p \leq k$, then this subset (actually the index set $\{i_1, i_2, \ldots, i_p\}$) is explicitly considered by A_k, so $v(A_k(I)) = m$ and $r_{A_k}(I) = 1$. Now consider the case when $p > k$. The set consisting of the largest k objects in the optimal solution will be considered explicitly as T by A_k. Let i_q be the first index in the optimal solution that is not added to this T by A_k. (If there is no such i_q, then A_k gives an optimal solution.) The object i_q is not one of the k largest of the objects in the optimal solution, which has sum m, so $s_{i_q} \leq m/(k+1)$. Since object i_q was rejected, the unfilled space in the knapsack is less than s_{i_q}. Thus $v(A_k(I)) + s_{i_q} > C \geq m$, and, writing r for the ratio $r_{A_k}(I)$, we have

$$r = \frac{m}{v(A_k(I))} < 1 + \frac{s_{i_q}}{v(A_k(I))} \leq 1 + \frac{m}{(k+1)v(A_k(I))} = 1 + \frac{r}{(k+1)}.$$

So $r < 1+r/(k+1)$ which simplifies to $r < 1+1/k$. Thus $R_{A_k}(m) \leq 1+1/k$ and $S_{A_k}(n) \leq 1+1/k$.

□

Corollary 9.14   Given any $\varepsilon > 0$, there is a polynomial-bounded algorithm A for the knapsack problem for which $R_A(m) \leq 1+\varepsilon$ and $S_A(n) \leq 1+\varepsilon$ for all m and n.

Even though approximation algorithms exist for which the ratio $r(I)$ can be made arbitrarily close to one, it is very unlikely that any approximation algorithm A can guarantee a constant bound on the absolute error, $opt(I) - v(A(I))$. It can be proved that if there is such an algorithm, then $P = NP$. (The proof is not very hard, and this problem is included in the exercises, but the reader may find it helpful to read Section 9.6.2 before tackling it.)

9.6
Graph Coloring

9.6.1 Some Basic Techniques

For the knapsack and bin packing problems we have found approximation algorithms that give fairly good results; the behavior ratio for any particular optimal value is bounded by a small constant. A number of heuristic algorithms have been developed for the graph coloring problem, but unfortunately they all can produce colorings that are very far from optimal. In fact, it has been shown that if there were an approximation algorithm for graph coloring that was guaranteed to use at most roughly twice the optimal number of colors, then it would be possible to obtain an optimal coloring in polynomial-bounded time, and that would imply $P = NP$. Thus getting near-optimal colorings is as hard as getting optimal ones. (We will prove a slightly weaker version of this statement in Section 9.6.2.)

We will start here with an easy heuristic strategy. It can produce poor colorings, but it is useful as a subroutine in more complex algorithms that use fewer colors. In the next section we will present such an algorithm.

Let $G = (V, E)$ where $V = \{v_1, \ldots, v_n\}$, and let the "colors" be positive integers. The *sequential coloring* strategy always colors the next vertex, say v_i, with the minimum acceptable color, that is, the minimum color not already assigned to a vertex adjacent to v_i.

Algorithm 9.3   Sequential Coloring

Input: $G = (V, E)$, a graph, where $V = \{v_1, \ldots, v_n\}$.

Output: A coloring of G.

```
for i := 1 to n do
    c := 1;
    while there is a vertex adjacent to v_i that is colored c
        do c := c+1 end;
    color v_i with c
end { for }
```

Algorithm 9.3 can easily be implemented so that its worst-case complexity is in $O(n^2)$.

The behavior of the sequential-coloring algorithm (henceforth abbreviated *SC*) on a given graph depends on the ordering of the vertices. For $k \geq 2$, define the graphs $G_k = (V_k, E_k)$, where $V_k = \{a_i, b_i \mid 1 \leq i \leq k\}$ and $E_k = \{a_i b_j \mid i \neq j\}$. See Fig. 9.9 for an illustration. If V is given in the order $a_1, \ldots, a_k, b_1, \ldots, b_k$, then *SC* will color all the a's with one color and all the b's with another, producing an optimal coloring. However, if the vertices are ordered $a_1, b_1, a_2, b_2, \ldots, a_k, b_k$, then *SC* needs a new color for each pair a_i and b_i, using a total of k colors. Thus $R_{SC}(2) = \infty$, and if we take $n = |V|$ as the size of a graph, $S_{SC}(n) \geq n/4$ for $n \geq 4$.

Let $\Delta(G)$ be the maximum degree of any vertex in G. The following theorem is easy to prove.

Theorem 9.15   The number of colors used by the sequential-coloring scheme is at most $\Delta(G)+1$.

Several more complicated graph coloring algorithms based on sequential coloring have additional features intended to prevent the poor behavior of *SC*. One such feature is to interchange two colors in the colored portion of the graph when so doing avoids the need for a new color. The interchange rule, formulated as follows, is illustrated in Fig. 9.10. Suppose that $v_1, \ldots, v_{p-1}$ have been colored using colors $1, 2, \ldots, c$ (where $c \geq 2$) and that v_p is adjacent to a vertex of each color. For each pair i and j with $1 \leq i < j \leq c$, let G_{ij} be the subgraph consisting of all vertices colored

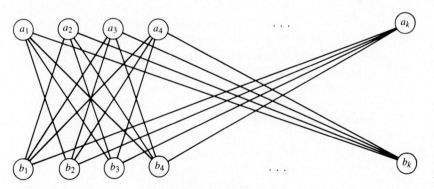

Figure 9.9 The graph G_k.

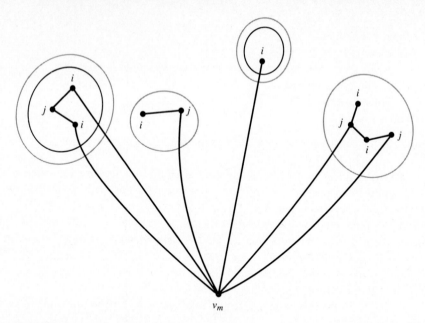

v_m

Figure 9.10 Switching colors. G_{ij} consists of the four connected components circled. S_i consists of the two components doubly circled.

i or j and all edges between these vertices. If there is a pair (i, j) such that in each connected component of G_{ij} the vertices adjacent to v_p are all of the same color, then an interchange will be done. Specifically, let S_i be the set of all vertices in connected components of G_{ij} where vertices adjacent to v_p are colored i; colors i and j are interchanged in S_i and v_p is colored with i. The algorithm then goes on to v_{p+1}.

The work needed to determine when to interchange colors and to carry out the interchange may add significantly to the time requirement of the algorithm, but doing the interchanges will produce better colorings than *SC* for many graphs. It will give optimal colorings for the graphs G_k where *SC* can do poorly (the reader should check this), and it can be shown that it will yield an optimal coloring for any graph G with $opt(G)$, i.e., $\chi(G)$, the chromatic number of G, equal to 1 or 2. However, for $k \geq 3$, there are graphs G_k' with $3k$ vertices and $\chi(G_k') = 3$ for which sequential coloring with interchanges (henceforth *SCI*) uses k colors: $G_k' = (V_k', E_k')$, where $V_k' = \{a_i, b_i, c_i \mid 1 \leq i \leq k\}$ and $E_k' = \{a_i b_j, a_i c_j, b_i c_j \mid i \neq j\}$. Thus $R_{SCI}(3) = \infty$, and $S_{SCI}(n) \geq n/9$ for most n.

The reader may observe that, if the vertices in G_k' are ordered $a_1, \ldots, a_k, b_1, \ldots, b_k, c_1, \ldots, c_k$, then *SCI* produces an optimal coloring. Thus another approach to the problem of improving the basic sequential-coloring strategy is to order the vertices in a special way before assigning colors. Some such techniques yield improvements in the colorings produced for many graphs, but again, in some cases they perform about as badly as *SC* and *SCI*.

9.6.2 Approximate Graph Coloring is Hard

No polynomial-bounded graph coloring algorithms are known for which the ratio of the number of colors used to the optimal number is bounded by a constant. In fact, guaranteeing a small constant bound on the ratio is *NP*-hard, that is, at least as hard as the *NP*-complete problems.

Theorem 9.16   If there is a polynomial time graph coloring algorithm that colors every graph G with fewer than $4\chi(G)/3$ colors, then the 3-colorability problem could be solved in polynomial time (and since 3-colorability is *NP*-complete, it would follow that $P = NP$).

Proof.   Suppose that G is an input for the 3-colorability problem, i.e., we want to know whether G can be colored with three colors. Let A be an approximate graph coloring algorithm described in the theorem, and let $v_A(G)$ be the number of colors used by A. If G is 3-colorable, then

$$v_A(G) < \frac{4}{3}\chi(G) \le \frac{4}{3}3 = 4, \text{ so } v_A(G) < 4.$$

Since the number of colors is an integer, A uses at most three colors. If G is not 3-colorable, A would have to use more than three colors. Thus A uses three colors on G if and only if G is 3-colorable, and we can use A to solve the 3-colorability problem in polynomial time.   □

The easy argument used in this proof shows that we are unlikely to find algorithms that use at most 33 percent of $\chi(G)$ extra colors when $\chi(G)$ is small. But we can prove a similar theorem even for graphs with $\chi(G)$ large. The proof uses a construction called the *composition* of two graphs. Informally, in the composition of G_1 and G_2 each vertex of G_1 is replaced by a copy of G_2. An edge xy in G_1 is replaced by edges between each vertex in x's copy of G_2 and each vertex in y's copy of G_2. (See Fig. 9.11 for an example.) The formal definition follows.

Definition   Let $G_1 = (V_1, E_1)$ and $G_2 = (V_2, E_2)$ be two graphs. The *composition* of G_1 and G_2, denoted $G_1[G_2]$, is the graph $G = (V, E)$, where $V = V_1 \times V_2$ and $E = \{(v_1, w_1)(v_2, w_2) \mid v_1 = v_2 \text{ and } w_1 w_2 \in E_2, \text{ or } v_1 v_2 \in E_1\}$.

It is easy to see that the number of vertices and edges in G is bounded by polynomials in the number of vertices and edges in G_1 and G_2, and that G can be constructed in polynomial-bounded time.

Theorem 9.17   If there is a polynomial time graph-coloring algorithm that uses fewer than $4\chi(G)/3$ colors for every graph G with $\chi(G) \ge k$, for some integer k, then the 3-colorability problem could be solved in polynomial time.

Proof.   Let A be an algorithm as described in the theorem. Let G be an input for the 3-colorability problem. Let C_k be the complete graph with k vertices, and let

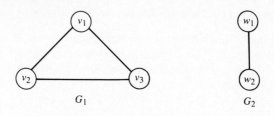

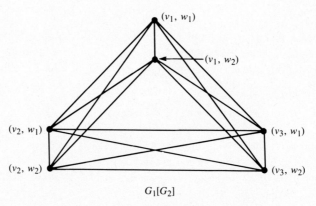

Figure 9.11   The composition of two graphs.

$H = C_k[G]$. H consists of k copies of G where each vertex in one copy is connected by an edge to each vertex in each other copy. Each copy of G can be colored with $\chi(G)$ colors, but because every vertex in one copy is adjacent to every vertex in each other copy, a new set of k colors is needed for each copy. So $\chi(H) = \chi(G)k$. Since this is at least k, A's performance guarantee holds for H. Now, run A on H, and let $v_A(H)$ denote the number of colors A uses. If G is 3-colorable, then

$$v_A(H) < \frac{4}{3}\chi(H) \leq \frac{4}{3}3k = 4k.$$

That is,

$$v_A(H) < 4k.$$

On the other hand, if G is not 3-colorable, then

$$v_A(H) \geq \chi(H) = \chi(G)k \geq 4k.$$

The running time of A is polynomial bounded in the size of H, and H can be constructed in polynomial time from G. Thus running A on H and checking whether the number of colors used is less than $4k$ answers the 3-colorability question for G in polynomial time.

$\square$

9.6.3   A Graph-coloring Algorithm

As usual, let $G = (V, E)$, and $n = |V|$. For many graph coloring heuristics, the worst ratio of the number of colors used to the optimal can be as bad as $\Theta(n)$; for some it is in $\Theta(n/\lg n)$. For a long time, no better algorithms were known. Now, however, there is one (due to A. Wigderson) that does somewhat better (though still, $R(3) = \infty$). The number of colors it uses is in $O(n^p)$ for $p < 1$ (but p depends on $\chi(G)$). If $\chi(G) = 3$, the algorithm uses at most $3\lceil \sqrt{n} \rceil$ colors.

Let $v \in V$. The neighborhood of v, denoted $N(v)$, is the set of vertices adjacent to v. The subgraph induced by v's neighbors, denoted $H(v)$, is the subgraph $(N(v), \{xy \mid xy \in E$ and $x,y \in N(v)\})$. Note that v is not in its neighborhood.

A key idea in the algorithm is that neighbors of vertices with high degree are colored first. While there are vertices with high degree, the neighborhood subgraphs

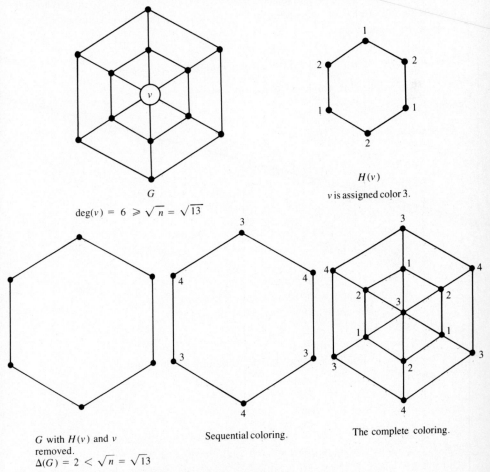

G

$\deg(v) = 6 \geqslant \sqrt{n} = \sqrt{13}$

$H(v)$

v is assigned color 3.

G with $H(v)$ and v removed.
$\Delta(G) = 2 < \sqrt{n} = \sqrt{13}$

Sequential coloring.

The complete coloring.

Figure 9.12   Example for Algorithm 9.4.

are colored recursively (with 2-colorable graphs as an easy boundary case). If all vertices have small degree, the graph is colored directly. The general algorithm and its analysis are not easy to follow, so we present and analyze Wigderson's nonrecursive algorithm for 3-colorable graphs first, and then briefly describe the general algorithm.

Neighborhoods and neighborhood subgraphs, $N(v)$ and $H(v)$, depend on the graph G. The algorithm discards vertices from G as they are colored; as G changes so do the neighborhoods of the vertices. $N(v)$ and $H(v)$ are always defined in terms of the current graph G.

The algorithm uses the following lemma, which is easy to prove.

Lemma 9.18   If G is k-colorable, then for any $v \in V$, $H(v)$ is $k-1$-colorable.

Since 2-colorable graphs can be colored (with only two colors) in polynomial time, the neighborhood of any vertex in a 3-colorable graph can be colored with two colors in polynomial time.

Algorithm 9.4   Approximate Coloring for 3-colorable Graphs

Input: G, a 3-colorable graph; n, the number of vertices in G.
Output: A coloring of G.

> **procedure** *Color3* (*G: Graph*);
> **var**
>   *c: integer*;   { the current color }
> **begin**
> $c := 1$;
>   **while** $\Delta(G) \geq \sqrt{n}$ **do**
>   let v be a vertex in G of maximum degree;
>   color $H(v)$ with colors c and $c+1$;
> $c := c+2$;
>   color v with the color c;
>   delete v and $N(v)$ from G, and delete all edges
>   incident with the deleted vertices
>   **end** { while };
>   { Now $\Delta(G) < \sqrt{n}$. }
>   use sequential coloring to color G beginning with color c;
> **end** { Color3 }

Example 9.5   *Color3* in action

Consider the graph in Fig. 9.12, in which most of the steps are explained. Note, however, that the sequential coloring step could have used three colors, not two. The coloring produced depends on the order in which the vertices are encountered. If after coloring the vertex at the top of the graph with 3, the next vertex encountered were the one at the bottom, the latter would have been colored 3 also (because these two vertices are not adjacent); then color 5 would have been needed.   ∎

Theorem 9.19   If G is 3-colorable, *Color3* produces a legal coloring.

Proof. By Lemma 9.18, each neighborhood is 2-colorable. The colors used for $N(v)$ are not used again, so they can cause no conflict. The color used for v is used again, but on the graph that remains after $N(v)$ is removed, so no other vertex assigned v's color is adjacent to v. The colors used for the sequential coloring of the graph with $\Delta < \sqrt{n}$ are not used again.   □

Theorem 9.20   If G is 3-colorable, *Color3* runs in polynomial-bounded time and uses at most $3\lceil\sqrt{n}\rceil$ colors.

Proof. First, the timing. Since each neighborhood is 2-colorable, each can be colored in polynomial time. (Use depth-first search, for example, and assign each vertex the color not assigned its parent.) The algorithm colors neighborhoods while $\Delta(G) \geq \sqrt{n}$. So for each v whose neighborhood is colored, $N(v)$ has at least $\sqrt{n}$ vertices. These vertices are discarded after they are colored, so the number of iterations of the **while** loop is at most $\sqrt{n}$. Sequential coloring is done once after the **while** loop; it (Algorithm 9.3) runs in polynomial time. Hence the total work is polynomial bounded.

Two new colors are used for each neighborhood colored in the **while** loop; hence $2\sqrt{n}$ colors at most are used by the loop for all the neighborhoods. (The color used for v is used for a later neighborhood or in the sequential coloring after the **while** loop; either way, that color will be counted.) When sequential coloring is used after the loop $\Delta(G) < \sqrt{n}$, and the number of colors used there is at most $\Delta(G)+1$ (by Theorem 9.15). If $\sqrt{n}$ is an integer, sequential coloring uses at most $\sqrt{n}$ colors, so the total is at most $3\sqrt{n}$. If $\sqrt{n}$ is not an integer, sequential coloring uses at most $\lceil\sqrt{n}\rceil$ colors, so the total is at most $3\lceil\sqrt{n}\rceil$.   □

At this point, the reader should be wondering what good this algorithm is, since determining whether a graph is 3-colorable is an *NP*-complete problem. In fact, *Color3* can produce a legal coloring for input graphs where $\chi(G) > 3$. After all, it can use as many as $3\lceil\sqrt{n}\rceil$ colors. *Color3* will get stuck if it tries to 2-color a neighborhood graph that is not 2-colorable. Such a failure can be detected and reported easily. Thus, *Color3* can easily be modified to return a Boolean variable *colored* that indicates whether it successfully colored its input graph. Theorems 9.19 and 9.20 can be generalized to state that *Color3* always runs in polynomial time, and that if it succeeds in coloring the input graph (which it is guaranteed to do if the graph is 3-colorable), it will produce a legal coloring using at most $3\lceil\sqrt{n}\rceil$ colors.

*The General Case

We will now consider the general coloring algorithm. Recall that the key idea was to color neighbors of vertices with high degree first. The neighborhood subgraphs are colored recursively. How small should $\Delta(G)$ be before we do a direct coloring rather than use recursion? This cutoff point is chosen to more or less balance the number of colors used by the recursive and nonrecursive parts. The value used is $n^{1-1/(k-1)}$, where k is a parameter to the algorithm that we might think of as a guess at $\chi(G)$. Let $p(k) = 1 - 1/(k-1)$. For a k-colorable graph G with n vertices, the recursive coloring algorithm, which we will call *Color*, runs in polynomial time and produces a legal coloring using at most $2k \lceil n^{p(k)} \rceil$ colors. (The proof is a more general, and harder, argument using the ideas of the proofs of Theorems 9.19 and 9.20.)

Once again we have the problem that we do not know if an arbitrary graph is k-colorable. We want a coloring algorithm that works well for any graph, whether or not we know its chromatic number. *Color* can be used to obtain such an algorithm. If k, the guess at $\chi(G)$, is too small, and *Color* cannot color G, it will fail on one of the "boundary" cases, i.e., when it tries to 2-color a graph that is not 2-colorable. So here too *Color* can be modified so that it returns a Boolean variable *colored* that indicates whether it colored its input graph successfully. *Color* is called repeatedly to find the minimum value of k for which it succeeds in coloring G. To find the minimum such k quickly, only powers of 2 are tried first. Then we use a binary-search-like scheme to check the values between two powers of 2. An outline of the scheme follows.

Algorithm 9.5 Approximate Graph Coloring

> $k := 1$;
> *colored* := *false*;
> **while not** *colored* **do**
> > $k := 2*k$;
> > *Color*($G, k, colored$)
> **end** { while };
> { The minimum k_0 for which *Color* succeeds is between $k/2$ and k. }
> do binary search on the integers $k/2, \ldots, k$ to find the smallest
> > one such that *Color* returns *colored* = *true*; let this be k_0;
> output the coloring produced by *Color*($G, k_0, colored$)

Theorem 9.21 The approximate graph coloring algorithm, Algorithm 9.5, runs in polynomial time and uses at most $2\chi(G) \lceil n^{1-1/(\chi(G)-1)} \rceil$ colors.

Proof. The number of calls to *Color* in Algorithm 9.5 (not counting the recursive calls from within *Color* itself) is roughly $2 \lg k_0$, so since *Color* runs in polynomial time, Algorithm 9.5 does also. For all $k \geq \chi(G)$, *Color*($G, k, colored$) will return *colored* = *true*, so $k_0 \leq \chi(G)$. So the number of colors used by Algorithm 9.5 is at most $2k_0 \lceil n^{p(k_0)} \rceil \leq 2\chi(G) \lceil n^{p(\chi(G))} \rceil \leq 2\chi(G) \lceil n^{1-1/(\chi(G)-1)} \rceil$. □

Exercises

Section 9.1: P *and* NP

9.1. Suppose that algorithms A_1 and A_2 have worst-case time bounds p and q, respectively. Suppose that algorithm A_3 consists of applying A_2 to the output of A_1. (The input to A_3 is the input to A_1.) Give a worst-case time bound for A_3.

9.2. Give a necessary and sufficient condition for a graph to be colorable with one color.

9.3. Write an algorithm to determine whether a graph $G = (V,E)$ is 2-colorable. The algorithm should run in $\Theta(\max(n, m))$ time where $n = |V|$ and $m = |E|$.

9.4. Show that each of the following decision problems is in *NP*. To do this, indicate what a "proposed solution" (in the sense of Section 9.1.3) would be, and tell what properties would be checked to determine if a proposed solution justifies a *yes* answer to the problem.

   a)   the bin packing problem.

   b)   the Hamilton circuit problem.

   c)   the CNF satisfiability problem.

*9.5. Give a dynamic programming solution for the subset sum problem. (There is a hint at the end of Section 9.1.4.)

Section 9.2: NP-*Complete Problems*

Note: For the exercises that ask you to show that one problem (Π_1) is polynomially reducible to another (Π_2), remember that this involves several steps: Define a transformation from Π_1 to Π_2, and show that the transformation satisfies both properties in the definition in Section 9.2.1.

9.6. The subset sum problem may be stated so that $s_1, \ldots, s_n$ and C are rational numbers. Show that this version of the problem is polynomial reducible to the version in the text and vice versa.

9.7. The *set intersection* problem is defined as follows:

   Given finite sets $A_1, A_2, \ldots, A_m$ and $B_1, B_2, \ldots, B_n$, is there a set T such that

$$|T \cap A_i| \geq 1 \ \text{ for } i = 1, 2, \ldots, m$$

and

$$|T \cap B_j| \leq 1 \ \text{ for } j = 1, 2, \ldots, n \ ?$$

Show that the set intersection problem is *NP*-complete by showing that it is in *NP* and that CNF-satisfiability is polynomially reducible to it.

9.8. Show that the Hamilton circuit problem for undirected graphs is reducible to the Hamilton circuit problem for directed graphs.

9.9. Suppose that we transform a directed graph $G = (V,E)$ into an undirected graph $G' = (V', E')$, where $V' = \{v^i \mid i = 1, 2 \text{ for } v \in V\}$ and $E' = \{v^1v^2 \mid v \in V\} \cup \{v^2w^1 \mid vw \in E\}$. Show by example that there is a directed graph G such that G does not have a Hamilton circuit but G' does.

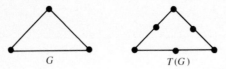

G $T(G)$

Figure 9.13   Transformation of a graph to a bipartite graph.

9.10. This problem considers an attempt at a polynomial transformation from one problem to another that *does not* work. Your problem is to find the flaw. A *bipartite graph* is a graph in which every cycle has even length. We attempt to show that the Hamilton circuit problem is reducible to the Hamilton circuit problem in bipartite graphs. We need a function T: {graphs} → {bipartite graphs} such that T can be computed in polynomial time and, for any graph G, G has a Hamilton cycle if and only if $T(G)$ has a Hamilton cycle. Let $T(G)$ be the bipartite graph obtained by inserting a new vertex in every edge. See Fig. 9.13 for an example. What is wrong with this transformation?

9.11. Show that the 3-colorability problem is reducible to the CNF-satisfiability problem. (This, of course, follows from Cook's theorem; give a direct transformation.)

9.12. The *clique problem* is:

Given a graph G and a positive integer k, does G have k mutually adjacent vertices? (A set of k mutually adjacent vertices is called a *k-clique*.)

Show that the clique problem is *NP*-complete by showing that it is in *NP* and then using the following polynomial transformation to reduce CNF-satisfiability to it. Suppose that $C_1, C_2, \ldots, C_p$ are the clauses in a CNF expression and let the literals in the ith clause be denoted $l_{i,1}, l_{i,2}, \ldots, l_{i,q_i}$. The expression is transformed to the graph with $V = \{(i, r) \mid 1 \le i \le p, 1 \le r \le q_i\}$, i.e., V has a vertex representing each occurrence of a literal in a clause, and $E = \{(i, r)(j, s) \mid i \ne j \text{ and } l_{ir} \ne \overline{l}_{js}\}$. In other words, there is an edge between two vertices representing literals in different clauses so long as it is possible for both of those literals to be assigned the value *true*. Let $k = p$.

9.13. A *vertex cover* for a graph $G = (V, E)$ is a subset V' of V such that every edge in E is incident with a vertex in V'. The *vertex cover problem* is:

Given G and a positive integer k, does G have a vertex cover with k vertices?

Show that the clique problem (see the previous exercise) is polynomial reducible to the vertex cover problem.

9.14. A *feedback vertex set* in a directed graph $G = (V, E)$ is a subset V' of V such that V' contains at least one vertex from each directed cycle in G. The *feedback vertex set problem* is:

Given a directed graph G and an integer k, does G have a feedback vertex set with at most k vertices?

Show that the vertex cover problem (see the previous exercise) is polynomial reducible to the feedback vertex set problem.

9.15. Consider the following problem.

An organization has 200 members and 17 committees. Each committee must meet for a full afternoon during the week of the organization's annual meeting. You are given a list of the members of each committee. Is it possible to schedule the committee meetings in five afternoons so that each member can attend the meeting of each committee of which he or she is a member?

Which of the problems discussed in Sections 9.1 and 9.2 and in the exercises up to this point most closely resembles this problem? Explain the correspondence.

9.16. Devise an algorithm to determine the chromatic number of graphs with the property that each vertex has degree at most 2 (i.e., is incident with at most two edges). The running time of your algorithm should be linear in the number of vertices in the graph.

*9.17. Devise a polynomial-bounded algorithm to determine whether a CNF expression with exactly two literals per clause is satisfiable. What is the worst-case complexity of your algorithm? (Hint: One possible solution constructs a digraph associated with the expression, then uses an algorithm from Chapter 4.)

9.18. Give a polynomial-bounded algorithm to determine whether a graph has a 4-clique. What is the worst-case complexity of your algorithm? (It is known that the largest clique possible in a planar graph has four vertices, so the k-clique problem for planar graphs is in P.)

9.19. Give necessary and sufficient conditions for a graph (undirected) with maximum degree 2 to have a Hamilton circuit. Outline an efficient algorithm to test the conditions.

9.20. Show that if the bin packing decision problem can be solved in polynomial time, then the optimization problem (find the smallest number of bins into which the objects can be packed) can also be solved in polynomial time.

Section 9.3: Approximation Algorithms

9.21. We can state the CNF-satisfiability problem as an optimization problem in the following form:

> Given a CNF expression E, find a truth assignment for the variables in E to make the maximum possible number of clauses true.

Describe the set FS_E and the function v_E for this problem.

9.22. Let $F = \{S_1, \ldots, S_n\}$ be a set of subsets of a set A such that $\bigcup_{i=1}^{n} S_i = A$. A *cover* of A is a subset of F, say $\{S_{i_1}, \ldots, S_{i_k}\}$ such that $\bigcup_{j=1}^{k} S_{i_j} = A$. ($F$ itself is a cover of A.) A *minimum cover* is a cover using the smallest possible number of sets. The *set cover* problem is:

> Given F as described above, find a minimum cover of A.

What is the set FS_F and the function v_F for this problem?

Section 9.4: Bin Packing

9.23. a) Construct an example for the bin packing problem in which the *Niff* algorithm uses three bins but the optimal number is two.

b) Construct an infinite sequence of examples I_t, where I_t has n_t objects for some $n_1 < n_2 < \cdots$, and $opt(I_t) = 2$ but *Niff* uses three bins.

*9.24. Show that Lemma 9.10 cannot be made stronger by constructing a sequence of examples such that for each $k \geq 2$ there is an input I with $opt(I) = k$ and *Niff* puts $k-1$ objects in extra bins.

9.25. Show that, if $2 \le opt(I) \le 4$, *Niff* uses at most $opt(I)+1$ bins.

9.26. Write a *nonincreasing best fit* algorithm for bin packing. What is the order of the worst-case running time?

9.27. a) Give an example in which the *nonincreasing best fit* (henceforth *Nibf*) strategy for bin packing produces a packing that is not optimal.

b) Give an example in which *Nibf* produces a different packing from *Niff*.

9.28. a) Write a *next fit* algorithm for bin packing.

b) Prove that $r_{NF}(I) < 2$ for all inputs I.

Section 9.5: The Knapsack and Subset Sum Problems

9.29. Show that the output of the greedy algorithm described in the text for the simplified knapsack problem (i.e., the subset sum problem) is always more than half the optimal. (Hint: Consider the two cases — the algorithm's result is greater than $C/2$, and the algorithm's result is at most $C/2$.)

9.30. Show that if the greedy algorithm described in the text for the simplified knapsack problem (i.e., the subset sum problem) did not explicitly consider the object with the largest size, it could give a result arbitrarily far from the optimal. (Hint: Construct an example with only two objects.)

*9.31. Devise an algorithm that, when given n and k such that $1 \le k \le n$, generates all subsets of $\{1, 2, \ldots, n\}$ containing at most k elements. The number of operations done between generating one subset and generating the next one should be in $O(k)$ and independent of n.

*9.32. Extend the approximation algorithms for the simplified knapsack problem and Theorems 9.12 and 9.13 to the general formulation of the problem (with profits as well as sizes).

Section 9.6: Graph Coloring

9.33. Describe data structures for representing the graph and the coloring in Algorithm 9.3 to achieve a fast implementation. What is the complexity of your implementation?

9.34. Prove Theorem 9.15.

9.35. Describe how the *SCI* strategy behaves on the graphs G_k defined in Section 9.6. In particular, how many times are pairs of colors interchanged?

9.36. Suppose that $G_1 = (V_1, E_1)$ and $G_2 = (V_2, E_2)$, where $|V_1| = n_1$, $|V_2| = n_2$, $|E_1| = m_1$, and $|E_2| = m_2$. How many vertices and edges are in $G = G_1[G_2]$?

9.37. Show that the graph in Fig. 9.12 is 3-colorable.

9.38. Lemma 9.18 states that if a graph is 3-colorable, then the neighborhood graph for each vertex is 2-colorable. The converse is: If the neighborhood of every vertex is 2-colorable, then the graph is 3-colorable. If the converse were true, there would be a polynomial algorithm for the 3-colorability problem (since it is easy to determine whether each neighborhood is 2-colorable). Show by constructing an example that the converse of Lemma 9.18 is not true.

9.39. Describe the coloring that would be produced by Wigderson's algorithm (*Color3*, Algorithm 9.4) for the graphs $G_k = (V_k, E_k)$, where $V_k = \{a_i, b_i \mid 1 \le i \le k\}$ and $E_k = \{a_i b_j \mid i \ne j\}$. (In Section 9.6.1 we observed that the sequential coloring algorithm may do a very poor job on these graphs.)

Additional Problems

9.40. For each of the following statements, indicate whether it is true, false, or unknown. ("Unknown" means it is currently not known if the statement is true or false.) Do not be too hasty.

  a)   The CNF-satisfiability problem is polynomially reducible to the traveling salesperson problem.

  b)   If $P \neq NP$, then no problem in NP can be solved in polynomial time.

  c)   2-CNF (the satisfiability problem, in which each clause has exactly two literals) is polynomially reducible to the CNF-satisfiability problem.

  d)   There cannot exist any (polynomial time) approximation algorithm for graph coloring that is guaranteed to use fewer than $2\chi(G)$ colors for all graphs G (where $\chi(G)$ is the chromatic number of G).

9.41. Consider the following optimization problem:

  Given $t_1, t_2, \ldots, t_n$, where all the t_i are positive integers, find a partition of these integers into two subsets that minimizes the larger sum.

  This may be thought of as the problem of scheduling jobs on two processors. Job i' takes time t_i. We want to finish the set of jobs at the earliest possible time.

  a)   Give a reasonable, but fairly simple, polynomial time approximation algorithm A for this problem. (How much time does your algorithm take?)

  b)   Give an example that shows that your algorithm does not always produce an optimal schedule.

  *c)   Say as much as you can about the quality of your algorithm's output, i.e., about the functions S_A and R_A.

9.42. Consider the following generalization of the previous problem:

  You have p bins, each with unbounded capacity, and are given integers $t_1, \ldots, t_n$. Pack the t_i into the bins so as to minimize the maximum bin level.

  Think of the bins as processors and of the t_i as the time requirements for n independent jobs. The problem is to assign jobs to processors to minimize the total time required.

  Give a polynomial time approximation algorithm A for this problem. Say as much as you can about the quality of its output.

9.43. Let $G = (V, E)$ be a graph. Consider the following attempt at an algorithm to find a minimum vertex cover C for G. (See Exercise 9.13 for the definition of a vertex cover.)

  Initially all edges are "unmarked" and C is empty;
  repeat
      choose a vertex v incident with the maximum number
          of unmarked edges;
      put v in C;
      mark all edges incident with v
  until there are no unmarked edges left

  Find an example in which this algorithm does not produce a minimum vertex cover.

*9.44. Suppose that we had a polynomial time subprogram TSP to solve the traveling salesperson decision problem (i.e., given a complete weighted graph and an integer k, it determines whether there is a tour of total weight at most k).

    a)   Show how to use the TSP subprogram to determine the weight of an optimal tour in polynomial time.

    b)   Show how to use the TSP subprogram to find an optimal tour in polynomial time.

*9.45. Show that if there were a polynomial time approximation algorithm for the knapsack problem that was guaranteed to fill the knapsack with objects whose total value differed from the optimal by a constant, then an optimal solution could be found in polynomial time. (In other words, if there is a polynomial time algorithm A and an integer k such that for all inputs I, $opt(I) - v_A(I) \leq k$, then an optimal solution could be found in polynomial time.)

*9.46. Suppose that there is a polynomial time algorithm for the CNF-satisfiability problem. Give a polynomial time algorithm that when given a CNF expression, finds a truth assignment for the variables that satisfies the expression, if one exists, or tells that the expression is not satisfiable, if that is the case.

Notes and References

The two papers that began the intensive study of *NP*-complete problems are Cook (1971) and Karp (1972). The latter outlines proofs of polynomial reducibility among many *NP*-complete problems. Both Stephen Cook and Richard Karp have won the ACM Turing Award, and their Turing Award lectures (1983 and 1986, respectively) present interesting overviews of computational complexity and their own views of the context of their work.

    A major source for more detail, formalism, applications, implications, approximation algorithms, etc., is Garey and Johnson (1979), an entire book on the subject of *NP*-completeness. The definition of *NP* given here uses an informal version of the definition of nondeterministic Turing machines given in Garey and Johnson. The latter also contains a proof of Cook's theorem, a list of several hundred *NP*-complete problems, and a long bibliography (thus we will not repeat most of the original references here).

    The approximation algorithms in Sections 9.4 through 9.6 are from Sahni (1975) (knapsack); Garey, Graham, and Ullman (1972) and Johnson (1972) (bin packing); and Johnson (1974) and Wigderson (1983) (graph coloring). (The faster approximation scheme mentioned for the knapsack problem is in Ibarra and Kim (1975).) There are more algorithms and references in Garey and Johnson (1979). Empirical studies of the expected behavior of the bin-packing heuristics are in Bentley *et al.* (1983). Approximation algorithms for scheduling problems are in Sahni (1976). Lawler (1985) is a book on the traveling salesman problem. Garey and Johnson (1979) has more theorems concerning the unlikelihood of obtaining good approximation algorithms for some problems.

10

Parallel Algorithms

10.1
Parallelism, the PRAM, and Other Models

10.1.1   Introduction

In most of this book, our model of computation has been a general-purpose, deterministic, random access computer that carried out one operation at a time. Several times we used specialized models to establish lower bounds for various problems; these were not general-purpose machines, but they too carried out one operation at a time. We will use the term *sequential algorithm* for the usual one-step-at-a-time algorithms that we have studied up to now. (Such algorithms are sometimes called *serial algorithms*.) In this chapter we will consider *parallel algorithms*, algorithms in which several operations may be executed at the same time in parallel, that is, algorithms for machines that have more than one processor working on one problem at the same time.

In recent years, as microprocessors have become cheaper and the technology for interconnecting them has improved, it has become possible and practical to build general-purpose parallel computers containing a very large number of processors. There has been a burst of activity in developing the hardware, the algorithms, and the theoretical models to make use of parallel computers. It seems clear that in the future more and more parallelism will be used. The purpose of this chapter is to introduce some of the concepts, formal models, techniques, and algorithms from the growing area of parallel computation.

Parallel algorithms are natural for many applications. In image processing (for example, in vision systems for robots) different parts of a scene may be processed simultaneously, i.e., in parallel. Parallelism can speed up computation for graphics displays. In search problems (e.g., bibliographic search, scanning news stories, and text editing) different parts of the database or text can be searched in parallel. Simulation programs often do the same computation to update the states of a large number of components in the system being simulated; these can be done in parallel for each simulated time step. Artificial intelligence applications (which include image processing and a lot of searching) can benefit from parallel computing. The fast Fourier transform is implemented on specialized parallel hardware. Algorithms for many combinatorial optimization problems (such as the optimization versions of some of the *NP*-complete problems described in Chapter 9) involve examining a large number of feasible solutions; some of the work can be done in parallel. Parallel computation can also speed up computation easily and substantially in many other application areas.

For the examples of parallel applications just mentioned, and for some of the algorithms studied earlier in this book, there seem to be fairly straightforward ways to break up the computation into parallel subcomputations. Many other well-known and widely used algorithms seem inherently sequential. Thus a lot of work has been done both to find parallel implementations of sequential algorithms where that approach is fruitful, and to develop entirely new techniques for parallel algorithm design.

If the number of processors in parallel computers were small, say two or six, then there would be a practical advantage in using them for some problems in which computation could be speeded up by some small constant factor. But such machines, and the algorithms for them, would not be very interesting in the context of this book, where we often ignore small constants. Parallel algorithms become interesting from a computational complexity point of view when the number of processors is very large. One of the parallel machines introduced in 1986 has 65,536 processors. The number of processors is larger than the input size for many of the actual cases for which a program is used. This is where we can get significant speedups and interesting algorithms.

How much can parallelism do for us? Suppose that a sequential algorithm for a problem does $W(n)$ operations in the worst case for input of size n, and that we have a machine with p processors. Then the best we can hope for from a parallel implementation of the algorithm is that it runs in $W(n)/p$ time, and we cannot necessarily achieve this maximum speedup in every case. Suppose that the problem is putting on socks and shoes, and that a processor is a pair of hands. A common sequential algorithm is: Put on the right sock, put on the right shoe, put on the left sock, put on the left shoe. If we have two processors we can assign one to each foot and accomplish the task in two time units instead of four. However, if we have four processors, we cannot cut the time down to one unit, because the socks must go on before the shoes.

Parallel machines have the potential to solve many problems much faster than before. However, many difficulties, both practical and concerning the theoretical models, remain to be resolved. There are several general-purpose and special-purpose formal models of parallel machines that correspond to various (real or theoretical) hardware designs. We will focus on one major class of models for general purpose parallel computers: the PRAM (pronounced "p ram"), or parallel random access machine. Although the PRAM model has some unrealistic features (which we will mention later), it is now widely used and serves as a good tool for introducing parallel computing.

We will not always give the most efficient algorithm known for a problem; our aim here is to present some techniques and algorithms that can be understood without great difficulty. Since this is a short, introductory chapter, much that is interesting and important in the study of parallel algorithms is left out. The notes and references at the end of the chapter suggest a few additional topics and sources for the reader who wishes to pursue the subject further.

10.1.2 The PRAM

A parallel random access machine (PRAM) consists of p general-purpose processors, $P_1, P_2, \ldots, P_p$, all of which are connected to a large shared, random access memory M. (See Figure 10.1.) The processors have a private, or local, memory for their own computation, but all communication among them takes place via the

Processors

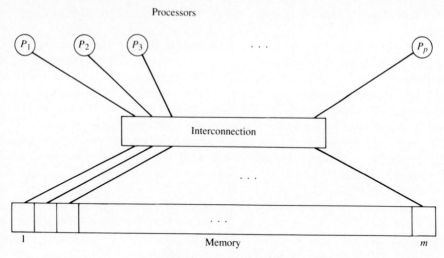

Figure 10.1 A PRAM.

shared memory. Unless we indicate otherwise, the input for an algorithm is assumed to be in the first n memory cells, and the output is to be placed in the first cell (or an initial sequence of cells). All memory cells that do not contain input are assumed to contain zero when a PRAM program starts. All the processors run the same program, but each processor "knows" its own index, and the program may instruct processors to do different things depending on their indexes. (Frequently a processor uses its own index as the index of the memory cell from which to read or into which to write.)

Each time step has three phases: the read phase, in which each processor may read from a memory cell; the computation phase, in which each processor may do some computation; and the write phase, in which each processor may write to a memory cell. The model allows processors to do lengthy computations in one step because for parallel algorithms we wish to focus on the complexity of the communication among processors.

PRAM processors are synchronized. That is, they all begin each step at the same time, and all the processors that write at any step write at the same time.

Any number of processors may read the same memory cell concurrently (i.e., at the same time). There are several variations of the PRAM depending on how write conflicts are handled. After looking at a few algorithms in which write conflicts are not a problem, we will consider the variations in Section 10.2. (There is also a variation of the PRAM model that prohibits concurrent reads, but we will not consider it here.)

Several programming languages for describing parallel algorithms exist, but we will use a mixture of English and our usual algorithm language. Several of our algorithms have **for** loops. Each processor can keep track of the loop index and do the appropriate incrementing and testing during the computation phases of its steps.

(Alternatively, the program may reside in a separate control unit that issues instructions to each processor at each step.) Several algorithms use arrays stored in the shared memory. We can assume that these are handled just as arrays in high-level languages are handled. That is, a compiler decides on some fixed arrangement of the arrays in memory following the input, and translates array references to instructions to compute specific memory addresses. For example, if the input occupies n cells, and *alpha* is the third k-element array, the compiler translates an instruction telling processor P_i to read *alpha*[j] into PRAM instructions to compute *index* := $n+2k+j$, and then to read $M[index]$. The address computation takes one PRAM step.

10.1.3 Other Models

Although the PRAM provides a good framework for developing and analyzing algorithms for parallel machines, the model would be difficult or expensive to provide in actual hardware. The PRAM assumes a complex communication network that allows all processors to access any memory cell at the same time, in one time step, and to write in any cell in one time step. Thus any processor can communicate with any other in two steps: One processor writes some data into a memory location in one step, and the other processor reads that location in the next step. Other parallel computation models do not have a shared memory, thus restricting communication between processors. A model that more closely resembles some actual hardware is the *hypercube*. A hypercube has 2^d processors for some d (the *dimension*), each connected to its neighbors. Figure 10.2(a) shows a hypercube of dimension 3. Each processor has its own memory and communicates with the other processors by sending messages. At each step each processor may do some computation, then send a message to one of its neighbors. To communicate with a nonneighbor, a processor may send a message that includes routing information indicating the ultimate destination; the message may take as many as d time steps to reach its destination.

In a hypercube with p processors, each processor is connected to $\lg p$ other processors. Another class of models, called *bounded degree networks*, restricts the connections further. In a bounded degree network, each processor is directly connected to at most d other processors, for some constant d (the *degree*). There are different designs for bounded degree networks; a 5×5 grid is illustrated in Fig. 10.2(b). Hypercubes and bounded degree networks are more realistic models than the PRAM, but algorithms for them can be harder to specify and analyze. The routing of messages among processors, an interesting problem in itself, is eliminated in the PRAM.

Since the PRAM, while not very practical, is conceptually easy to work with when developing algorithms, a lot of effort has gone to finding efficient ways to simulate PRAM computations on other parallel models, particularly models without shared memory. For example, each PRAM step can be simulated in approximately $O(\lg p)$ steps on a bounded degree network. Thus we can develop algorithms for the PRAM and know that these algorithms can be translated to algorithms for actual machines. The translation may even be done automatically by a translator program.

In Chapter 9, we defined the class of problems P to help distinguish between tractable and intractable problems. P contains problems that can be solved in

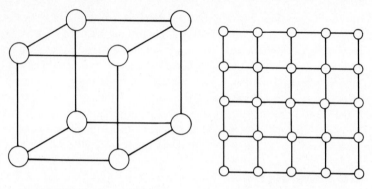

(a) A hypercube (dimension = 3).   (b) A bounded degree network (degree = 4).

Figure 10.2 Other parallel architectures. (a) A hypercube (dimension = 3). (b) A bounded degree network (degree = 4).

polynomially bounded time. For parallel computation, too, we classify problems according to their use of resources: time and processors. *NC* is defined as the class of problems that can be solved by a parallel algorithm with p (the number of processors) bounded by a polynomial in the input size, and the number of time steps bounded by a polynomial in the log of the input size. More succinctly, if the input size is n, then $p(n) \in O(n^k)$ for some constant k, and the worst-case time, $T(n)$ is in $O((\lg n)^m)$ for some constant m. The time bound, sometimes referred to as "poly-log time" because it is a polynomial in the log of n, is quite small, but we expect parallel algorithms to run very fast. The bound on the number of processors is not so small. Even for small $k > 1$, it may be impractical to use n^k processors for moderately large input. The reasons for using a polynomial bound in the definition of *NC* are similar to the reasons for using a polynomial bound on time to define the class P. First, the class of problems that can be solved in poly-log time using a polynomially bounded number of processors is independent of the specific parallel computation model used (among a large class of models considered "reasonable"). Thus *NC* is independent of whether we are using a PRAM or a bounded degree network. Second, if a problem *cannot* be solved fast with a polynomial number of processors, then that is a strong statement about how hard the problem is. In fact, for most of the algorithms we will look at, $p \in O(n)$.

10.2
Some PRAM Algorithms and the Handling of Write Conflicts

10.2.1   The Binary Fan-in Technique

Consider the problem of finding the largest key in a list of n keys. We have seen two algorithms for this problem: Algorithm 1.3 and the tournament method described in Section 3.3.2. In Algorithm 1.3, we proceeded sequentially through the list, comparing *max*, the largest key found so far, to each remaining key. After each comparison, *max* may change; we cannot do the next comparison in parallel because we do not know which value to use. In the tournament method, however, elements are paired off and compared in "rounds." In succeeding rounds, the winners from the preceding round are paired off and compared. (See Fig. 3.1.) The largest key is found in $\lceil \lg n \rceil$ rounds. All of the comparisons in one round can be performed at the same time. Thus the tournament method naturally gives us a parallel algorithm.

In a tournament, the number of keys under consideration at each round decreases by half, so the number of processors needed at each round also decreases by half. To keep the description of the algorithm short and clear, however, we specify the same instructions for all processors at each time step. The extra work being done may be confusing; it is helpful to look at Fig. 10.3 first. The figure shows the work that actually contributes to the answer. A slanted line represents a *read* operation. A vertical line represents data that has been saved in a processor's local memory; each processor always saves the largest key it has seen so far. A twisted line represents a *write* operation; a processor writes (the largest key it has seen) in the memory cell with the same number as the processor (i.e., P_i writes in $M[i]$). A dot represents a comparison (and some "bookkeeping" computation). If a *read* line comes into P_i from P_j, that means P_i reads from $M[j]$, since that is where P_j wrote. Figure 10.4 shows a complete example of the activity of all the processors. The shaded parts correspond to Fig. 10.3 and show the computations that affect the answer.

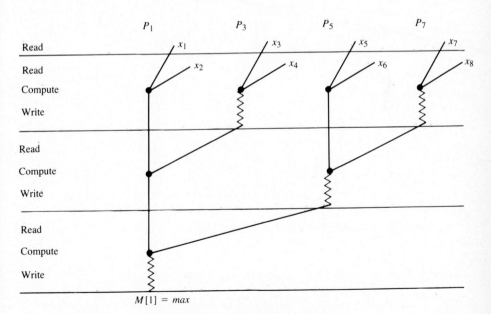

Figure 10.3 A parallel tournament.

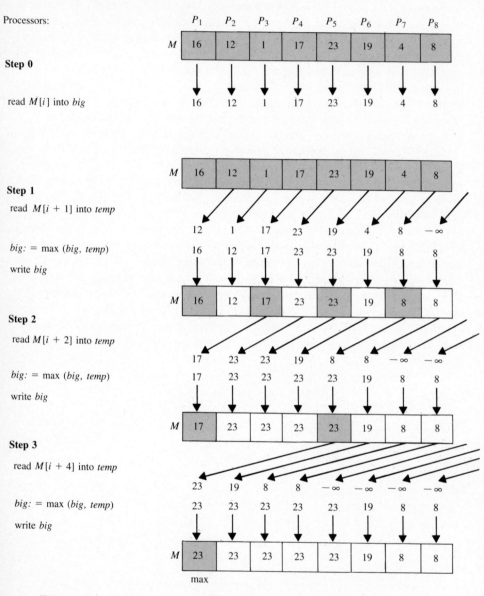

Figure 10.4   A tournament example showing the activity of all the processors.

Algorithm 10.1   A Parallel Tournament for Finding the Largest Key

Input: n keys $x_1, x_2, \ldots, x_n$, initially in memory cells $M[1], M[2], \ldots, M[n]$.

Output: The largest key will be left in $M[1]$.

Comment: Each processor carries out the algorithm using its own number for *i*. Each processor has a local variable *big* in which it saves the largest key it has seen so far, a variable *temp*, and a variable *incr* that is used to compute the cell number for its next *read*. Since *n* may not be a power of 2, the algorithm initializes cells $M[n+1], \ldots, M[2n]$ with some small value, because some of these cells will enter the tournament. (Since all memory cells in a PRAM that do not contain input are assumed to contain zero, this initialization would not be needed if all the keys were known to be nonnegative.)

```
read M[i] into big;
incr := 1;
write −∞ { some very small value } into M[n+i];
for step := 1 to ⌈lgn⌉ do
    read M[i+incr] into temp;
    big := max(big, temp);
    incr := 2*incr;
    write big into M[i]
end { for }
```

Analyzing the algorithm is easy. The initialization before the **for** loop takes one read/compute/write step, and each iteration of the **for** loop is also one read/compute/write step; the total is $\lceil \lg n \rceil + 1$ steps. So Algorithm 10.1 uses *n* processors and $\Theta(\lg n)$ time. (It actually needed only $n/2$ processors, but writing it for $n/2$ processors would involve slightly messier indexing.)

The tournament, or binary fan-in, scheme used by Algorithm 10.1 can be applied to several other problems as well, so it is worth doing a formal proof of the correctness of this algorithm. We want to show (by induction on *t*) that after the *t*th iteration of the **for** loop, $M[1]$ contains the maximum of $x_1, \ldots, x_{2^t}$. Thus when the loop terminates after $\lceil \lg n \rceil$ iterations, $M[1]$ will contain the maximum of $x_1, \ldots, x_{2^{\lceil \lg n \rceil}}$. Since *n* may not be a power of 2, we let $x_j = -\infty$ (or some very small value) for $j > n$. To establish our claim about the contents of $M[1]$ after the *t*th iteration, we need to prove a more general statement that also tells us the contents of other cells.

Theorem 10.1   At the end of the *t*th iteration of the **for** loop, each cell $M[i]$, for $1 \le i \le n$, contains the maximum of $x_i, \ldots, x_{i+2^t-1}$, and $incr = 2^t$. (Thus when $t = \lceil \lg n \rceil$ and $i = 1$, the desired conclusion follows.)

Proof, by induction on t.   For the basis of the induction, let $t = 0$. The theorem claims that before the **for** loop is executed, $M[i]$, for $1 \le i \le n$, contains the maximum of $x_i, \ldots, x_i$, i.e., x_i, which is true because that is the input. Also, from the initialization step, $incr = 1 = 2^0$.

Now let $t > 0$; let us examine the *t*th iteration of the loop. By the induction assumption, at the end of the $t-1$st iteration, $incr = 2^{t-1}$. Since *big* was written into $M[i]$ at the end of the $t-1$st iteration, on the *t*th iteration the new value for *big* is:

$$big := \max\{M[i], M[i+2^{t-1}]\}.$$

We can use the induction hypothesis for the contents of $M[i]$, but not always for $M[i+2^{t-1}]$. The latter may not be one of the first n cells; some of the processors access cells with index greater than n (though never greater than $2n$). The induction hypothesis does not tell us what is in these cells, but a simple observation does. For $n<j\leq2n$, $M[j]$ always contains $-\infty$ because, after the initialization preceding the loop, no processor ever writes into these cells. Thus it is always true that $M[j]$ contains the maximum of $x_j,\dots,x_{2n}$. So, returning to the computation of big, we see that

$M[i]$ contains the maximum of $x_i,\dots,x_{i+2^{t-1}-1}$,

and that

$M[i+2^{t-1}]$ contains the maximum of $x_{i+2^{t-1}},\dots,x_{i+2^{t-1}+2^{t-1}-1}$, that is, the maximum of $x_{i+2^{t-1}},\dots,x_{i+2^t-1}$ (or the maximum of $x_{i+2^{t-1}},\dots,x_{2n}$ if $i+2^t-1>2n$).

So, at the tth iteration, big is assigned the maximum of $x_i,\dots,x_{i+2^t-1}$, and this value is written in $M[i]$. Also, at the tth iteration, $incr$ is doubled, so now $incr=2^t$. This establishes the induction claim for t and completes the proof. $\square$

Note that Algorithm 10.1 overwrites the input data. If this is not desirable, it is a simple matter to copy the input (in one parallel step) to a scratch area in memory and do the computation there.

With only slight modification to Algorithm 10.1, the binary fan-in scheme can be used to find the minimum of n keys, to compute the Boolean **or** or Boolean **and** of n bits, and to compute the sum of n keys, each in $O(\lg n)$ steps, without any write conflicts.

10.2.2 Some Easily Parallelizable Algorithms

Consider the problem of multiplying two $n\times n$ matrices A and B. The usual matrix multiplication algorithm has a natural parallel version. Recall the formula for the entries of the product matrix C:

$$c_{ij} = \sum_{k=1}^{n} a_{ik} b_{kj} \quad \text{for } 1 \leq i, j \leq n.$$

Since concurrent reads are allowed, we can simply assign one processor to each element of the product, thus using n^2 processors. Each processor P_{ij} can compute its c_{ij} in $2n$ steps. (There are n terms to add, and each term needs two reads.) With more processors we can do better. Clearly, the binary fan-in scheme used in Algorithm 10.1 can be used to add n integers in $\lceil\lg n\rceil+1$ steps. The work done would "look" the same as in Fig. 10.3, but the dots would represent additions instead of comparisons. Thus the matrix product can be computed in $O(\lg n)$ time with $\Theta(n^3)$ processors. (Of course, for this problem the output does not all go in $M[1]$; we assume that n^2 cells of memory are designated for the elements of C.)

Many dynamic programming algorithms can be speeded up easily by doing computation in parallel. Recall that dynamic programming solutions usually involve computing elements of a table. Often the elements in one row (or column or diagonal) depend only on entries in earlier rows (columns or diagonals). Thus with n processors, all elements in one row of an $n \times n$ table can be computed in parallel, cutting the running time of the algorithm by a factor of n.

10.2.3 Handling Write Conflicts

PRAM models vary according to how they handle write conflicts. The CREW (Concurrent Read, Exclusive Write) PRAM model requires that only one processor write in a particular cell at any one step; an algorithm that would have more than one processor write to one cell at the same time is an illegal algorithm. In the Common-Write model, it is legal for several processors to write in the same cell at the same time if and only if they all write the same value. In the Priority-Write model, if several processors attempt to write in the same memory cell at the same time, the processor with the smallest index succeeds.[1] The models differ in how fast they can solve various problems. To illustrate the difference, we will consider the problem of computing the Boolean **or** function on n bits.

Using the binary fan-in scheme of Algorithm 10.1, each processor performs an **or** operation on a pair of bits at each round, and the problem is solved in $O(\lg n)$ time. This method will work on all three of the models mentioned because there are no write conflicts; the processors write the results of their operations in different memory cells. It has been shown that for the CREW model $\Theta(\lg n)$ time is required to compute the **or** of n bits (even if more than n processors are used). Consider the following algorithm, using n processors:

Algorithm 10.2 Computing the **or** of n Bits

Input: Bits $x_1, \ldots, x_n$ in $M[1], \ldots, M[n]$.

Output: $x_1 \vee \cdots \vee x_n$ in $M[1]$.

> P_i reads x_i from $M[i]$;
> If $x_i = 1$, then P_i writes 1 in $M[1]$.

Since all the processors that write, write the same value, this is a legal program for the Common-Write and Priority-Write models. Thus the **or** of n bits can be computed in one step on these models with n processors.

---

[1] Warning: The abbreviations used for the various models in research papers are not consistent. EREW and CREW are used consistently for Exclusive Read, Exclusive Write and Concurrent Read, Exclusive Write, respectively, but CRCW (Concurrent Read, Concurrent Write) sometimes means Common-Write, sometimes Priority-Write, and sometimes other variants. To avoid ambiguity, we will spell out the rule for write conflicts.

10.2.4 A Fast Algorithm for Finding Max

If we use the Common-Write or Priority-Write PRAMs, we can obtain an algorithm for finding the maximum of n numbers in less time than the binary fan-in method. This algorithm uses $n(n-1)/2$ processors. The strategy is to compare all pairs of keys in parallel, then communicate the results through the shared memory. We use an array *loser* that occupies memory cells $M[n+1], \ldots, M[2n]$. Initially, all entries in this array are zero. If x_i "loses" a comparison, then *loser*$[i]$ will be assigned the value 1.

Algorithm 10.3   Finding the Largest of n Keys

Input: n keys $x_1, x_2, \ldots, x_n$, initially in memory cells $M[1], M[2], \ldots, M[n]$ ($n > 2$).

Output: The largest key will be left in $M[1]$.

Comment: For clarity, the processors will be numbered $P_{i,j}$, for $1 \le i < j \le n$. Figure 10.5 illustrates the algorithm.

    Step 1
        $P_{i,j}$ reads x_i (from $M[i]$).
    Step 2
        $P_{i,j}$ reads x_j (from $M[j]$).
        $P_{i,j}$ compares x_i and x_j.
        Let k be the index of the smaller key.
        (If the keys are equal, let k be the smaller index.)

Initial memory contents ($n = 4$).

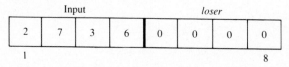

After Step 2

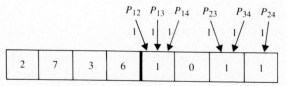

After Step 3

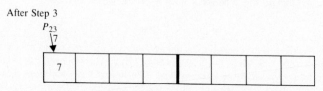

Figure 10.5   Example for the fast max-finding algorithm.

$P_{i,j}$ writes 1 in *loser*$[k]$.
{ At this point, every key other than the largest has lost a comparison. }
Step 3
$P_{i,i+1}$ reads *loser*$[i]$ (and $P_{1,n}$ reads *loser*$[n]$);
Any processor that read a 0 writes x_i in $M[1]$. ($P_{1,n}$ would write x_n.)
{ $P_{i,i+1}$ already has x_i in its local memory; $P_{1,n}$ has x_n. }

This algorithm does only three steps. However, the number of processors is in $\Theta(n^2)$. If the number of processors is limited to n, the largest key can be found in $O(\lg\lg n)$ time by an algorithm that uses Algorithm 10.3 on small groups of keys repeatedly. (See the notes and references at the end of the chapter.)

This algorithm shows that, if common writes are allowed, the binary fan-in scheme is not the fastest way to find the maximum key. In the matrix mulitplication example in Section 10.2.2, we suggested using binary fan-in to add n integers in $O(\lg n)$ time. The reader may now wonder if addition can also be done in constant time on Common-Write PRAMs. In Section 10.5.2 we will show that it cannot. Thus adding n integers is a harder problem than finding the maximum of n integers.

10.3
Merging and Sorting

10.3.1   Introduction

It is not difficult to find ways to speed up some of the sorting algorithms we studied in Chapter 2 by doing some of the operations in parallel. The reader should be able to find $\Theta(n)$-time parallel implementations of, for example, Insertion Sort and Mergesort (to sort n keys). In this section we will present a parallel sorting algorithm based on Mergesort that does roughly $(\lg n)^2/2$ PRAM steps using n processors.

The algorithm presented here gives a dramatic improvement over $\Theta(n\lg n)$ sequential sorting. For example, a list of 1000 keys can be sorted in 55 parallel steps; a sequential algorithm does about 10,000 comparisons. It is not the fastest parallel sorting algorithm known; parallel sorting can be done in $O(\lg n)$ time. However, the algorithm we describe is very easy to understand, it uses few (n) processors, and the number of steps is a small multiple of $(\lg n)^2$.

As usual, we will assume that we wish to sort into nondecreasing order.

10.3.2   Parallel Merging

As shown in Section 2.3.6, we can merge two sorted lists of $n/2$ keys each by doing at most $n-1$ comparisons. The merging algorithm we used there (Algorithm 2.6) seems essentially sequential; the two keys compared at one step depend on the result of the previous comparison. Here we use a different approach to merge in $\lg n$ parallel steps. Since we intend to use the merge algorithm in a Mergesort-style sorting algorithm in which we will always merge two lists of equal length, we will write the merge algorithm for lists of equal length. It is an easy exercise to generalize the algorithm and its analysis to lists of different sizes.

Suppose that the two sorted lists are in the first n memory cells $M[1], \ldots, M[n/2]$ and $M[n/2+1], \ldots, M[n]$. For clarity, we will refer to the first list as $X = (x_1, x_2, \ldots, x_{n/2})$ and to the second as $Y = (y_1, y_2, \ldots, y_{n/2})$. Each of the n processors is assigned to one key and has the task of determining where that key belongs in the merged list. Assume for the moment that all the keys are distinct. A processor assigned to a key in X, say x_i, does a binary search in Y to determine how many keys of Y are smaller than x_i. Specifically, the processor finds the smallest j such that $x_i < y_j$. Then, since x_i is greater than exactly $i-1$ keys in X and $j-1$ keys in Y, its proper position in the merged list is in $M[i+j-1]$. (If the processor finds that x_i is greater than all the keys in Y, then clearly, x_i belongs in position $n/2+i$.) Similarly, processors assigned to keys in Y do a binary search in X to find the correct position for their key. After the binary searches are completed, each processor writes its key in the correct position. (See Fig. 10.6.)

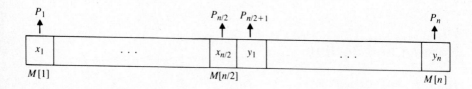

(a) Assignment of processors to keys.

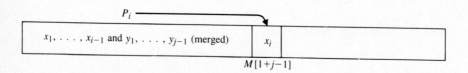

(b) Binary search steps: P_i finds j such that $y_{j-1} < x_i < y_j$.

(c) Output step.

Figure 10.6   Parallel merging.

If there are duplicate keys, the algorithm as described will not quite work. It will work if we simply require that a key from X that is equal to a key from Y be treated as if it were smaller than the key from Y.

Algorithm 10.4   Parallel Merging

Input: Two sorted lists of $n/2$ keys each, in the first n cells of memory.

Output: The merged list, in the first n cells of memory.

Comment: Each processor P_i has a local variable x (if $i \le n/2$) or y (if $i > n/2$) and other local variables for conducting its binary search. Each processor has a local variable *position* that will indicate where to write its key.

   Initialization:

   P_i reads $M[i]$ into x (if $i \le n/2$) or into y (if $i > n/2$).

   P_i does initialization for its binary search.

Binary search steps:

   Processors P_i, for $1 \le i \le n/2$, do binary search in $M[n/2+1], \ldots, M[n]$ to find the smallest j such that $x < M[n/2+j]$, and assign $i+j-1$ to *position*. If there is no such j, P_i assigns $n/2+i$ to *position*.

   Processors $P_{n/2+i}$, for $1 \le i \le n/2$, do binary search in $M[1], \ldots, M[n/2]$ to find the smallest j such that $y < M[j]$, and assign $i+j-1$ to *position*. If there is no such j, P_i assigns $n/2+i$ to *position*.

Output step:

   Each P_i (for $1 \le i \le n$) writes its key (x or y) in $M[position]$.

Theorem 10.2   The parallel merge algorithm does $\lfloor \lg n \rfloor + 1$ steps to merge two sorted lists, with $n/2$ keys each, using n processors.

Proof.   The initialization is one PRAM step. The binary searches are all done in lists of $n/2$ keys, so they take $\lfloor \lg n/2 \rfloor + 1 = \lfloor \lg n \rfloor$ read/computation steps. Since the binary searches do not involve any writing to the shared memory, the output can be done in the last binary search step. Thus the total is $\lfloor \lg n \rfloor + 1$.   □

Note that since there are no write conflicts, the merge algorithm works on all the variations of the PRAM we have described.

10.3.3 Sorting

Suppose that we have a list of n keys to be sorted. Recall the strategy of Mergesort:

   Break the list into two halves.
   Sort the two halves (recursively).
   Merge the two sorted halves.

If we "unravel" the recursion, we see that the algorithm merges small sorted lists (one key each) first, then merges slightly larger lists (two keys each), then larger lists, and so on until finally it merges two lists of size (roughly) $n/2$. The recursive algorithm merges some larger lists before doing all the small lists (since it completely sorts the first half of the keys before even beginning on the second half). To write a systematic, iterative parallel algorithm, we merge all the pairs of lists of size 1 in the first pass (in parallel), then merge all the pairs of lists of size 2 in the next pass, and so on. Clearly we use $\lceil \lg n \rceil$ merge passes. The assignment of processors to their tasks is very easy (though the indexing in the algorithm obscures it a little). Whenever two sublists occupying, say, $M[i], \ldots, M[j]$ are being merged, processors $P_i, \ldots, P_j$ do the merge using Algorithm 10.4. Figure 10.7 illustrates one pass.

Algorithm 10.5   Sorting by Merging

Input: A list of n keys in $M[1], \ldots, M[n]$.

Output: The n keys sorted in nondecreasing order in $M[1], \ldots, M[n]$.

Comment: The indexing in the algorithm is easier if the number of keys is a power of 2, so the first step will "pad" the input with large keys at the end. We still use only n processors.

> P_i writes ∞ (some large key) in $M[n+i]$;
> **for** $t := 1$ **to** $\lceil \lg n \rceil$ **do**
> $k := 2^{t-1}$;   { the size of the lists being merged }
> $P_i, \ldots, P_{i+2k-1}$ merge the two sorted lists
>   of size k beginning at $M[i]$;
> **end** { for }

Theorem 10.3   Algorithm 10.5 sorts n keys in $(\lceil \lg n \rceil + 1)(\lceil \lg n \rceil + 2)/2$ steps. Hence parallel sorting can be done in $O((\lg n)^2)$ time with n processors.

Proof.   At the tth merge pass, each pair of sublists being merged has a total of 2^t keys, so, by Theorem 10.2, the tth merge pass does $t+1$ steps. In total, all the merge passes do

$$\sum_{t=1}^{\lceil \lg n \rceil} (t+1) = \frac{(\lceil \lg n \rceil + 1)(\lceil \lg n \rceil + 2)}{2} - 1$$

steps, and there is one initialization step.   □

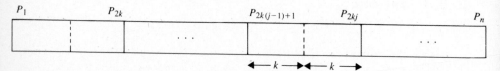

Figure 10.7 Assignment of processors for one merge pass. Processors $P_{2k(j-1)+1}, \ldots, P_{2kj}$ merge the jth pair of sublists of size k.

10.4
A Parallel Connected Component Algorithm

10.4.1   Strategy and Techniques

Let $G = (V, E)$ be a graph (undirected) with $|V| = n$ and $|E| = m$. To keep the notation simple, let $V = \{1, 2, \ldots, n\}$. In Chapter 4 we studied a sequential algorithm to find the connected components of G. It used depth-first search and ran in $\Theta(\max\{n, m\})$ time. Although depth-first search may seem inherently sequential, there has been recent progress in developing fast parallel algorithms to construct depth first search trees. However, it is not necessary to do depth first search to find connected components. In this section we present a parallel algorithm that finds connected components in $O(\lg n)$ time using $\max\{2n, 2m\}$ processors. The algorithm will have write conflicts, so, among the variations of the PRAM we have described, the Priority-Write PRAM is the one that must be used here. However, a weaker model will do. In the Arbitrary-Write model, when several processors try to write in the same memory cell at the same time, an arbitrary one of them succeeds. An algorithm for this model must work correctly no matter which processor "wins" the write conflict. The connected component algorithm will work on the Arbitrary-Write model.

The connected component algorithm is more complicated than any of the other parallel algorithms we have looked at so far. We will give a high-level description of the algorithm, then show how its various parts can be implemented on a PRAM.

The algorithm begins with each vertex in a separate tree, then repeatedly combines trees that are in the same connected component and shortens the trees. Ultimately each connected component is converted to a (directed) tree of depth 1, with all the vertices pointing to the root. Such a tree is called a *star*. See Fig. 10.8 for an illustration. The trees are represented by an array *parent*, such that *parent*[v] is the parent of vertex v. (The *parent* of a root is the root itself.) Once the connected components have been converted to stars, we can determine whether two vertices are in the same component in constant time by comparing their parents.

To shorten trees, the algorithm uses a technique called *shortcutting*, which is also useful in other parallel graph algorithms. Shortcutting simply changes the *parent* of each vertex to make the vertex point directly to its grandparent. That is, the shortcutting operation on vertex v is

$$parent[v] := parent[parent[v]].$$

Shortcutting is applied in parallel to all vertices. To see the speed with which this operation can cut down long paths, consider a simple chain of vertices as in Fig. 10.9(a), where $parent[v] = v-1$ (and $parent[1] = 1$). Figure 10.9(b) shows the *parent* pointers after the first and second applications of the shortcutting operation. If we started with n vertices in the chain, after $\lceil \lg n \rceil$ applications of shortcutting, all vertices would have the same parent.

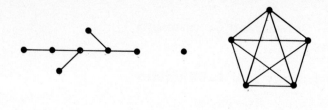

(a) A graph.

(b) Its components as stars.

Figure 10.8   Connected components turned into stars.

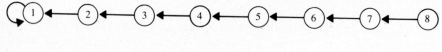

(a) A chain of vertices.

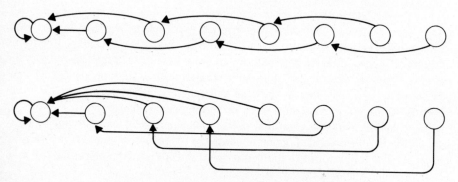

(b) After two applications of shortcutting. All vertices
will point to the root after the next shortcut operation.

Figure 10.9   The effect of shortcutting in a simple example.

Shortcutting never joins two separate trees. We need another operation, called *hooking*, to connect trees. *Hook(i, j)* means attach *i*'s root to the parent of *j*. *Hook(i, j)* is applied only if *i* is a root or the child of a root; thus the operation is

$$parent[parent[i]] := parent[j].$$

There are two versions of hooking in the algorithm. The first is

Conditional hooking: If the parent of *i* is a root, *j* is adjacent to *i* (in *G*), and *j*'s parent is less than *i*'s parent, then *Hook(i, j)*.

The requirement that we hook to the smaller of the two parents helps avoid the introduction of cycles. Conditional hooking is illustrated in Fig. 10.10(b). The second hooking operation is

Star hooking: If *i* belongs to a star, *j* is adjacent to *i* (in *G*), and *j* is not in *i*'s star, then hook *Hook(i, j)*.

Star hooking is illustrated in Fig. 10.10(c).

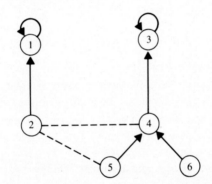

(a) Trees before hooking. (Dotted lines are edges in *G*.)

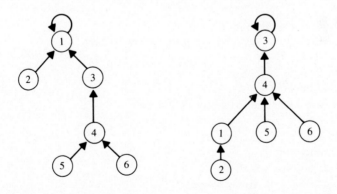

(b) Conditional hooking with *i*=4, *j*=2.   (c) Star hooking with *i*=2, *j*=5.

Figure 10.10   Illustrations of hooking.

At any one time, for a particular vertex i, there may be several vertices j that satisfy the conditions for conditional hooking or star hooking, but only one value can be stored as the new parent of i's root. In the parallel algorithm, different processors will be trying the different choices for j, and several processors may try to write in *parent*[*parent*[*i*]] at the same time. Only one succeeds in writing, but the algorithm will work properly no matter which one succeeds. By running long enough, the algorithm eventually hooks together all trees that are part of one connected component.

10.4.2   The Algorithm

The algorithm begins with each vertex of G in a separate tree. It repeatedly does hooking and shortcutting until the desired structure is achieved. We first give a high-level description. Unfortunately, Algorithm 10.6 does not quite work. Rather than make it more complicated, we will solve the problem by slightly modifying the input graph and the initialization step. After presenting the algorithm, we will explain the problem and the solution.

Algorithm 10.6   Finding Connected Components of a Graph

Input: A graph G (undirected).

Output: A forest of directed trees of depth 1, represented by the array *parent*, indexed by the vertices. Each tree contains the vertices of one connected component.

Comment: An instruction specified for a vertex v is performed in parallel for all vertices. The hooking steps are performed in parallel for all edges ij in G (and *only* for pairs i and j such that ij is an edge). Each edge, say xy, is processed twice (in parallel), once with x in the role of i, and once with y in the role of i.

```
{ Initialization }
parent[v] := v;
repeat
    { Conditional hooking }
    if i's parent is a root and parent[i] > parent[j] then
        Hook(i, j)
    end { if };
    { Star hooking }
    if i is in a star and j is not in i's star then
        Hook(i, j)
    end { if };
    { Shortcutting }
    if v is not in a star then
        parent[v] := parent[parent[v]]
    end { if }
until the shortcutting step did not produce any changes.
```

One of the facts used in the proof that the algorithm works correctly is that conditional hooking and star hooking do not produce new stars. But unfortunately, on the first pass through the loop, they do. A single vertex hooked to a star yields a star. Then, since stars can be created by conditional hooking, the star hooking step might hook two stars to each other in both directions, thus creating a cycle. These problems are eliminated by attaching a "dummy" vertex $v+n$ to each vertex v in G and initializing $parent[v+n]$ to v. Thus every tree always has at least two vertices. (Since the dummy vertices are not adjacent to any vertices in G, they affect nothing in the algorithm except whether or not a tree is a star.) Figure 10.11 illustrates the action of the algorithm. The proof that the algorithm works correctly includes many details expressed in several lemmas, which follow. The algorithm itself is not very hard to understand if a few examples are worked through, so the reader should examine Fig. 10.11 carefully before proceeding.

Assuming that the dummy vertices have been added and their parents are initialized as described in the previous paragraph, the following theorems show that the algorithm works.

Theorem 10.4   At any time during execution of Algorithm 10.6, the structure defined by the *parent* pointers is a forest.

Theorem 10.5   When Algorithm 10.6 terminates, the forest defined by the *parent* pointers consists only of stars, and the vertices in each star are exactly the vertices of a connected component of G (and the "dummy" vertices associated with them).

The proofs of the theorems use several lemmas.

Lemma 10.6   Conditional hooking and star hooking never create new stars.

Proof.   When the root of a tree with at least two nodes is attached to another tree, the new tree will have depth at least 2.   □

Lemma 10.7   The star hooking step never hooks a star onto another star.

Proof.   Suppose it does. Then, at the beginning of the star hooking step there were two stars, S_1 and S_2, containing vertices i and j, respectively, such that ij is an edge in G. Since conditional hooking does not create stars (Lemma 10.6), S_1 and S_2 were stars at the beginning of the conditional hooking step. They satisfy the conditions for conditional hooking, so the one with the larger root would have been hooked in the conditional hooking step, and it would no longer be a star.   □

Lemma 10.8   Once a vertex becomes a leaf, it will always be a leaf.

Proof.   It is clear that shortcutting does not change a leaf to a nonleaf. The only way a leaf could become a nonleaf is by having something hooked to it, but the hooking steps always hook to the parent of j, for some j. The parent of a vertex

The graph

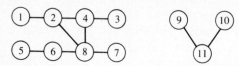

The initial forest. (Each "dummy" vertex $v + n$ is denoted by v'.)

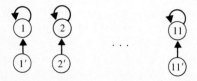

After conditional hooking. (The processors that succeeded in writing are P_{21}, P_{42}, P_{65}, P_{87}, and $P_{11,10}$.)

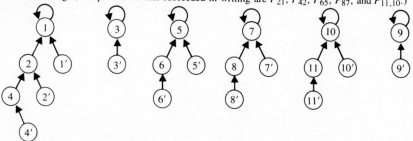

After star hooking. (P_{34} and $P_{9,11}$ wrote.)

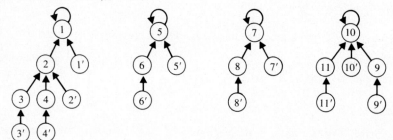

After shortcutting.

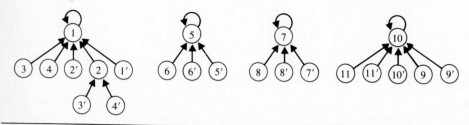

Figure 10.11   Illustration of the connected component algorithm.

After conditional hooking. (P_{82} succeeded in writing.)

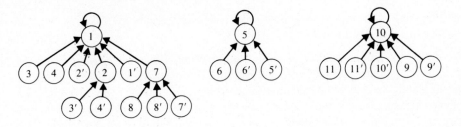

After star hooking. (P_{68} wrote.)

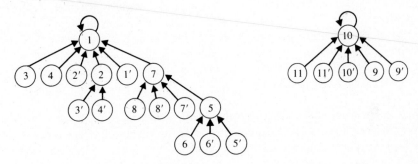

After shortcutting.

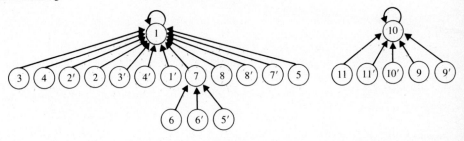

There will be no hooking on the next iteration. Shortcutting will turn the first tree into a star.

On the last iteration there will be no changes and the algorithm will stop.

could not have been a leaf. (For the case in which j is its own parent, i.e., j is a root, recall that every tree has at least two vertices, so a root is never a leaf.) $\square$

Lemma 10.9   After each shortcut step and each conditional hooking step, if $i < parent[i]$, then i is a leaf.

Proof.   The condition holds (vacuously) after the initialization because no vertex is less than its parent. Since a leaf never becomes a nonleaf (Lemma 10.8), no step in the algorithm can "undo" the required property for any vertex $i < parent[i]$ unless i's parent changes. So we must check what happens when a vertex is given a new parent. Conditional hooking hooks $parent[i]$ to $parent[j]$ only if $parent[i] > parent[j]$, so the lemma is not violated by this change. Star hooking may hook a star with root x onto a vertex y where $x < y$. (So the lemma makes no claim for the star hooking step.) Now, at the beginning of the shortcutting step, consider any chain in some tree of the form $\ldots \rightarrow x \rightarrow y \rightarrow z \rightarrow \ldots$. If $x < y$, then either x is already a leaf or it was the root of a star that just got hooked to y, and it will be a leaf when the shortcutting is done on its children. If $x > y$, then either y is the root of a star that was just hooked to z, so x is a leaf, or $y > z$, so $x > z$. In all cases, after shortcutting produces $x \rightarrow z$, either x is a leaf or $x > z$. $\square$

Proof of Theorem 10.4   The algorithm starts with trees; we have to show that no step introduces a cycle. It is clear that shortcutting cannot introduce a cycle. Since star hooking always hooks a star to a nonstar (Lemma 10.7), it does not introduce a cycle. Suppose that a cycle were formed in conditional hooking. Let j be the largest vertex in the cycle, and let i be the vertex in the cycle that points to j. Then $i < parent[i] = j$. But by Lemma 10.9, i must be a leaf, so it cannot be in a cycle. $\square$

Lemma 10.10   Any star that exists at the end of the star hooking step must be an entire connected component.

Proof.   By Lemma 10.7, the star was a star at the beginning of the star hooking step. But if any vertex in the star were adjacent (in G) to a vertex in any other tree, the star hooking step would have hooked the star to another tree, and it would no longer be a star. $\square$

Proof of Theorem 10.5   Since the vertices of G start out in disjoint trees, and two trees are hooked only if they contain vertices i and j that are adjacent, all the vertices in any one tree at any time are in the same connected component. The algorithm stops when shortcutting produces no changes. This can happen only when there are no vertices of distance 2 from their roots; i.e., when all vertices are in stars at the end of the star hooking step. By Lemma 10.10, each such star is an entire component. $\square$

10.4.3   PRAM Implementation of the Algorithm

Now we consider how the instructions in the algorithm are assigned to processors, as well as how many PRAM steps are needed to carry out each instruction.

Some processors have two "names." When we must do some operation for each vertex (say shortcutting), we will use the first $2n$ processors, referring to them as P_v. When we must do an operation for each edge, we use the first $2m$ processors, referring to them by the names P_{ij}. Since operations on vertices and operations on edges are done in different instructions, each processor will be doing only one thing at a time.

The PRAM algorithm assumes that the input is in the form of a list of edges in the graph G. Each edge appears twice in the list; that is, if ij is an edge, the pairs (i, j) and (j, i) are in the input list. Each of the $2m$ processors reads a (distinct) memory cell containing an edge, and from then on considers itself to be the processor for that particular (oriented) edge; i.e., it "knows" that its alternative name is P_{ij} if it read (i, j).

The form of the input is not critical to the speed of the algorithm. If the input were provided as an adjacency matrix, we would have n^2 processors read the matrix entries in the first step. Those that read a zero would just quit without doing any more work for edges.

We present the algorithm again with more implementation detail. The important observation to make here is that each step of the algorithm can be implemented in a constant number of PRAM steps.

Algorithm 10.7   Finding the Connected Components of a Graph

Input: A list of edges in the graph, each edge listed in both orientations; n, the number of vertices.

Output: A forest of directed trees of depth 1, represented by the array *parent*, indexed by the vertices. Each tree contains the vertices of one connected component.

Comment: A Boolean array *star* is used to tell if a vertex is in a star; *star*[v] is true if and only if v is in a star. The instructions for computing *star* are shown after the main algorithm. The shared Boolean variable *noChange* tells whether or not the shortcutting step has made any changes at each iteration of the loop.

    { Initialization }
    each edge processor reads an edge;
    each processor reads n;
    P_v and P_{2v} write v in *parent*[v] and *parent*[$2v$], respectively, where $1 \leq v \leq n$;

repeat

    { Conditional hooking }
    P_{ij} reads *parent*[i], *parent*[*parent*[i]], and *parent*[j];
    if *parent*[i] = *parent*[*parent*[i]] { i.e., if *parent*[i] is a root }
        and *parent*[i] > *parent*[j] **then**
            write *parent*[j] in *parent*[*parent*[i]]
    end { if };

{ Star hooking }
P_v computes and writes $star[v]$ { as described below };
P_{ij} reads $parent[i]$, $parent[j]$, and $star[i]$;
if $star[i]$ **and** $parent[i] \neq parent[j]$ { i.e., j is not in i's star } **then**
      write $parent[j]$ in $parent[parent[i]]$
end { if };

{ Shortcutting }
P_v writes *true* in *noChange*;
P_v reads $parent[v]$ and $parent[parent[v]]$;
if $parent[parent[v]] \neq parent[v]$ **then**
    P_v writes $parent[parent[v]]$ in $parent[v]$;
    P_v writes *false* in *noChange*
end { if }

until *noChange* { All processors read *noChange* to determine whether
they should stop. }

A vertex is not in a star if and only if one of the following conditions holds:

1.   Its parent is not its grandparent,
2.   It is the grandparent, but not the parent, of some other vertex, or
3.   Its parent has a nontrivial grandchild.

Figure 10.12 illustrates all three cases and the computation of *star*. The computation
is described in the following algorithm.

Algorithm 10.8   Determining if a Vertex is in a Star

Comment: These steps are carried out by P_v (for $1 \leq v \leq 2n$).

write *true* in $star[v]$;
read $parent[v]$ and $parent[parent[v]]$;
if $parent[v] \neq parent[parent[v]]$ **then**
    write *false* in $star[v]$;
    write *false* in $star[parent[parent[v]]]$
end { if };
read $star[parent[v]]$;
if not $star[parent[v]]$ **then**
    write *false* in $star[v]$.
end { if }

The computation of $star[v]$ and each of the steps in Algorithm 10.7's main
loop can be carried out in constant time by a PRAM, so the number of iterations of
the loop determines the order of the running time. All that remains is to show that
the number of iterations is in $O(\lg n)$.

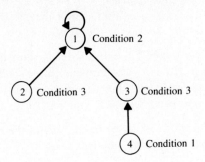

(a) How to tell each vertex is not in a star.

Initial values of *star*

T	T	T	T

If *v* has a nontrivial grandparent, *star* [*v*] := *false* (condition 1).

T	T	T	F

If *v* has a nontrivial grandparent, *star* [*v*'s grandparent] := *false* (condition 2).

F	T	T	F

(In general at this point, if the tree is not a star, only children of the root may still have *star* = *true*.)

If *star* [*v*'s parent] is *false*, then *star* [*v*] := *false*.

F	F	F	F

(b) The computation.

Figure 10.12 Computation of *star*.

Lemma 10.11   Let d be the depth of a nonstar tree before the shortcutting step. After shortcutting, its depth is at most $2d/3$.

Proof.   Actually, the depth is cut roughly in half after each shortcutting step. It is easy to show by induction that if d is even, the depth after shortcutting will be exactly $d/2$, and that if d is odd, the depth after shortcutting will be $(d+1)/2$. The worst case is when $d = 3$. The new depth will be 2, which is $2d/3$.   □

For any connected component C, let $d_C(t)$ be the sum of the depths of all the trees in C at the end of the tth iteration of the loop (for $t \geq 0$).

Lemma 10.12   For any connected component whose vertices do not form a star at the beginning of the tth iteration of the loop (for $t \geq 1$), $d_C(t) \leq 2d_C(t-1)/3$.

Proof.   Consider what happens to the trees of C during the tth iteration. Since a tree is never hooked to a leaf, the depth of a tree that results from hooking is at most the sum of the depths of the two trees that were hooked. After shortcutting, each tree is at most two-thirds as deep as it was before, so the sum of the depths is at most two-thirds what it was before.   □

Theorem 10.13   Algorithm 10.6 runs in $O(\lg n)$ time in the worst case.

Proof.   From Lemma 10.12, for any connected component C, we have

$$d_C(t-1) \geq \frac{3}{2} d_C(t).$$

Iterating this recurrence gives

$$d_C(0) \geq \left[\frac{3}{2}\right]^t d_C(t).$$

At the end of the 0th iteration (i.e., after the initialization), there are n trees, each of depth 1, so $d_C(0) \leq n$. Let T be the number of the first iteration after which the vertices of C are in one star. Then $d_C(T) = 1$. So

$$n \geq d_C(0) \geq \left[\frac{3}{2}\right]^T d_C(T) = \left[\frac{3}{2}\right]^T,$$

i.e.,

$$n \geq \left[\frac{3}{2}\right]^T.$$

So

$$T \leq \lg_{3/2} n.$$

Since T is an integer, we conclude that after $\lfloor \lg_{3/2} n \rfloor$ iterations, each component is a star. The algorithm does just one more iteration, in which nothing changes, so the total number of iterations, and the running time of the algorithm, is in $O(\lg n)$.   □

10.5
Lower Bounds

In this section we present a lower-bound argument for parallel computation. Most of this section is about Boolean functions, that is, computations in which the input consists of n bits. However, we will also get a lower bound for addition of integers. Before we attack the lower bound, let us first consider how to compute Boolean functions quickly.

10.5.1 Computing Boolean Functions

Let f be a function from $\{0,1\}^n$ into $\{0,1\}$. The Boolean **or** and **and** functions are examples. We will show that any such function can be computed in $O(\lg n)$ time (on any of the versions of the PRAM defined in Section 10.2.3). The key idea is to use the binary fan-in scheme described in Section 10.2.1 to encode the bits into one integer. Then any processor can read the integer and compute the function value. (If it seems strange that all the actual computation of the function will be done by one processor in one step, and all the rest of the time is used to communicate the n input bits to that processor, recall that models for parallel computation tend to focus on the communication between processors.)

We define the function *Encode* on n bits $b_1, b_2, \ldots, b_n$ as follows:

$$Encode\,(b_1, b_2, \ldots, b_n) = b_n b_{n-1} \cdots b_2 b_1 = \sum_{i=1}^{n} b_i\, 2^{i-1}$$

Encode is computed by the following PRAM algorithm.

Algorithm 10.9   Encoding n Bits as an Integer

Input: Bits $b_1, b_2, \ldots, b_n$, initially in the first n memory cells.

Output: The integer $b_n b_{n-1} \cdots b_2 b_1$ in $M[1]$.

Comment: Each processor uses local variables x (to accumulate part of the integer), *temp*, and *incr* (for determining which processor's output to read next and how much the value read needs to be "shifted"). As usual, i is the processor number. (Some processors will read from memory cells with indexes larger than n; as usual, we assume that they contain zeros.)

```
Read M[i] into x;
incr := 1;
for step := 1 to ⌈lg n⌉ do
    read M[i+incr] into temp;
    x := x + 2^incr * temp;
    incr := 2 * incr;
    write x into M[i]
end { for };
```

This algorithm clearly takes $\lceil \lg n \rceil + 1$ time steps. The proof that the algorithm actually works, i.e., that after the loop terminates, the contents of $M[1]$ will be the integer $b_n b_{n-1} \cdots b_2 b_1$ is similar to the proof of correctness for Algorithm 10.1. We leave the details as an exercise.

Algorithm 10.10   Computing a Boolean Function on n Bits

Input: Bits $b_1, b_2, \ldots, b_n$, initially in the first n memory cells.

Output: $f(b_1, b_2, \ldots, b_n)$ in $M[1]$, where f is some specific Boolean function.

> *Encode* { using Algorithm 10.9 };
> P_1 reads $M[1]$;
> P_1 computes $f(b_1, b_2, \ldots, b_n)$
>   { by decoding the integer from $M[1]$ into its component bits if necessary };
> P_1 writes the result in $M[1]$.

We summarize with the following theorem.

Theorem 10.14   A PRAM can compute any Boolean function in $\lceil \lg n \rceil + 2$ steps using n processors.

10.5.2   A Lower Bound for Computing a Function on n Bits

The *Encode* function defined in Section 10.5.1 is a hardest-to-compute function on n bits. That is, no other function can require time of higher order because, once all the input bits have been encoded into one integer, any other function can be computed by one processor in one step after reading the encoding. In this section we will derive a lower bound on the number of steps required to compute *Encode*.

Several of our earlier lower-bound arguments used decision trees. The idea underlying those arguments was that there had to be enough branching in the tree, enough decisions, to distinguish inputs that should generate different outputs. A similar idea is used here. A PRAM for *Encode* must do enough steps to distinguish between all possible 2^n inputs because each input is encoded as a distinct integer; i.e., each input generates a distinct output. Since the output is written in $M[1]$, a PRAM must do enough steps so that any of 2^n different values could be written in $M[1]$. Of course, for one particular input, a PRAM always writes exactly one specific value in $M[1]$ at any step. Here we consider the space of all inputs; we count all the different values a PRAM could write for all possible inputs.

The value in a memory cell depends on what the processors write (or do not write). What a processor writes depends on the "state" of the processor at the beginning of a step and what it reads from memory on that step. Think of the state of a processor as everything internal to the processor that affects its action (e.g., the values of all the variables in its local memory and its own index). The proof of the lower bound counts how many different states processors can be in, and how many values could be written in memory cells, after each step.

Theorem 10.15   (P. Beame) Any Priority-Write PRAM that computes *Encode* must do at least $\lg n + 1 - \lg\lg(4p)$ steps (where p is the number of processors).

Proof.   We want to find answers to the following two questions.

How many different values can be in any particular memory cell $M[i]$ after t steps?

How many different states can any particular processor P_i be in after t steps?

We define two sequences of numbers,

$$r_0 = 1 \qquad s_0 = 2$$
$$r_{t+1} = r_t s_t \qquad s_{t+1} = pr_{t+1} + s_t \tag{10.1}$$

where p is the number of processors. Here are the first few values in the sequences.

t	r_t	s_t
0	1	2
1	2	$2p + 2$
2	$2(2p + 2)$	$2p(2p + 2) + 2p + 2$

Lemma 10.16   The number of distinct states a processor may be in after t steps (considering all possible inputs) is at most r_t. The number of distinct values that could be in a memory cell after t steps (considering all possible inputs) is at most s_t.

Proof.   We prove the lemma by induction on t. For $t = 0$ (that is, before the PRAM has executed any instructions), each processor can be in only one state, its initial state. Each memory cell contains one of two possible values: 0 and 1. Since $r_0 = 1$ and $s_0 = 2$, the basis for the induction is established.

Now, for $t \geq 0$, assume that after t steps a processor can be in any one of at most r_t steps, and that a memory cell can have one of at most s_t values. The new state of a processor after step $t+1$ depends on the old state (the state after step t) and the value read from memory by that processor on step $t+1$. Thus the number of possible states after step $t+1$ is at most $r_t s_t$, which is r_{t+1}. On step $t+1$ any processor can write in a particular memory cell, and a processor can write a different value for each state it could be in. That gives pr_{t+1} possible values, but it is also possible that no processor writes in the cell on this step, so any of the s_t values that could have been there before may still be in the cell after step $t+1$. Thus the total number of possible values in a memory cell at the end of step $t+1$ is $pr_{t+1} + s_t$, which is s_{t+1}.   □

Lemma 10.17   For $t \geq 1$, $s_{t+1} \leq s_t^2$.

Proof.   Using Eq. 10.1,

$$s_{t+1} = pr_{t+1} + s_t = pr_t s_t + s_t = s_t(pr_t + 1) \leq s_t(pr_t + g \, s_{t-1}) = s_t^2. \qquad \square$$

Lemma 10.18   For $t \geq 1$, $s_t \leq (4p)^{2^{t-1}}$.

Proof.   For $t = 1$, $s_1 = pr_0 s_0 + s_0 = p \times 1 \times 2 + 2 = 2p + 2 \leq 4p$ since $p \geq 1$.   For $t \geq 1$,

$$s_{t+1} \leq s_t^2 \leq \left[(4p)^{2^{t-1}}\right]^2 = (4p)^{2^t}. \qquad \square$$

Now, continuing with the proof of the theorem, we observe that if any PRAM algorithm computes *Encode* in T steps, then $s_T \geq 2^n$, because 2^n distinct outputs could appear in $M[1]$ when the algorithm terminates. So

$$2^n \leq s_T \leq (4p)^{2^{T-1}},$$

and, taking logs,

$$n \leq 2^{T-1} \lg(4p).$$

Taking logs again,

$$\lg n \leq T - 1 + \lg\lg(4p).$$

Therefore

$$T \geq \lg n + 1 - \lg\lg(4p). \qquad \square$$

Corollary 10.19   Any CREW PRAM or Common-Write PRAM that computes *Encode* must do at least $\lg n + 1 - \lg\lg(4p)$ steps. Hence if p is bounded by any polynomial in n, the parallel complexity of *Encode* is in $\Theta(\lg n)$.

Proof.   Any program for either of these models is also a valid program for the Priority-Write model, so the lower bound in the theorem applies.

If p is bounded by a polynomial in n, then $\lg\lg(4p)$ is in $\Theta(\lg\lg n)$, and $\lg n + 1 - \lg\lg(4p)$ is in $\Theta(\lg n)$. $\qquad \square$

And, finally, here is our lower bound on parallel addition of integers.

Corollary 10.20   Any PRAM algorithm (CREW, Common-Write, or Priority-Write) to compute the sum of n integers with at most n bits does at least $\lg n - \lg\lg(4p)$ steps. (Hence if p is bounded by any polynomial in n, the parallel complexity of addition of n integers is in $\Theta(\lg n)$.)

Proof. Let *Add* be any PRAM algorithm to compute the sum of n integers. With one more step, we can compute *Encode* as follows:

Read b_i from $M[i]$;
Compute $b_i 2^{i-1}$;
Write $b_i 2^{i-1}$ into $M[i]$;
Add.

If *Add* were faster than the corollary claims, we would have a faster way to compute *Encode*, contradicting Theorem 10.15.   □

Exercises

Section 10.2: Some PRAM Algorithms and the Handling of Write Conflicts

10.1. For Algorithm 10.1, what does P_2 compute in the first three iterations of the loop?

10.2. Write a CREW PRAM algorithm to compute the sum of n integers in $O(\lg n)$ time.

10.3. Show that the Boolean **and** of n bits can be computed in constant time by a Common-Write or Priority-Write PRAM.

10.4. Show that the Boolean matrix product of two $n \times n$ Boolean matrices can be computed in constant time by a Priority-Write or Common-Write PRAM. (The number of processors should be bounded by a polynomial in n.)

*10.5. Using the lower bound given in the text for computing the **or** of n bits on a CREW PRAM, show that computing the maximum of n integers requires $\Omega(\lg n)$ time on a CREW PRAM.

10.6. Using the lower bound given in the text for computing the **or** of n bits on a CREW PRAM, show that Boolean matrix multiplication requires $\Omega(\lg n)$ time on a CREW PRAM.

10.7. Modify Algorithm 10.1 so that it outputs an index of the largest key instead of the largest key itself. (The modified algorithm should not have write conflicts, and it should still do $\Theta(\lg n)$ steps.)

10.8. Would Algorithm 10.3 work correctly if we did not specify how k should be chosen when a processor compares two equal keys? Justify your answer with an argument or a counterexample.

10.9. Modify Algorithm 10.3 so that it outputs an index of the largest key instead of the largest key itself. (The modified algorithm should do only a constant number of steps.)

Section 10.3: Merging and Sorting

10.10. Give a PRAM implementation of Insertion Sort for n keys that runs in $O(n)$ time steps. (You may use any PRAM variation, but specify which one.)

10.11. Modify the parallel merge algorithm (Algorithm 10.4) to merge two sorted lists of n and m keys, respectively. How many steps does the revised algorithm do?

*10.12. Describe an algorithm to sort n keys in $O(\lg n)$ steps on a CREW PRAM. The number of processors may be greater than n, but it should be bounded by a polynomial in n.

(Hint: Exercise 10.2 implies that you can compute the sum of n bits in $O(\lg n)$ time. Use that along with the trick used in the constant-time max algorithm (Algorithm 10.3).)

*10.13. Give an algorithm to merge two sorted lists of n keys each in constant time on a CREW PRAM. The number of processors may be greater than n, but it should be bounded by a polynomial in n.

Section 10.4: A Parallel Connected Component Algorithm

10.14. The connected component algorithm does not tell us the number of connected components in the input graph. Show that the number of connected components can be determined in $O(\lg n)$ time.

10.15. In the example in Fig. 10.11, when more than one processor tried to write in one memory cell at the same time, we made an arbitrary choice as to which one succeeded. Redo the example making a different valid choice at each step at which there was a write conflict.

Section 10.5: Lower Bounds

10.16. Prove that Algorithm 10.9 computes *Encode*. Hint: Show by induction that after the tth iteration, $M[i]$ contains

$$b_i + 2b_{i+1} + \ldots + 2^{2^t-1}b_{i+2^t-1}.$$

Additional Problems

10.17. The n-bit unary representation of an integer k is a sequence of k ones followed by $n-k$ zeroes. For each of the following problems you should use at most n processors.

 a) Show that a PRAM (CREW, Common-Write, or Priority-Write) can read an integer k between 0 and n from $M[1]$ and convert k to its unary representation in one step. (The output is to go in cells $M[1], \ldots, M[n]$.)

 b) Show that a Priority-Write PRAM with n processors can read the unary representation of an integer k from cells $M[1], \ldots, M[n]$ and write k in $M[1]$ in one step.

 c) Show that a CREW PRAM can solve the problem in (b) in two steps.

 d) Show that, in a constant number of steps, a PRAM (any variation) can convert the n-bit unary representation of an integer (given in cells $M[1], \ldots, M[n]$) to the $\lfloor \lg n \rfloor + 1$ bit binary representation of the same integer.

*10.18. Suppose that you have a sorted list of n keys in memory and p processors, where p is small compared to n. Give a CREW PRAM algorithm for searching the list for a key x. How many steps does your algorithm do? (Hint: use a generalization of binary search. You may find part (c) of the previous exercise helpful. Your search algorithm should do $\Theta(\lg(n+1)/\lg(p+1))$ steps in the worst case.)

10.19. Go through the earlier chapters of this book and pick out any algorithm that has a natural parallel version. Write the parallel algorithm and tell how many processors and time steps it uses. (Choose an algorithm for which the running time of the parallel version is of lower order than the sequential version.)

10.20. Make a list of the problems covered in this chapter that are in the class *NC* (defined in Section 10.1). Are any algorithms in this chapter not *NC* algorithms?

Notes and References

The PRAM model was presented (in slightly different form) in Fortune and Wyllie (1978) and Goldschlager (1978). There are now many dozens of papers using this model.

Section 10.3 is from Shiloach and Vishkin (1981). Their paper gives algorithms for sorting (and several other problems) in which the number of processors is smaller than the number of keys. It also contains the $O(\lg\lg n)$ algorithm for finding the largest of n keys mentioned in Section 10.2.4 and a solution to Exercise 10.17(c).

Section 10.4 is based on Awerbuch and Shiloach (1983) and Shiloach and Vishkin (1982). Awerbuch and Shiloach also contains a parallel algorithm for finding minimum spanning forests. Section 10.5 is from Beame (1986).

For those who wish to read further, the Bibliography includes a sampling of other papers: Cook, Dwork and Reischuk (1986), on upper and lower bounds for several problems considered in Sections 10.2 and 10.3; Chandra, Stockmeyer and Vishkin (1984), on a number of interesting problems and the relations between their parallel complexity; Kruskal (1983) and Snir (1985), on parallel searching (including the solution to Exercise 10.18); Batcher (1969), on sorting networks; Landau and Vishkin (1986), on approximate string matching; and Tarjan and Vishkin (1985), on biconnected components of graphs. Akl (1985) is a book on parallel sorting (using various models of parallel computation); Richards (1986), a bibliography on parallel sorting, contains nearly 400 entries. Quinn and Deo (1984) is a survey of parallel graph algorithms. Quinn (1987) is a text on parallel algorithms.

Bibliography

Aho, Alfred V., and M. J. Corasick, "Efficient string matching: An aid to bibliographic search," *Communications of the ACM*, vol. 18, no. 6, pp. 333-340, June, 1975.

Aho, Alfred V., John E. Hopcroft, and Jeffery D. Ullman, *The Design and Analysis of Computer Algorithms*, Addison-Wesley, Reading, Mass., 1974.

Aho, Alfred V., John E. Hopcroft, and Jeffrey D. Ullman, *Data Structures and Algorithms*, Addison-Wesley, Reading, Mass., 1983.

Akl, Selim, *Parallel Sorting*, Academic Press, Orlando, Fla., 1985.

Angluin, Dana, "The four Russians' algorithm for Boolean matrix multiplication is optimal in its class," *SIGACT News*, vol. 8, no. 1, pp. 29-33, January–March, 1976.

Apostolico, Alberto, and Raffaele Giancarlo, "The Boyer-Moore-Galil string searching strategies revisited," *SIAM Journal on Computing*, vol. 15, no. 1, pp. 98-105, February, 1986.

Arlazarov, V. L., E. A. Dinic, M. A. Kronrod, and I. A. Faradzev, "On economical construction of the transitive closure of a directed graph," *Soviet Mathematics, Doklady*, vol. 11, no. 5, pp. 1209-1210, 1970.

Awerbuch, Baruch, and Yossi Shiloach, "New connectivity and MSF algorithms for ultracomputer and PRAM," *Proceedings of the IEEE International Conference on Parallel Processing*, pp. 175-179, 1983.

Batcher, Kenneth E., "Sorting networks and their applications," *Proceedings of the AFIPS Spring Joint Computer Conference*, pp. 307-314, 1968.

Beame, Paul, "Limits on the power of concurrent-write parallel machines," *Proceedings of the Eighteenth Annual ACM Symposium on Theory of Computing*, pp. 169-176, 1986.

Bellman, Richard E., *Dynamic Programming*, Princeton University Press, Princeton, N. J., 1957.

Bellman, Richard E., and Stuart E. Dreyfus, *Applied Dynamic Programming*, Princeton University Press, Princeton, N.J., 1962.

Bentley, Jon L., *Writing Efficient Programs*, Prentice-Hall, Englewood Cliffs, N.J., 1982.

Bentley, Jon L., *Programming Pearls*, Addison-Wesley, Reading, Mass., 1986.

Bentley, Jon L., "Programming Pearls," *Communications of the ACM*, monthly column.

Bentley, Jon L., David Johnson, F. T. Leighton, and Catherine C. McGeoch, "An experimental study of bin packing," *Proceedings of the 21st Annual Allerton Conference on Communication, Control, and Computing*, pp. 51-60, 1983.

Blum, Manuel, Robert W. Floyd, Vaughan Pratt, Ronald L. Rivest, and Robert E. Tarjan, "Time bounds for selection," *Journal of Computer and System Sciences*, vol. 7, no. 4, pp. 448-461, August, 1973.

Borodin, Allan, and Ian Munro, *Computational Complexity of Algebraic and Numeric Problems*, American Elsevier, New York, 1975.

Boyer, Robert S., and J. Strother Moore, "A fast string searching algorithm," *Communications of the ACM*, vol. 20, no. 10, pp. 762-772, October, 1977.

Brassard, Gilles, "Crusade for a better notation," *SIGACT News*, vol. 17, no. 1, pp. 60-64, Summer, 1985.

Brigham, E. Oran, *The Fast Fourier Transform*, Prentice-Hall, Englewood Cliffs, N.J., 1974.

Chandra, Ashok K., Larry J. Stockmeyer, and Uzi Vishkin, "Constant depth reducibility," *SIAM Journal on Computing*, vol. 13, no. 2, pp. 423-439, May, 1984.

Cook, Stephen A., "The complexity of theorem proving procedures," in *Proceedings of the Third Annual ACM Symposium on Theory of Computing*, pp. 151-158, 1971.

Cook, Stephen A., "An overview of computational complexity," *Communications of the ACM*, vol. 26, no. 6, pp. 400-408, June, 1983. 1982 Turing Award Lecture.

Cook, Stephen, Cynthia Dwork, and Rudiger Reischuk, "Upper and lower time bounds for parallel random access machines without simultaneous writes," *SIAM Journal on Computing*, vol. 15, no. 1, pp. 87-97, February, 1986.

Cooley, J. W., and J. W. Tukey, "An algorithm for the machine calculation of complex Fourier series," *Mathematics of Computation*, vol. 19, pp. 297-301, 1965.

Cooper, Doug, *Standard Pascal User Reference Manual*, W. W. Norton & Co., New York, 1983.

Coppersmith, Donald, and Shmuel Winograd, "Matrix multiplication via arithmetic progressions," *Proceedings of the 19th Annual ACM Symposium on Theory of Computing*, pp. 1-6, 1987.

Deo, Narsingh, *Graph Theory with Applications to Engineering and Computer Science*, Prentice-Hall, Englewood Cliffs, N.J., 1974.

Dijkstra, Edsger W., "A note on two problems in connexion with graphs," *Numerische Mathematik*, vol. 1, pp. 269-271, 1959.

Dijkstra, Edsger W., *A Discipline of Programming*, Prentice-Hall, Englewood Cliffs, N.J., 1976.

Even, Shimon, *Algorithmic Combinatorics*, Macmillan, New York, 1973.

Even, Shimon, *Graph Algorithms*, Computer Science Press, Inc., Rockville, Md., 1979.

Fischer, Michael J., "Efficiency of equivalence algorithms," in *Complexity of Computer Computations*, eds. R.E. Miller and J.W. Thatcher, pp. 153-167, Plenum Press, New York, 1972.

Fischer, Michael J., and A. R. Meyer, "Boolean matrix multiplication and transitive closure," in *Conference Record, IEEE 12th Annual Symposium on Switching and Automata Theory*, pp. 129-131, 1971.

Floyd, Robert W., "Algorithm 97: Shortest path," *Communications of the ACM*, vol. 5, no. 6, p. 345, June, 1962.

Floyd, Robert W., "Algorithm 245: Treesort 3," *Communications of the ACM*, vol. 7, no. 12, p. 701, December, 1964.

Ford, Lester R. Jr., and D. R. Fulkerson, *Flows in Networks*, Princeton University Press, Princeton, N.J., 1962.

Fortune, Steven, and James Wyllie, "Parallelism in random access machines," *Proceedings of the Tenth Annual ACM Symposium on Theory of Computing*, pp. 114-118, 1978.

Gabow, H. N., "Two algorithms for generating weighted spanning trees in order," *SIAM Journal on Computing*, vol. 6, no. 1, pp. 139-150, 1977.

Galil, Zvi, "Real-time algorithms for string-matching and palindrome recognition," in *Proceedings of the Eighth Annual ACM Symposium on Theory of Computing*, pp. 161-173, 1976.

Galil, Zvi, "On improving the worst case running time of the Boyer-Moore string matching algorithm," *Communications of the ACM*, vol. 22, no. 9, pp. 505-508, September, 1979.

Galler, B. A., and Michael J. Fischer, "An improved equivalence algorithm," *Communications of the ACM*, vol. 7, no. 5, pp. 301-303, May, 1964.

Gardner, Martin, *Wheels, Life, and Other Mathematical Amusements*, W. H. Freeman, San Francisco, 1983.

Garey, Michael R., Ronald L. Graham, and Jeffery D. Ullman, "Worst-case analysis of memory allocation algorithms," in *Proceedings of the Fourth Annual ACM Symposium on Theory of Computing*, pp. 143-150, 1972.

Garey, Michael R., and David S. Johnson, "The complexity of near-optimal graph coloring," *Journal of the ACM*, vol. 23, no. 1, pp. 43-49, January, 1976.

Garey, Michael R., and David S. Johnson, *Computers and Intractability: A Guide to the Theory of NP-completeness*, W. H. Freeman, San Francisco, 1979.

Gentleman, W. M., and G. Sande, "Fast Fourier transforms - for fun and profit," in *Proceedings, Fall Joint Computer Conference*, pp. 563-578, 1966.

Gibbons, Alan, *Algorithmic Graph Theory*, Cambridge University Press, Cambridge, 1985.

Goldschlager, Leslie M., "A unified approach to models of synchronous parallel machines," *Proceedings of the Tenth Annual ACM Symposium on Theory of Computing*, pp. 89-94, 1978.

Gries, David, *The Science of Programming*, Springer-Verlag, New York, 1981.

Guibas, Leo J., and Andrew M. Odlyzko, "A new proof of the linearity of the Boyer-Moore string searching algorithms," *Proceedings of the 18th Annual IEEE Symposium on Foundations of Computer Science*, pp. 189-195, 1977.

Hall, Patrick A. V., and Geoff R. Dowling, "Approximate string matching," *ACM Computing Surveys*, vol. 12, no. 4, pp. 381-402, December, 1980.

Hantler, S. L., and J. C. King, "An introduction to proving the correctness of programs," *ACM Computing Surveys*, vol. 8, no. 3, pp. 331-353, September, 1976.

Hoare, C. A. R., "Quicksort," *Computer Journal*, vol. 5, no. 1, pp. 10-15, April 1962.

Hopcroft, John E., and Robert E. Tarjan, "Dividing a graph into triconnected components," *SIAM Journal on Computing*, vol. 2, no. 3, pp. 135-158, September, 1973a.

Hopcroft, John E., and Robert E. Tarjan, "Algorithm 447: Efficient algorithms for graph manipulation," *Communications of the ACM*, vol. 16, no. 6, pp. 372-378, June, 1973b.

Hopcroft, John E., and Robert E. Tarjan, "Efficient planarity testing," *Journal of the ACM*, vol. 21, no. 4, pp. 549-568, October, 1974.

Hopcroft, John E., and Jeffery D. Ullman, "Set merging algorithms," *SIAM Journal on Computing*, vol. 2, no. 4, pp. 294-303, December, 1973.

Hyafil, Laurent, "Bounds for selection," *SIAM Journal on Computing*, vol. 5, no. 1, pp. 109-114, March, 1976.

Ibarra, Oscar H., and Chul E. Kim, "Fast approximation algorithms for the knapsack and sum of subset problems," *Journal of the ACM*, vol. 22, no. 4, pp. 463-468, October, 1975.

Johnson, David S., "Fast allocation algorithms," in *Proceedings of the Thirteenth Annual Symposium on Switching and Automata Theory*, pp. 144-154, 1972.

Johnson, David S., "Approximation algorithms for combinatorial problems," in *Proceedings of the Fifth Annual ACM Symposium on Theory of Computing*, pp. 38-49, 1973.

Johnson, David S., "Worst-case behavior of graph coloring algorithms," in *Proceedings of the Fifth Southeastern Conference on Combinatorics, Graph Theory, and Computing*, pp. 513-528, Utilitas Mathematica Publishing, Winnipeg, Canada, 1974.

Karp, Richard M., "Reducibility among combinatorial problems," in *Complexity of Computer Computations*, eds. R.E. Miller and J.W. Thatcher, pp. 85-104, Plenum Press, New York, 1972.

Karp, Richard M., "Combinatorics, complexity, and randomness," *Communications of the ACM*, vol. 29, no. 2, pp. 98-109, February 1986. 1985 Turing Award Lecture.

King, K. N., and Barbara Smith-Thomas, "An optimal algorithm for sink-finding," *Information Processing Letters*, vol. 14, no. 3, pp. 109-111, May 1982.

Knuth, Donald E., *The Art of Computer Programming, Volume I: Fundamental Algorithms*, Addison-Wesley, Reading, Mass., 1968.

Knuth, Donald E., *The Art of Computer Programming, Volume III: Sorting and Searching*, Addison-Wesley, Reading, Mass., 1973.

Knuth, Donald E., "Big omicron and big omega and big theta," *SIGACT News*, vol. 8, no. 2, pp. 18-24, April-June, 1976.

Knuth, Donald E., "The complexity of songs," *Communications of the ACM*, vol. 27, no. 4, pp. 344-346, April, 1984.

Knuth, Donald E., James H. Morris, Jr., and Vaughn R. Pratt, "Fast pattern matching in strings," *SIAM Journal on Computing*, vol. 6, no. 2, pp. 323-350, June, 1977.

Kronsjo, Lydia, *Computational Complexity of Sequential and Parallel Algorithms*, John Wiley & Sons, Chichester, U.K., 1985.

Kruse, Robert L., *Data Structures and Program Design, Second Edition*, Prentice-Hall, Englewood Cliffs, N.J., 1987.

Kruskal, Clyde P., "Searching, merging, and sorting in parallel computation," *IEEE Transactions on Computers*, vol. C-32, no. 10, pp. 942-946, October, 1983.

Kruskal, J. B. Jr., "On the shortest spanning subtree of a graph and the traveling salesman problem," *Proceedings of the AMS*, vol. 7, no. 1, pp. 48-50, 1956.

Landau, Gad M., and Uzi Vishkin, "Introducing efficient parallelism into approximate string matching and a new serial algorithm," *Proceedings of the 18th Annual ACM Symposium on Theory of Computing*, pp. 220-230, 1986.

Lawler, Eugene L., J. K. Lenstra, A. H. G. Rinnooy Kan, and David B. Schmoys (editors), *The Traveling Salesman Problem*, Wiley-Interscience, Chichester, U.K., 1985.

Lueker, George S., "Some techniques for solving recurrences," *Computing Surveys*, vol. 12, no. 4, pp. 419-436, December 1980.

Prim, R. C., "Shortest connection networks and some generalizations," *Bell System Technical Journal*, vol. 36, pp. 1389-1401, 1957.

Purdom, Paul W. Jr., and Cynthia A. Brown, *The Analysis of Algorithms*, Holt, Rinehart and Winston, New York, 1985.

Quinn, Michael J., *Designing Efficient Algorithms for Parallel Computers*, McGraw-Hill, New York, 1987.

Quinn, Michael J., and Narsingh Deo, "Parallel graph algorithms," *ACM Computing Surveys*, vol. 16, no. 3, pp. 319-348, September, 1984.

Rabin, Michael O., "Complexity of computations," *Communications of the ACM*, vol. 20, no. 9, pp. 625-633, September, 1977. 1976 Turing Award Lecture.

Reingold, Edward M., "On the optimality of some set merging algorithms," *Journal of the ACM*, vol. 19, no. 4, pp. 649-659, October, 1972.

Reingold, Edward M., Jurg Nievergelt, and Narsingh Deo, *Combinatorial Algorithms: Theory and Practice*, Prentice-Hall, Englewood Cliffs, N.J., 1977.

Reingold, Edward M., and A. I. Stocks, "Simple proofs of lower bounds for polynomial evaluation," in *Complexity of Computer Computations*, eds. R.E. Miller and J.W. Thatcher, pp. 21-30, Plenum Press, New York, 1972.

Richards, Dana, "Parallel sorting — a bibliography," *SIGACT News*, vol. 18, no. 1, pp. 28-48, Summer, 1986.

Sahni, Sartaj K., "Approximate algorithms for the 0/1 knapsack problem," *Journal of the ACM*, vol. 22, no. 1, pp. 115-124, January, 1975.

Sahni, Sartaj K., "Algorithms for scheduling independent tasks," *Journal of the ACM*, vol. 23, no. 1, pp. 116-127, January, 1976.

Savage, John E., "An algorithm for the computation of linear forms," *SIAM Journal on Computing*, vol. 3, no. 2, pp. 150-158, June, 1974.

Schonhage, A., M. Paterson, and Nicholas Pippenger, "Finding the median," *Journal of Computer and System Sciences*, vol. 13, no. 2, pp. 184-199, October, 1976.

Sedgewick, Robert, "Quicksort with equal keys," *SIAM Journal on Computing*, vol. 6, no. 2, pp. 240-267, June, 1977.

Sedgewick, Robert, "Implementing Quicksort programs," *Communications of the ACM*, vol. 21, no. 10, pp. 847-857, October, 1978.

Sedgewick, Robert, *Algorithms*, Addison-Wesley, Reading, Mass., 1983.

Sharir, M., "A strong-connectivity algorithm and its application in data flow analysis," *Computers and Mathematics with Applications*, vol. 7, no. 1, pp. 67-72, 1981.

Shell, Donald L., "A high-speed sorting procedure," *Communications of the ACM*, vol. 2, no. 7, pp. 30-32, July, 1959.

Shiloach, Yossi, and Uzi Vishkin, "Finding the maximum, merging and sorting in a parallel computation model," *Journal of Algorithms*, vol. 2, pp. 88-102, 1981.

Shiloach, Yossi, and Uzi Vishkin, "An $O(\log n)$ parallel connectivity algorithm," *Journal of Algorithms*, vol. 3, pp. 57-67, 1982.

Smit, G. de V., "A comparison of three string matching algorithms," *Software - Practice and Experience*, vol. 12, pp. 57-66, 1982.

Snir, Marc, "On parallel searching," *SIAM Journal on Computing*, vol. 14, no. 3, pp. 688-708, August, 1985.

Strassen, Volker, "Gaussian elimination is not optimal," *Numerische Mathematik*, vol. 13, pp. 354-356, 1969.

Tarjan, Robert E., "Depth-first search and linear graph algorithms," *SIAM Journal on Computing*, vol. 1, no. 2, pp. 146-160, June, 1972.

Tarjan, Robert E., "On the efficiency of a good but not linear set union algorithm," *Journal of the ACM*, vol. 22, no. 2, pp. 215-225, 1975.

Tarjan, Robert E., and Uzi Vishkin, "An efficient parallel biconnectivity algorithm," *SIAM Journal on Computing*, vol. 14, no. 4, pp. 862-874, November, 1985.

Tenenbaum, Aaron M., and Moshe J. Augenstein, *Data Structures Using Pascal, Second Edition*, Prentice-Hall, Englewood Cliffs, N.J., 1986.

van Leeuwen, Jan, and Robert E. Tarjan, "Worst-case analysis of set union algorithms," *Journal of the ACM*, vol. 31, no. 2, pp. 245-281, April, 1984.

Wagner, Robert A., and Michael J. Fischer, "The string-to-string correction problem," *Journal of the ACM*, vol. 21, no. 1, pp. 168-178, January, 1974.

Wainwright, Roger L., "A class of sorting algorithms based on quicksort," *Communications of the ACM*, vol. 28, no. 4, pp. 396-403, April 1985.

Warshall, Stephen, "A theorem on Boolean matrices," *Journal of the ACM*, vol. 9, no. 1, pp. 11-12, January, 1962.

Wegner, P., "Modification of Aho and Ullman's correctness proof of Warshall's algorithm," *SIGACT News*, vol. 6, no. 1, pp. 32-35, 1974.

Weide, B., "A survey of analysis techniques for discrete algorithms," *ACM Computing Surveys*, vol. 9, no. 4, pp. 291-313, December, 1977.

Wigderson, Avi, "Improving the performance guarantee for approximate graph coloring," *Journal of the ACM*, vol. 30, no. 4, pp. 729-735, October, 1983.

Williams, J. W. J., "Algorithm 232: Heapsort," *Communications of the ACM*, vol. 7, no. 6, pp. 347-348, June, 1964.

Winograd, Shmuel, "On the number of multiplications necessary to compute certain functions," *Journal of Pure and Applied Math*, vol. 23, pp. 165-179, 1970.

Wirth, Niklaus, *Programming in Modula-2, Second edition*, Springer-Verlag, Berlin, 1983.

Yao, Andrew C., "On the average behavior of set merging algorithms," in *Proceedings of the Eighth Annual ACM Symposium on Theory of Computing*, pp. 192-195, 1976.

Yao, Frances, "Speed-up in dynamic programming," *SIAM Journal on Algebraic and Discrete Methods*, vol. 3, no. 4, pp. 532-540, December, 1982.

Index

Page numbers followed by n indicate material in footnotes.